CONSERVING EUROPE'S WILDLIFE

Law and Policy of the Natura 2000 Network of Protected Areas

Andrew L. R. Jackson

CRC Press
Taylor & Francis Group
Boca Raton London New York

CRC Press is an imprint of the
Taylor & Francis Group, an **informa** business

CRC Press
Taylor & Francis Group
6000 Broken Sound Parkway NW, Suite 300
Boca Raton, FL 33487-2742

First issued in paperback 2020

ISBN-13: 978-1-138-20365-5 (hbk)
ISBN-13: 978-0-367-50854-8 (pbk)

Visit the Taylor & Francis Web site at
http://www.taylorandfrancis.com

and the CRC Press Web site at
http://www.crcpress.com

British Library Cataloguing-in-Publication Data
A catalogue record for this book is available from the British Library

Library of Congress Cataloging-in-Publication Data
Names: Jackson, Andrew (Law teacher), author.
Title: Conserving Europe's wildlife : law and policy of the Natura 2000 Network of protected areas / Andrew Jackson.
Description: Abingdon, Oxon [UK] ; New York : Routledge, 2018. | Includes bibliographical references and index.
Identifiers: LCCN 2017037643 | ISBN 9781138203655 (hardback) | ISBN 9781315471211 (ebook)
Subjects: LCSH: Wildlife conservation—Law and legislation—European Union countries. | Natura 2000 (Network) | Habitat conservation—Law and legislation—European Union countries. | Biodiversity conservation—Law and legislation—European Union countries.
Classification: LCC KJE6255 .J33 2018 | DDC 346.2404/69516—dc23
LC record available at https://lccn.loc.gov/2017037643

Typeset in Bembo
by Swales & Willis Ltd, Exeter, Devon, UK

CONSERVING EUROPE'S WILDLIFE

The Natura 2000 network of protected areas is the centrepiece of European Union nature policy, currently covering almost one-fifth of the EU's entire land territory plus large marine areas. This vast EU-wide network, which aims to conserve Europe's most valuable and threatened species and habitats, has major impacts on land use throughout all Member States of the EU. This book critically assesses the origins and implementation of the Natura 2000 network, established under the Birds Directive of 1979 and the Habitats Directive of 1992.

Based on original archival research and interviews with key participants, the book records a detailed history of the origins and negotiation of Natura 2000 policy and law, with the history of EU environmental policy provided as a framework. An historical institutionalist approach is adopted, which emphasises the importance of understanding legal and policy development as processes that unfold over time. Three phases in the history of EU environmental policy are identified and described, and the history of EU nature policy is placed within the context of these three phases. Informed by this history, the author presents a comprehensive summary and assessment of the law and policy that protects Natura 2000 sites at EU level, and reviews the nature conservation outcomes for the targeted species and habitats. The book reveals how a knowledge of the history of Natura 2000 enriches our understanding of key issues such as conflicts in establishing and conserving the Natura 2000 network, EU integration in the field of nature conservation, and the future of EU nature policy.

Andrew L. R. Jackson is an Assistant Professor in environmental and planning law, UCD Sutherland School of Law, University College Dublin, Ireland. He is a graduate of Oxford University (BA, Law) and of Cambridge University (LLM). He also holds an MSc in Biodiversity and Conservation from Trinity College Dublin (TCD) and a PhD from TCD on EU environmental policy and law. He previously worked for the international law firm Slaughter and May in London and Paris; for the UK government's legal service, in the Department for Environment, Food and Rural Affairs (Defra); and with NGOs Friends of the Irish Environment and An Taisce, the National Trust for Ireland, where he was Natural Environment Officer and in-house Solicitor. Dr Jackson has been involved in public interest environmental and planning cases for many years, including before the Irish, English and EU courts.

Routledge Research in International Environmental Law

The Precautionary Principle in Marine Environmental Law
With Special Reference to High Risk Vessels
Bénédicte Sage-Fuller

International Environmental Law and Distributive Justice
The Equitable Distribution of CDM Projects under the Kyoto Protocol
Tomilola Akanle Eni-Ibukun

Environmental Governance in Europe and Asia
A Comparative Study of Institutional and Legislative Frameworks
Jona Razzaque

Climate Change, Forests and REDD
Lessons for Institutional Design
Edited by Joyeeta Gupta, Nicolien van der Grijp and Onno Kuik

International Environmental Law and the Conservation of Coral Reefs
Edward J. Goodwin

Forthcoming:

International Environmental Law and the International Court of Justice
Aleksandra Cavoski

www.routledge.com/Routledge-Research-in-International-Environmental-Law/book-series/INTENVLAW

To Suzanne, Saoirse, Róisín, Cillian and Ailbhe

CONTENTS

PREFACE, ACKNOWLEDGMENTS AND NOTE ON THE TEXT

Growing up in Scotland in the 1970s and 1980s was predominantly an outdoor pursuit. Countless hours were spent rock-pooling, or fishing for sticklebacks with jam jars, or hunched over a murky pool of water in an old sink in the garden, studying mosquito larvae and other undesirables. I should start, therefore, by thanking my parents, Jack and Sheilah, for nurturing in me a love of nature from my earliest years, and for forgiving occasional home invasions by winged intruders from my beloved sink.

My specific interest in EU nature conservation law and policy goes back to my time working as a government lawyer – in the mid-2000s – in the Countryside and Nature Conservation division of the legal service of the UK's Department for Environment, Food and Rural Affairs (Defra), where I advised on various aspects of environmental law, including nature conservation. Two insights stayed with me on leaving Defra: first, that my work as an environmental lawyer would benefit from a deeper knowledge of environmental science; and, second, that institutional memory is of inestimable value – its presence a major boon and its absence a gaping loss. Having been particularly taken by the legal strength and impacts of the Birds and Habitats Directives during my time with Defra, the latter insight led to my becoming interested in the history of EU nature law and policy. My sense then – which has only grown since – was that a knowledge of this history would be invaluable in seeking to understand and apply the law. This subsequently became the subject of my doctoral research at Trinity College Dublin, funded by what is now the Irish Research Council.

The present book is a significantly revised and updated version of my doctoral thesis. I thank my doctoral supervisor Professor David Taylor, now of the National University of Singapore. I owe a debt of gratitude to the interviewees and correspondents who participated in the present research – too numerous to mention individually here – for giving so generously of their time, information and expertise, and for encouraging me to publish this book. Many thanks are due also to DG Environment of the European Commission for allowing me access to their archive – particular thanks to René Deprez and Micheál Ó Bríain in this regard. Additional special thanks to Micheál for generously providing the cover image for this book. Thanks also to archivists in the other EU institutions, who were extraordinarily helpful and responsive in providing me with information.

Sincere thanks to Amy Johnston, Tim Hardwick and all at Routledge, who were unfailingly patient in allowing me time to complete this project. Many thanks also to Megan Symons of Swales & Willis for her expert production assistance.

I thank my colleagues at University College Dublin for creating a supportive atmosphere conducive to research. Specific thanks to the UCD College of Social Sciences and Law Research Fund for support.

For support and encouragement throughout, and for comments on the manuscript, I thank Professor Suzanne Kingston of University College Dublin.

Please note that the terms European Economic Community, European Community, and European Union (and the related acronyms EEC, EC, EU) – used over the years to refer to the supranational bodies now known as the European Union – have been used interchangeably in this book. For ease, I have tended to favour using "the EU" even where this is not strictly correct according to the time period being discussed. I have endeavoured to state the law as at 31 July 2017.

ABBREVIATIONS

ACE	Actions Communautaires pour l'Environnement (Community Action relating to the Environment)
ACNAT	Actions Communautaires pour la Conservation de la Nature (Community Action relating to Nature Conservation)
AFP	Agence France-Presse
ANCGE	Association Nationale des Chasseurs de Gibiers d'Eau (National Association of Waterfowl Hunters, France)
BANC	British Association of Nature Conservationists
CAP	Common Agricultural Policy of the European Union
CEC	Council of the European Communities
CEU	Council of the European Union
CIC	Conseil International de la Chasse et de la Conservation du Gibier (International Council for Game and Wildlife Conservation)
CITES	Convention on International Trade in Endangered Species of Wild Fauna and Flora
CJEU	Court of Justice of the European Union
COG	Chiefs of Government
COREPER	Committee of Permanent Representatives of the Governments of the Member States of the European Union
CORINE	Coordination of Information on the Environment
DBV	Deutscher Bund für Vogelschutz (German Association for the Protection of Birds)
Defra	Department for Environment, Food and Rural Affairs (UK)
DG	Directorate General (of the European Commission)
DG III	Directorate General for Industrial Affairs, Science and Technology
DG XI	Directorate General for Environment, Consumer Protection and Nuclear Safety; later became DG Environment

EC	European Community
ECB	European Central Bank
ECJ	European Court of Justice (now known as the Court of Justice of the European Union, CJEU)
ECPS	Environment and Consumer Protection Service of the European Commission (later became DG XI, then DG Environment)
EEA	European Environment Agency
EEAS	European External Action Service
EEB	European Environmental Bureau
EEC	European Economic Community
EIA	Environmental Impact Assessment
ELO	European Landowners' Organization
ESC	European Economic and Social Committee
ETC-BD	European Topic Centre on Biological Diversity
EU	European Union
EUI	European University Institute
FACE	Fédération des Associations de Chasse et Conservation de la Faune Sauvage de l'UE (Federation of Associations for Hunting and Conservation of the European Union; formerly Federation of Hunting Associations of the EEC)
HAEU	Historical Archives of the European Union
IBA	Important Bird Areas
ICBP	International Council for Bird Preservation
IEEP	Institute for European Environmental Policy
IELA	Irish Environmental Law Association
IMPEL	European Union Network for the Implementation and Enforcement of Environmental Law
IROPI	Imperative Reasons of Overriding Public Interest – a concept from Article 6(4) of the Habitats Directive
IRSNB	Institut Royal des Sciences Naturelles de Belgique (Royal Belgian Institute of Natural Sciences)
IUCN	International Union for Conservation of Nature
IWRB	International Waterfowl Research Bureau
JNCC	Joint Nature Conservation Committee (adviser to UK government)
LIFE	L'Instrument Financier pour l'Environnement (EU Financing Instrument for the Environment)
LIPU	Liga Italiana Protezione Uccelli (Italian League for Bird Protection)
MEP	Member of the European Parliament
MIT	Massachusetts Institute of Technology
MP	Member of Parliament
MPI	Max-Planck-Institut für Verhaltensphysiologie
NABU	Naturschutzbund Deutschland (Nature and Biodiversity Conservation Union, Germany)

NCC	Nature Conservancy Council (UK government agency from 1973 to 1991)
NGO	Non Governmental Organisation
NVTBVV	Nederlandse Vereniging Tot Bescherming Van Vogels (Dutch Society for the Protection of Birds)
OJ	Official Journal of the European Union
pSCI	Proposed Site of Community Importance under the Habitats Directive
REFIT	EU Regulatory Fitness and Performance Programme
RGI	Renewables Grid Initiative
RSPB	Royal Society for the Protection of Birds
SAC	Special Area of Conservation under the Habitats Directive
SCI	Site of Community Importance under the Habitats Directive
SDC	Sustainable Development Commission (UK)
SEA	Single European Act
SFF	Secrétariat de la Faune et de la Flore (Fauna and Flora Secretariat, a unit of France's National Museum of Natural History)
SPA	Special Protection Area under the Birds Directive
SSSI	Site of Special Scientific Interest (site protection designation in Scotland, England, and Wales)
TEU	Treaty on European Union
TFEU	Treaty on the Functioning of the European Union
UK	United Kingdom
UNCBD	UN Convention on Biological Diversity
UNEP	United Nations Environment Programme
US	United States of America
WEBS	Working Group of European Bird Protection Societies
WTO	World Trade Organization
WWF	World Wildlife Fund

EU CASES

Case	*Stage*	*Chapter (reference): page(s)*
Case 240/83 *ADBHU*, EU:C:1985:59	Judgment	Chapter 11 (CJEU, 1985): p. 260
Case 236/85 *Commission* v. *Netherlands*, EU:C:1987:436	Judgment	Chapter 5 (CJEU, 1987d): p. 90
Case 247/85 *Commission* v. *Belgium*, EU:C:1986:458	AG's Opinion	Chapter 5 (CJEU, 1986a): p. 90
Case 247/85 *Commission* v. *Belgium*, EU:C:1987:339	Judgment	Chapter 5 (CJEU, 1987a): p. 90
Case 252/85 *Commission* v. *France*, EU:C:1988:202	Judgment	Chapter 5 (CJEU, 1988): p. 90
Case 262/85 *Commission* v. *Italy*, EU:C:1987:340	AG's Opinion	Chapter 5 (CJEU, 1986b): p. 90
Case 262/85 *Commission* v. *Italy*, EU:C:1987:340	Judgment	Chapter 5 (CJEU, 1987b): p. 90
Case 412/85 *Commission* v. *Germany*, EU:C:1987:370	Judgment	Chapter 5 (CJEU, 1987c): p. 90
Case 247/87 *Star Fruit* v. *Commission*, EU:C:1989:58	Judgment	Chapter 11 (CJEU, 1989): p. 277
C-339/87 *Commission* v. *Netherlands*, EU:C:1990:119	Judgment	Chapter 8 (CJEU, 1990b): p. 180
C-288/88 *Commission* v. *Germany*, EU:C:1990:273	Judgment	Chapter 8 (CJEU, 1990a): p. 180
Case 57/89 R *Commission* v. *Germany (Leybucht Dykes)*, EU:C:1989:334	Order	Chapter 8 (CJEU, 1989): p. 163
Case 57/89 *Commission* v. *Germany (Leybucht Dykes)*, EU:C:1990:432	AG's Opinion	Chapter 4 (CJEU, 1990): p. 66 Chapter 8 (CJEU, 1990c): pp. 180–181 Chapter 9 (CJEU, 1990): p. 196 Chapter 11 (CJEU, 1990): p. 271

(continued)

(continued)

Case	*Stage*	*Chapter (reference): page(s)*
Case 57/89 *Commission* v. *Germany (Leybucht Dykes)*, EU:C:1991:89	Judgment	Chapter 6 (Leybucht): p. 107 Chapter 8 (Leybucht): p. 182 Chapter 9 (CJEU, 1991; and Leybucht): pp. 196–201, 206, 213–214 Chapter 10 (CJEU, 1991; and Leybucht): pp. 229, 231, 234, 248 Chapter 11 (CJEU, 1991; and Leybucht): pp. 271–272, 278–279, 281–282, 284
C-157/89 *Commission* v. *Italy*, EU:C:1990:385	AG's Opinion	Chapter 8 (CJEU, 1990d): p. 180
C-355/90 *Commission* v. *Spain (Santoña marshes)*, EU:C:1993:229	AG's Opinion	Chapter 6 (CJEU, 1993): p. 119 Chapter 8 (CJEU, 1993): pp. 163, 180
C-355/90 *Commission* v. *Spain (Santoña marshes)*, EU:C:1993:331	Judgment	Chapter 10 (CJEU 1993): p. 229
C-44/95 *Lappel Bank*, EU:C:1996:297	Judgment	Chapter 1 (CJEU, 1996): pp. 5, 8, 13 Chapter 10 (CJEU, 1996; and Lappel): pp. 229–230, 241, 248 Chapter 11 (CJEU, 1996; and Lappel): pp. 273, 281–282
C-3/96 *Commission* v. *Netherlands*, EU:C:1998:238	Judgment	Chapter 4 (CJEU, 1998): p. 69 Chapter 10 (CJEU, 1998): pp. 229–230 Chapter 11 (CJEU, 1998): p. 262
C-96/98 *Commission* v. *France*, EU:C:1999:580	Judgment	Chapter 10 (CJEU, 1999): p. 248
C-371/98 *First Corporate Shipping*, EU:C:2000:600	Judgment	Chapter 1 (CJEU, 2000): pp. 5, 8 Chapter 8 (CJEU, 2000): p. 168 Chapter 10 (CJEU, 2000b): p. 233 Chapter 11 (CJEU, 2000b): p. 273
C-374/98 *Commission* v. *France*, EU:C:2000:670	Judgment	Chapter 10 (CJEU, 2000a): p. 231 Chapter 11 (CJEU, 2000a): pp. 267, 273
C-127/02 *Waddenzee*, EU:C:2004:482	Judgment	Chapter 10 (CJEU, 2004): pp. 235–236, 248 Chapter 11 (CJEU, 2004): p. 273
C-117/03 *Dragaggi*, EU:C:2005:16	Judgment	Chapter 10 (CJEU, 2005a): p. 238
C-6/04 *Commission* v. *UK*, EU:C:2005:626	Judgment	Chapter 9 (CJEU, 2005): p. 214 Chapter 10 (CJEU, 2005b): pp. 228, 235 Chapter 11 (CJEU, 2005): p. 273

C-239/04 *Commission* v. *Portugal*, EU:C:2006:665	Judgment	Chapter 10 (CJEU, 2006c): p. 236
C-418/04 *Commission* v. *Ireland*, EU:C:2007:780	Judgment	Chapter 4 (CJEU, 2007): p. 74 Chapter 10 (CJEU, 2007): pp. 230, 241 Chapter 11 (CJEU, 2007): p. 290
C-191/05 *Commission* v. *Portugal*, EU:C:2006:150	AG's Opinion	Chapter 10 (CJEU, 2006d): p. 230
C-191/05 *Commission* v. *Portugal*, EU:C:2006:472	Judgment	Chapter 10 (CJEU, 2006a): p. 230
C-244/05 *Bund Naturschutz*, EU:C:2006:579	Judgment	Chapter 10 (CJEU, 2006b): p. 239 Chapter 11 (CJEU, 2006): p. 273
C-503/06 R *Commission* v. *Italy*, EU:C:2006:800	Order	Chapter 8 (CJEU, 2006): p. 163
C-535/07 *Commission* v. *Austria*, EU:C:2010:602	Judgment	Chapter 10 (CJEU, 2010b): pp. 235, 248
C-241/08 *Commission* v. *France*, EU:C:2010:114	Judgment	Chapter 10 (CJEU, 2010a): p. 235
C-308/08 *Commission* v. *Spain*, EU:C:2010:281	Judgment	Chapter 10 (CJEU, 2010c): p. 249
C-240/09 *LZ I*, EU:C:2011:125	Judgment	Chapter 10 (CJEU, 2011a): p. 240
C-404/09 *Commission* v. *Spain*, EU:C:2011:768	Judgment	Chapter 10 (CJEU, 2011b): pp. 235, 249
C-43/10 *Nomarchiaki*, EU:C:2012:560	Judgment	Chapter 10 (CJEU, 2012a): p. 249 Chapter 11 (CJEU, 2012a): p. 291
C-182/10 *Solvay*, EU:C:2012:82	Judgment	Chapter 10 (CJEU, 2012c): p. 249 Chapter 11 (CJEU, 2012c; and Solvay): pp. 276, 278, 291
C-340/10 *Commission* v. *Cyprus*, EU:C:2012:143	Judgment	Chapter 8 (CJEU, 2012): p. 183 Chapter 10 (CJEU, 2012b): pp. 239, 248 Chapter 11 (CJEU, 2012b): p. 291
C-258/11 *Sweetman*, EU:C:2012:743	AG's Opinion	Chapter 11 (CJEU, 2012d): p. 276
C-258/11 *Sweetman*, EU:C:2013:220	Judgment	Chapter 10 (CJEU, 2013): pp. 234–236, 238 Chapter 11 (CJEU, 2013): p. 273
C-301/12 *Cascina Tre Pini*, EU:C:2014:214	Judgment	Chapter 10 (CJEU, 2014a): pp. 231, 233, 248
C-521/12 *Briels*, EU:C:2014:330	Judgment	Chapter 10 (CJEU, 2014b): p. 237 Chapter 11 (CJEU, 2014): p. 273
C-141/14 *Commission* v. *Bulgaria*, EU:C:2016:8	Judgment	Chapter 10 (CJEU, 2016c): pp. 231, 235
C-399/14 *Grüne Liga Sachsen*, EU:C:2016:10	Judgment	Chapter 10 (CJEU, 2016d): pp. 235–236, 240–241, 249

(continued)

(continued)

Case	*Stage*	*Chapter (reference): page(s)*
C-461/14 *Commission* v. *Spain*, EU:C:2016:895	Judgment	Chapter 10 (CJEU, 2016f): p. 248
C-504/14 *Commission* v. *Greece*, EU:C:2016:847	Judgment	Chapter 10 (CJEU, 2016b): pp. 235, 249
C-243/15 *LZ II*, EU:C:2016:838	Judgment	Chapter 10 (CJEU, 2016a): p. 240 Chapter 11 (CJEU, 2016a): p. 273
C-387/15 *Orleans*, EU:C:2016:583	Judgment	Chapter 10 (CJEU, 2016e): p. 237 Chapter 11 (CJEU, 2016b): p. 273
C-142/16 *Commission* v. *Germany*, EU:C:2017:301	Judgment	Chapter 10 (CJEU, 2017a): p. 249 Chapter 11 (CJEU, 2017): p. 273
C-281/16 *VHL*, EU:C:2017:476	AG's Opinion	Chapter 10 (CJEU, 2017b): pp. 233, 248
C-441/17 R, *Commission* v. *Poland*, EU:C:2017:622	Order	Chapter 10 (CJEU, 2017c): p. 224

1

INTRODUCTION

> [The creation of an EU network of protected nature conservation areas] would constitute an awesome challenge to the sensitivities of member nations. It is one thing for Brussels to dictate the content of sausages, but quite another for it to control the uses to which the motherland is put.
>
> *(Shoard, 1990)*

EU environmental policy has in the past been described as a flanking policy to other apparently more central concerns such as the single market (e.g., European Commission, 1987, pp. 54–55). However, environmental policy has long since stepped out of the shadow of the single market, and has shown itself of high political salience. The establishment and conservation of the EU's vast network of protected areas for wildlife – known as Natura 2000 – is a good example of this. Regulating land use at EU level has indeed proven controversial, as Shoard (1990) predicted.

For example, when asked for his views regarding the EU immediately before and after assuming the US presidency in 2017, Donald Trump focused on his failure to obtain permission to build a 2.8km-long, 4m-high, 20m-wide wall on a Natura 2000 beach in Ireland, a project aimed at safeguarding a Trump golf course from coastal erosion. As the *Washington Post* commented,

> To Trump, the bloc's environmental protection regulations were a threat to his exquisitely manicured fairways and putting greens. 'I found it to be a very unpleasant experience,' he told British and German interviewers [in January 2017] after bringing up the wall dispute, unbidden, when asked his opinion of the EU. 'A very bad experience,' he emphasized weeks later as he raised the issue at his first White House news conference.
>
> *(Witte, 2017)*

Trump returned to the theme *again* in May 2017 during his first visit to Europe as US president, reportedly telling the Belgian prime minister that his golf course experience in Ireland "did not give him a very good image of the European Union" (Demonty et al., 2017). "The bureaucratic battle over a golf-course sea wall makes for an unlikely inflection point in geopolitical history", notes the *Washington Post* (Witte, 2017). And yet, EU nature conservation policy has been a sensitive and much-debated topic since its earliest origins, touching as it does on fundamental issues such as land rights and land use control.

To a developer such as Trump, the legislation and institutions that operate to conserve the Natura 2000 network may be something of a culture shock (see Demonty et al., 2017). As this book details, the EU's nature conservation laws set demanding requirements for Member States, which in turn have been strictly interpreted by the Court of Justice of the EU (CJEU) (see Chapter 10). There can be little doubt that these laws are amongst the most powerful and far-reaching the EU has ever passed in the environmental sphere. This raises interesting questions about the origins and history of these laws, their development, and the interplay between the key actors involved.

This book critically assesses the origins, development and implementation of the EU's Natura 2000 network of protected areas, established under the Birds Directive (1979) and the Habitats Directive (1992). Natura 2000 is the flagship of EU nature policy, currently covering almost one fifth of the EU's entire land territory plus large marine areas. This huge EU-wide network of protected areas, which aims to conserve Europe's most valuable and threatened species and habitats, has major impacts on land use across the EU, as Donald Trump and many others have discovered.

Based on original archival research and interviews with key participants,[1] the book records a detailed history of the origins and negotiation of Natura 2000 policy and law. As such, the book adopts an historical institutionalist approach – which emphasises the importance of understanding legal and policy development as processes which unfold over time (Pierson, 1996) – in seeking to understand current EU nature conservation policy in the light of its history and the institutional interplay that brought it into being. Three phases in the history of EU environmental policy and law are identified. EU nature conservation policy and law are then placed in the context of these three phases. Under the umbrella of the first phase (1958 to 1987), the origins and negotiation of the Birds Directive and the road to the Habitats Directive are assessed. The second phase (1987 to 1992) covers the negotiation and adoption of the Habitats Directive. The third phase (1992 to present day) covers the implementation of the Natura 2000 network.

The book provides a detailed history and assessment of Natura 2000 policy and law in the EU, against the backdrop of the general history of EU environmental policy. The book then applies an historical institutionalist analysis to address the following four central issues. What insights can this detailed history and the attendant historical institutionalist approach contribute to understanding:

1. Conflicts in establishing the Natura 2000 network?
2. The resolution of acute conflicts between human land uses and Natura 2000 sites?
3. The process of EU integration in the nature policy field?
4. The future of EU nature policy?

As regards the negotiation phases of the Birds and Habitats Directives, the history presented here focuses primarily on the negotiation of provisions relating to the designation and protection of Natura 2000 sites. In the context of the Habitats Directive, other interesting issues such as the drafting and negotiation of the annexes of species and habitats requiring Natura 2000 sites, and the negotiation of provisions relating to the protection of land outside the Natura 2000 network – e.g., corridors and stepping stones – are beyond the scope of this book.

Introduction to EU nature conservation policy and Natura 2000

At the global level, in 2002 the Parties to the Convention on Biological Diversity pledged to achieve by 2010 a significant reduction in the then current rate of biodiversity loss (UNCBD, 2002). The EU adopted a more ambitious target, seeking to halt biodiversity loss in the EU by 2010 (European Council, 2001), but both the EU's target and the global target were missed (European Commission, 2010b; UNCBD, 2010). The revised global aim, adopted in October 2010, was to take "effective and urgent action to halt the loss of biodiversity in order to ensure that by 2020 ecosystems are resilient and continue to provide essential services, thereby securing the planet's variety of life, and contributing to human well-being, and poverty eradication" (UNCBD, 2010, paragraph 12). The EU adopted an even more ambitious headline target for 2020: halting the loss of biodiversity and the degradation of ecosystem services in the EU by 2020, and restoring them in so far as feasible, while stepping up the EU contribution to averting global biodiversity loss – and included in addition a vision for the better protection of biodiversity in the EU by 2050 (European Commission, 2011, p. 2).

The Birds Directive (1979) and the Habitats Directive (1992) are the "cornerstones of the EU's biodiversity policy" (European Commission, 2010a, p. 17). The protection provided by the Directives is divided between species protection measures and site protection measures (European Commission, 2007). In general terms, the species protection measures require the EU's Member States to protect certain species wherever they are found in the wild in the Member States' territories, while the site protection measures require Member States to identify, designate and protect conservation areas for certain habitat types and for the habitats of certain species: namely, Special Protection Areas (SPAs) under the Birds Directive, and Special Areas of Conservation (SACs) under the Habitats Directive. The network of SACs and SPAs – which is intended to comprise an ecologically coherent whole

(Article 3(1) of the Habitats Directive, 1992) – is collectively known as Natura 2000, and is the "centrepiece" of EU biodiversity policy (European Commission, 2009).[2]

Notwithstanding the difficulties that have been encountered in establishing the Natura 2000 network, and the fact that the network is not yet complete (European Commission, 2017),[3] the network's importance and impacts should not be underestimated. In 2017, 25 years after the Habitats Directive was adopted, the Natura 2000 network now covers more than 27,500 sites in 28 countries, comprising about 18% of the EU's land territory plus large marine areas (European Commission, 2017),[4] making it the largest network of protected areas in the world (European Commission, 2010b). Recent research has revealed that the network is delivering conservation benefits,[5] notwithstanding a poor picture for biodiversity overall in the EU (EEA, 2015a).

Of course, it is not merely the network's extent that gives Natura 2000 its significance in terms of land use control. Rather, it is the combined effect of this extent, the strict legal protection afforded to Natura 2000 sites by EU law, and the fact that protection for the sites can be enforced by the European Commission, a supranational authority. When one combines these elements, conflicts with human land uses are inevitable. As Claus Stuffmann, the European Commission's senior negotiator in respect of both the Birds and Habitats Directives, reflects,

> areas placed under national protection status can always be victims of arbitration when taking account only national interests. It is quite a different matter, however, when this arbitration must comply with criteria established at Community level . . . Designating a special area of conservation under the Community directives is not merely another label, but it also constitutes considerable additional legal protection.
>
> *(Stuffmann, 1994, p. 205)*

This has arguably been one contributory factor to the delays experienced in establishing the Natura 2000 network: Coffey and Shaw (2001, p. 22) refer to the Habitats Directive as "one of the most notorious items of EC law, as far as implementation by the Member States is concerned", and it is notable that the Habitats and Birds Directives consistently account for a relatively high percentage of the European Commission's infringement caseload in the environmental sector (e.g., see European Commission, 2016b).[6]

An historical approach to understanding EU nature law and policy

Whilst there are existing overviews of the history of EU nature conservation policy (e.g., see Krämer, 1993; Van Hoorick, 1997), and detailed accounts of the history of the Birds Directive by Meyer (2010, 2011, 2014), there is no existing detailed account of the history of the Habitats Directive, the Natura 2000 network, or the origin and negotiation of the rules applying to protect that network, notwithstanding the network's far-reaching impacts on day-to-day land use planning in the EU.[7] This book aims to fill this gap, mindful of Jordan's (1998, p. 235) emphasis on the need for research to "link broader political and economic currents with the detailed nitty gritty of policy development and implementation". This book therefore provides as a backdrop an overview of the history of EU environmental policy, against which the detailed development of EU nature policy played out. As this book aims to show, a knowledge of this history serves to illuminate the key issues set out below.

Conflicts in establishing the Natura 2000 network

Whilst the establishment of the Natura 2000 network represents a remarkable achievement, there have nevertheless been difficulties regarding the network's implementation, including delays in establishing the network, resistance at national level, and, in some cases, inadequate management of and protection for sites. Krämer (1993, p. 25) provides an important insight into these problems. It is

> the difference between the European Community (EC) and the United States that has the greatest implication for nature protection within Europe. The Federal Government of the United States owns one-third of all United States territory, including land that provides vital habitat for wildlife. For example, the United States Federal Government owns ninety percent of the habitat of the spotted owl. Consequently, the Federal Government is authorized to protect vast areas of wildlife habitat, including that of the spotted owl, without needing or requiring intervention by the states. In contrast, the Member States and private parties own all the land within the EC . . . Even when Member States adopt nature protection measures, widespread ownership of private land within the Member States may limit the effectiveness of the measures. Consequently, all nature protection measures within the EC are subject to a greater level of competing interests than are many habitat protection measures within the United States.
>
> *(Krämer, 1993, p. 25)*

Engelen et al. (2008, pp. 4–5) cite several additional reasons for the problems encountered in establishing the Natura 2000 network, including the "purely scientific legitimation underlying Natura 2000". That is, as a matter of EU law, Natura 2000 sites must be chosen solely on the basis of scientific criteria, and no account can be taken of socio-economic considerations in selecting sites and their boundaries (CJEU, 1996, 2000). Such an approach, Engelen et al. (2008, p. 5) argue, seems "to invite a technocratic, top-down mode of policy-making that is increasingly being rejected by local constituencies as too insensitive to local interests, too paternalistic for modern tastes and too elitist for modern democracies".[8] Low political priority at the national level, coupled with a tight implementation schedule, are said to have added to the problems, limiting "the extent to which local interests can collectively decide on the implementation process" (Engelen et al., 2008, p. 5; Paavola, 2004). Moreover, a "lack of communication, information, and recognition has further added to the national and local resistance to the implementation of Natura 2000" (Engelen et al., 2008, p. 5). One might cite here, for example, a certain wariness about what the designation of a Natura 2000 site will mean in terms of future restrictions on activities (Baldock, 1992), and resistance based on the perception that site designations are being "imposed" (Leibenath, 2008, p. 242). Allied to this, in certain countries, is Euroscepticism (Björkell, 2008) – i.e., a specific objection grounded on the belief that site designations are being imposed by 'the EU' – and strong landowning traditions, "emphasising the sovereignty of the landowner" (Björkell, 2008, p. 110; Laffan and O'Mahony, 2008).

It suffices to note here that the literature abounds with examples of conflicts during the implementation of Natura 2000, and that many of these studies cite as reasons some or all of the points mentioned above. For example, Hiedanpää (2002) and Björkell (2008) address the implementation of Natura 2000 in Finland, called by some "the largest environmental conflict in Finland's history", the main reason being "a lack of legitimacy in its implementation process" (Björkell, 2008, p. 110). Bogaert and Leroy (2008) review implementation in Flanders, Belgium, which ran into "severe difficulties" (Engelen et al., 2008, p. 17). Alphandéry and Fortier (2001) and Pinton (2001, 2008) review the situation in France, the latter recording that the process of drawing up an inventory of Natura 2000 sites caused

> a scientific and political crisis . . . Opposition reached its peak in March 1996, when the administration presented its final list, including 1,316 sites covering . . . 13% of the country. The opposition movement . . . a powerful alliance of hunters, foresters, farmers and fishermen, fiercely objected to this list. Their action led to a suspension of the inventory procedure for several months in 1997 . . . As a result of the consultations with non-scientific stakeholders, the total area of sites proposed to the EU dropped from the initial figure of 13% of the country in 1996 to 7.6% at the end of 2002.
>
> *(Pinton, 2008, pp. 211–212)*[9]

Leibenath (2008, p. 237) considers the implementation of Natura 2000 in Germany, "a tale of neglect, delays and lawsuits". The main reason, he explains, "can be found in real or assumed conflicts with landowners and land-users, who were lobbying vigorously against Natura 2000. As the Länder [German states] had to bear the financial and legal consequences of selecting an area for Natura 2000, they followed a conflict-averse strategy in the first place and based their first tranche of [Natura 2000 sites] on political instead of ecological considerations", leading to EU legal proceedings against Germany (Leibenath, 2008, p. 239).[10] Raoul Beunen and colleagues have considered the situation in the Netherlands over the course of several publications, with an article in 2013 perceiving a "depressing" story of Natura 2000 implementation in the Netherlands, leading, the authors report, to a "negative attitude toward nature conservation and [a] crisis in Dutch nature conservation policy" (Beunen, 2006; Beunen et al., 2009; Beunen, 2010; Beunen and de Vries, 2011; Beunen et al., 2013, p. 280). In Poland, Chmielewski and Glogowska (2015) report many cases of conflicts but that stakeholders see Natura 2000 not merely as a restraint but also an opportunity.

Ferranti et al. (2010) consider implementation in Italy (and the Netherlands), noting that "the technical-cultural delay in the regional Natura 2000 implementation process [in Italy] is partly due to the lack of information and training by the government and partly to the fragmentation of responsibilities in the field of nature conservation, lack of communication between the authorities involved and low competence in nature conservation matters" (Ferranti et al., 2010, p. 303). Laffan and O'Mahony (2008, p. 177) address implementation in Ireland, where the Habitats Directive "triggered a high degree of political opposition from key domestic veto players that stymied the efforts of the Irish authorities to implement the directive on time and in a correct manner". Apostolopoulou and Pantis (2009, p. 221) examine Natura 2000 implementation in Greece, revealing, they report, a "national strategy . . . compromised by absence of conservation policy history, lack of state capacity, uncommunicated biological knowledge and lack of public participation". The UK has similarly experienced various conflicts during the implementation of Natura 2000 (e.g., see Ledoux et al., 2000; Morris, 2011; Fairbrass and Jordan, 2001). Finally, multiple country reviews such as those of Laffan (2004), WWF (2006), Bouwma et al. (2008), Rauschmayer et al. (2009) and Kati et al. (2015) (who consider strengths and weaknesses) identify similar messages to those detailed above. Additional reasons could also be cited such as, for example, inadequate funding for nature conservation in the EU, particularly when compared with the funding available for policies that have tended to have negative impacts on biodiversity (e.g., the common agricultural policy (Gammell, 1999)).[11]

In summary, Engelen et al. (2008, p. 5) argue that "the EU's evident legislative success in the field of nature conservation policies upstream has increasingly been frustrated by highly politicized decision-making downstream, which has resulted in delays, deadlocks, and even out-and-out resistance".

Acute conflicts in conserving Natura 2000 sites

In negotiating the Habitats Directive the Member States managed to introduce a derogation provision – Article 6(4) – which provides:

> If, in spite of a negative assessment of the implications for the [Natura 2000] site and in the absence of alternative solutions, a plan or project must nevertheless be carried out for imperative reasons of overriding public interest, including those of a social or economic nature, the Member State shall take all compensatory measures necessary to ensure that the overall coherence of Natura 2000 is protected. It shall inform the Commission of the compensatory measures adopted.

> Where the site concerned hosts a priority natural habitat type and/or a priority species, the only considerations which may be raised are those relating to human health or public safety, to beneficial consequences of primary importance for the environment or, further to an opinion from the Commission, to other imperative reasons of overriding public interest.

Thus, whilst the Member States cannot take account of socio-economic considerations at the time of choosing Natura 2000 sites for designation (CJEU, 1996, 2000), it is thereafter possible for national authorities to allow plans or projects that have been the subject of a negative impact assessment to proceed, but only where three conditions are met: (1) an "absence of alternative solutions" must be demonstrated; (2) "imperative reasons of overriding public interest" for proceeding with the plan or project must be demonstrated (and these reasons are limited in cases involving priority species or habitats, and must in certain cases be confirmed by an opinion from the European Commission);[12] and (3) the Member State in question must take "all compensatory measures necessary to ensure that the overall coherence of Natura 2000 is protected".

This derogation therefore represents something of a fault line:[13] it is the sole legal means by which the EU's Member States can allow damaging (or potentially damaging) plans/projects to proceed in respect of Natura 2000 sites. Plans or projects that national authorities seek to approve via this provision may be considered as *acute* conflicts because they pit the goal of conserving the EU's most threatened species and habitats against plans/projects that must proceed, according to national governments, for imperative reasons of overriding public interest. Given the 'high stakes' nature of these conflicts, cases relying on this Article 6(4) derogation make for a potentially fruitful area of study linked to the issue introduced immediately below – EU integration. Is this derogation provision an example of the Member States retaining an important degree of control over EU nature conservation policy? Is this a "counter-force" to supranational integration in the biodiversity field (Alter, 2009, p. 16)? What light can an analysis of the history of the Natura 2000 network shed on these questions?

Understanding EU integration in the field of nature conservation

As Jordan et al. (1999) argued almost two decades ago, "After forty years of sustained activity, the 'holy grail' of integration theory is still to establish whether states, either individually or collectively, control the process of [EU] integration or whether it has escaped their clutches in several important respects". This debate about the nature of EU integration fell somewhat out of fashion in academic circles for a time, for a variety of reasons,[14] with the focus of research attention shifting to "less grandiose issues" (Bulmer, 2015, p. 289) such as governance (multi-level governance, networks, etc.) and Europeanisation (i.e., the impact of EU membership on Member States). However, the central intergovernmental-supranational debate "continues to inform research" (Bache et al., 2011, p. 18), never went away in national politics,[15] and has returned centre stage by virtue of the "new intergovernmentalism" (Bickerton et al., 2015).

In terms of the basic intergovernmental vs. supranational divide: in intergovernmental perspectives, explain Bache et al. (2011, p. 17),

> European integration is a process whereby the governments of states voluntarily enter into agreements to work together to solve common problems. Some constraints operate on the autonomy of national governments, but they remain in control of the process. The alternative [supranational] perspective suggests that although governments started the process, integration soon took on a life of its own that went beyond the control of the governments.

The few theoretical accounts that exist regarding the EU integration process in the biodiversity field have tended to favour the idea that the EU's Member States lost control of EU biodiversity policy in certain respects, highlighting a potential contributory factor to the conflicts described above. For example, some would argue that the Habitats and Birds Directives have had, in certain important respects, rather different implications to those envisaged by the Member States at the time the instruments were negotiated (Stone Sweet, 2004; Fairbrass and Jordan, 2001), and that this is in part a result of decisions of the CJEU, which has interpreted the instruments more expansively and less flexibly than Member States might have intended and hoped.

More generally, having examined several examples from EU environmental policy, including nature policy, Stone Sweet (2004, p. 232) concludes that, "Through its rulings, [the CJEU] has acted – relatively systematically – to reduce the domain of national autonomy, to expand supranational modes of governance to the detriment of intergovernmental modes, and to create conditions for the gradual Europeanization of national administration and judging". This argument, emphasising supranationalism, forms part of Sandholtz and Stone Sweet's (2012) broader theory of EU integration, an updated version of neo-functionalism which, they claim, "offers a causal explanation of the development of EU institutions and the expansion of their authority" and is "the most theoretically viable and empirically productive general theory of European integration" (Sandholtz and Stone Sweet, 2012, p. 28). In stark contrast, more recently Bickerton et al. (2015) have argued that, since the Maastricht Treaty in 1992, EU integration has been taking place in the *absence* of traditional supranational policy-making, a phase they term "the new intergovernmentalism".

Theoretical claims such as these can usefully be tested in the context of an individual EU policy area such as nature policy. A promising method for this is historical institutionalism (Pierson, 1996), which (as noted above) emphasises the importance of understanding political development as a process that unfolds over time. As Alter (2009, p. 16) comments, for example,

> neo-functionalist theory brings with it theoretical baggage which is often unsupported by evidence. Historical institutionalism . . . offers a more open framework that deals better with contingency, and it captures important contextual elements and counter-forces that scholars who are looking only for forces supporting integration will miss. By focusing on counter-forces and surprising puzzles, we may well reveal dynamics that ultimately shape long term change, and thus that better account for important systemic transformations we have observed in Europe.

An historical institutionalist analysis has in the past been applied to EU biodiversity policy, notably by Fairbrass and Jordan (2001),[16] who argue that "a comparison of UK Government aims with long-term political outcomes in the sphere of EU biodiversity policy (*c.*1970–2000) reveals evidence of firm state control in the short-term, but of significant unintended consequences in the medium to long-term. These findings raise doubts about the explanatory power of intergovernmental accounts of EU environmental policy making" (Fairbrass and Jordan, 2001, p. 1). The UK government, they argue, switched between a "strategic manager" and a "passive observer" of EU biodiversity policy over the "full policy cycle":

> The two main instances of strategic management were the initial decision to adopt the Birds Directive (on the assumption that it was modelled on a supposedly superior UK model of conservation) and its subsequent amendment via the Habitats Directive [i.e., the application of Article 6(2) to (4) to sites protected under the Birds Directive, discussed in Chapters 9 and 10]. However, thereafter, the UK government was condemned to watch as the Commission, the [CJEU] and interest groups exploited EU biodiversity policy to suit their own political purposes.
>
> *(Fairbrass and Jordan, 2001, p. 18)*

Fairbrass and Jordan's (2001) analysis thus reaches a similar conclusion to Stone Sweet's (2004) account: namely, an important feature of EU nature policy has arguably been "unintended consequences" from the Member States' perspective. As noted above, Stone Sweet (2004) uses this conclusion as evidence in support of his 'supranational governance' (neo-functionalist) account of EU integration. In contrast, Fairbrass and Jordan (2001, p. 1) argue that their study casts doubt on "the explanatory power of intergovernmental accounts of EU environmental policy making", citing Moravcsik's (1998) liberal intergovernmentalism as a leading example, which accounts are often juxtaposed as rivals to neo-functionalist/ supranational governance theories of EU integration.

There would therefore appear to be something of a consensus around the idea that the Member States in some sense lost control of EU nature policy (if they ever had control of it in the first place), and that the policy has had unintended consequences from the (or at least from certain) Member States' perspective. This book tests this consensus, using historical institutionalism's long-term perspective.

What seems potentially missing from Fairbrass and Jordan's (2001) historical institutionalist analysis of EU biodiversity policy, in light of the history detailed in this book, is a consideration of counter-forces to EU integration and "surprising puzzles" in biodiversity policy (Alter, 2009, p. 16). Further, Fairbrass and Jordan's (2001) account can no longer be regarded as covering the full policy cycle of EU biodiversity policy, given that many years have passed since their paper was published, and fundamentally their analysis is limited to the UK. This book seeks to address these points, presenting an historical institutionalist analysis of EU nature policy that goes beyond the limits of Fairbrass and Jordan's (2001) account, both temporally and geographically.

The future of EU nature policy

In recent years certain Member States expressed a wish to renegotiate the Habitats and Birds Directives, implicitly or explicitly to weaken the protections that apply in respect of Natura 2000 sites. For example, in 2009 Jan Peter Balkenende, then Prime Minister of the Netherlands, wrote to José Manuel Barroso, then President of the European Commission, asking the Commission to "draft proposals to bring [the Habitats and Birds Directives] up to date", on the basis that "Natura 2000 fails to strike a balance between ecological value, economic interests and other uses" (Balkenende, 2009).[17] Balkenende framed his argument around the need to update these "relatively old directives" to "take account of the increasingly topical need to strike a balance between ecological interests and economic and human interests, and to address the impact of climate change on our living environment". This reference to climate change is interesting because it arguably serves to muddy the waters: i.e., this could simply be an opportunistic argument aimed at avoiding the Dutch government's stance being characterised as anti-environment.[18] Similar arguments were made in the UK in 2007, when the report of a consultation exercise recorded that the UK's Sustainable Development Commission had "heard from a number of stakeholders who believe that the [Habitats and Birds] Directives no longer reflect modern conservation priorities in light of the challenges presented by climate change" (SDC, 2007, pp. 130–131). Again, there is at least the possibility that climate change and the related need to promote renewable energy projects and grid development[19] could be used as convenient excuses to seek a weakening across the board of the protections applicable to Natura 2000 sites (Jackson, 2011).

More recently, in 2011 the UK's then Chancellor George Osborne said "we will make sure that gold plating of EU rules on things like Habitats aren't placing ridiculous costs on British businesses . . . If we burden them with endless social and environmental goals – however worthy in their own right – then not only will we not achieve those goals, but the businesses will fail, jobs will be lost, and our country will be poorer" (Osborne, 2011).[20] This led Sir John Lawton, then Vice-President of the Royal Society for the Protection of Birds (RSPB), to call Osborne a "bloody idiot" on wildlife protection (Jones, 2012b). Later still, David Cameron, then Prime Minister, stated that "we [the UK] need to examine whether the balance is right in so many areas where the European Union has legislated including on the environment . . . Nothing should be off the table" (Cameron, 2013).

Following national interventions such as these, the European Commission announced in 2013 that Natura 2000 (and hence the Habitats and Birds Directives) was to be the subject of a legislative "Fitness Check" as part of the European Commission's REFIT "Fit for Growth" programme to "prepare for future legislative initiatives to reduce regulatory burden" (European Commission, 2013, p. 7). When Jean-Claude Juncker took over from Barroso as Commission President in 2014, he effectively pre-empted the outcome of the Fitness Check by mandating the new EU Environment Commissioner Vella to "carry out an in-depth evaluation of the Birds and Habitats directives and assess the potential for merging them into a more modern piece of legislation" (Juncker, 2014, p. 4). Many environmentalists feared that "more modern" was a euphemism for deregulation and weakening, and there then began a high profile campaign by environmentalists to save the directives. The subsequent history is discussed in Chapter 10.

Whilst "crystal ball gazing in the social sciences is always risky" (Peterson, 2015, p. 202), the argument here will be that a knowledge of the detailed history of EU nature policy allows us to understand the past and current trajectories and themes of that policy, as well as the institutional and other factors that frame and channel the policy in particular directions. In this way, a knowledge of the history arguably informs our understanding of what may be key themes in the future of the policy.

Notes

1 With a view to ensuring historical accuracy insofar as possible, interviews were triangulated with one another and with available archival materials. Archival materials were sourced principally from the European Commission, the Council of the EU, the European Parliament, the Council of Europe, the University of Pittsburgh's online Archive of European Integration, and the EUI's Historical Archives of the European Union. Where a conflict arose, the documentary archive tended to be preferred on the basis of its contemporaneity. Interviewees and correspondents (numbering 36 in total) came from the following groups (at the time the Birds and Habitats Directives were negotiated): EU institutions (Commission and European Parliament); EU Member State governments; NGO representatives; and the Council of Europe. The present author reads English and French, which created a natural limit in terms of documentary sources. However, the overwhelming majority of documents in the relevant archives were in either English or French. Contemporaneous media articles were sourced principally from what is now the Nexis database, which provided extensive access to English language news but relatively scant access to other language sources. Notably, however, the relevant archive of the European Commission contained a large proportion of materials from or relating to the UK. That is, during the Habitats Directive negotiations members of the European Commission's negotiating team collected, received and retained news articles and other relevant documents from the national level (e.g., reports of parliamentary hearings, etc.), and these were largely from the UK. This may have been because several of the individuals closely involved with the pre-adoption history of the Habitats Directive (within and outside the Commission) were British (see Chapters 6–9), as well as the fact that the UK NGOs were particularly active during the Directive's passage, and a British individual headed up the main EU-wide NGO campaign in support of the Directive proposal. Despite best efforts, it was not possible to secure interviewees/correspondents who participated in the negotiations regarding the Birds and Habitats

Directives from every relevant Member State administration, but this was offset to some extent by archival materials from the Council and Commission, which revealed each Member State's stated views.

2 The central objective of the Habitats Directive (1992, Article 2(2)) is to maintain or restore, at a "favourable conservation status", wild fauna and flora of EU interest. On this subject, see Mehtälä and Vuorisalo (2007).

3 The terrestrial part of the Natura 2000 network is largely complete but there are still "important gaps" for the marine environment (European Commission, 2016a, p. 33).

4 The EEA (2015b, p. 8) states that about 4% of the EU's seas are within Natura 2000.

5 E.g., see the literature cited on pp. 37–38 and 84–85 of European Commission (2016a).

6 That is, legal cases brought by the European Commission against Member States for alleged breaches of EU environmental law.

7 Jones (2012a) touches on the early history of the Habitats Directive.

8 Van den Belt (2008, p. 228) offers an interesting counter-point to such views, citing the case of France: "One might hold that science . . . has no monopoly on defining which sites, habitats and species are worth protecting. Dr. Pinton suggests that local stakeholders questioned the credibility of the science used in making the preliminary inventories of the nature heritage in France and that they advanced the claim that their own 'local knowledge' (or practical knowledge) was more adequate. But how credible are such claims made by local stakeholders, given that they are obviously raised for opportunistic reasons in the context of a fierce power struggle?"

9 Plus 2.6% within the SPA network under the Birds Directive, bringing the total to 10.2% at the end of 2002 (Pinton, 2008). This has since increased to bring 12.84% of France's land territory within the Natura 2000 network (close to the original figure of 13%, which was strongly resisted) plus large marine areas (European Commission, 2017).

10 For more on the situation in Germany, see Stoll-Kleemann (2001) and Winter (2005).

11 For example, over the period 2014–2020, just 0.32% of the EU's Multiannual Financial Framework spending will be dedicated to nature conservation and other environmental aims via the EU's LIFE funding instrument, while 38% will go to the CAP (European Parliament, 2015, 2016).

12 Priority habitats/species are the subset of EU protected habitats/species that are given the highest priority in view of their danger of disappearance or because of the particular responsibility the EU bears for their conservation. In respect of such priority habitats/species, public interest arguments are limited, under the Article 6(4) derogation, to reasons of "human health or public safety, to beneficial consequences of primary importance for the environment or, further to an opinion from the Commission, to other imperative reasons of overriding public interest".

13 Holder (1997, p. 1475) describes the *Lappel Bank* case (CJEU, 1996), which concerned site selection rather than site protection, as taking place "at the seam between economics and environmental protection".

14 Bickerton et al. (2015, pp. 12–13) cite several potential reasons: a pragmatic approach seeking to investigate parts of the EU rather than its nature as a whole; a reluctance to explore the 'methods of integration' debate to avoid challenging the field's "broad normative commitment to further EU integration"; and a view that "these 'ontological' debates about the 'nature of the beast' have all been answered".

15 As evidenced, for example, by the UK government's (2012) 'Balance of Competences' review, part of then Prime Minister David Cameron's strategy to renegotiate the UK's relationship with the EU before the Brexit referendum (Cameron, 2013).

16 Also see Fairbrass (2000).

17 Barroso (2009) provided a detailed response, which began: "Natura 2000 does not threaten the balance between economic and natural development".

18 Nevertheless, Dutch MP Diederik Samsom called the letter an "intervention to undermine nature protection" (Dutch News, 2010).

19 See RGI (2011).

20 On the subject of so-called gold-plating, see Etherington (2006).

References

Alphandéry, P. and Fortier, A. 2001. Can a territorial policy be based on science alone? The system for creating the Natura 2000 network in France. *Sociologia Ruralis*, 41, 311–328.

Alter, K. J. 2009. *The European Court's political power*. Oxford: Oxford University Press.

Apostolopoulou, E. and Pantis, J. D. 2009. Conceptual gaps in the national strategy for the implementation of the European Natura 2000 conservation policy in Greece. *Biological Conservation*, 142, 221–237.

Bache, I., George, S. and Bulmer, S. 2011. *Politics in the European Union*. Oxford: Oxford University Press.

Baldock, D. 1992. The status of Special Protection Areas for the protection of wild birds. *Journal of Environmental Law*, 4, 139–144.

Balkenende, J. P. 2009. Letter dated 13 July 2009 from the then Prime Minister of the Netherlands to J. M. Barroso, President of the European Commission, attaching an annex entitled 'Dutch position paper on Natura 2000'.

Barroso, J. M. 2009. Letter dated 26 October 2009 from Commission President Barroso to Jan Peter Balkenende, Prime Minister of the Netherlands, attaching a response to the 'Dutch position paper on Natura 2000'.

Beunen, R. 2006. European nature conservation legislation and spatial planning: for better or for worse? *Journal of Environmental Planning and Management*, 49, 605–619.

Beunen, R. 2010. The governance of nature: how nature conservation ambitions have been dashed in planning practices. PhD, Wageningen University.

Beunen, R. and De Vries, J. R. 2011. The governance of Natura 2000 sites: the importance of initial choices in the organisation of planning processes. *Journal of Environmental Planning and Management*, 54, 1041–1059.

Beunen, R., Van Assche, K. and Duineveld, M. 2013. Performing failure in conservation policy: the implementation of European Union directives in the Netherlands. *Land Use Policy*, 31, 280–288.

Beunen, R., Van Der Knaap, W. G. M. and Biesbroek, G. R. 2009. Implementation and integration of EU environmental directives. Experiences from the Netherlands. *Environmental Policy and Governance*, 19, 57–69.

Bickerton, C., Hodson, D. and Puetter, U. 2015. The new intergovernmentalism and the study of European integration. *In:* Bickerton, C., Hodson, D. and Puetter, U. (eds) *The new intergovernmentalism: states and supranational actors in the post-Maastricht era*. Oxford: Oxford University Press.

Birds Directive 1979. Council Directive 79/409/EEC of 2 April 1979 on the conservation of wild birds, *OJ L 103*, 25 April 1979, p. 1, since codified as Directive 2009/147/EC of the European Parliament and of the Council of 30 November 2009 on the conservation of wild birds. *OJ L 20*, 26 January 2010, p. 7.

Björkell, S. 2008. Resistance to top-down conservation policy and the search for new participatory models. *In:* Keulartz, J. and Leistra, G. (eds) *Legitimacy in European nature conservation policy: case studies in multilevel governance*. The Netherlands: Springer.

Bogaert, D. and Leroy, P. 2008. Endangered legitimacy: Nature policy in Flanders – the sloppy management of participation. *In:* Keulartz, J. and Leistra, G. (eds) *Legitimacy in European nature conservation policy: case studies in multilevel governance*. The Netherlands: Springer.

Bouwma, I. M., Kamphorst, D. A., Beunen, R. and Van Apeldoorn, R. C. 2008. Natura 2000 benchmark, a comparative analysis of the discussion on Natura 2000 management issues. *WOTNM-Rapport*. Wageningen: Wettelijke Onderzoekstaken Natuur and Milieu.

Bulmer, S. 2015. Understanding the new intergovernmentalism: pre- and post-Maastricht EU studies. *In:* Bickerton, C., Hodson, D. and Puetter, U. (eds) *The new intergovernmentalism: states and supranational actors in the post-Maastricht era.* Oxford: Oxford University Press.

Cameron, D. 2013. David Cameron's EU speech – full text. *The Guardian*, 23 January 2013.

Chmielewski, W. and Glogowska, M. 2015. Implementation of the Natura 2000 Network in Poland – an Opportunity or a Threat to Sustainable Development of Rural Areas? Study on Local Stakeholders' Perception. *Eastern European Countryside*, 21, 153–169.

CJEU 1996. Judgment of the Court of Justice of the EU of 11 July 1996, Lappel Bank, C-44/95, EU:C:1996:297.

CJEU 2000. Judgment of the Court of Justice of the EU of 7 November 2000, First Corporate Shipping, C-371/98, EU:C:2000:600.

Coffey, C. and Shaw, K. 2001. The Habitats Directive 92/43: an inventory of interactions with other institutions. London: Institute for European Environmental Policy.

Demonty, B., Dubuisson, M. and Kuczkiewicz, J. 2017. Donald Trump, comme vous ne l'avez pas vu. *Le Soir*, 25 May 2017.

Dutch News. 2010. MPs ask PM to explain Natura 2000 claims. *Dutch News*, 11 January 2010.

EEA 2015a. The European environment – status and outlook 2015: synthesis report. European Environment Agency.

EEA 2015b. State of nature in the EU: results from reporting under the nature directives 2007–2012. European Environment Agency. EEA Technical report No 2/2015.

Engelen, E., Keulartz, J. and Leistra, G. 2008. European nature conservation policy making: from substantive to procedural sources of legitimacy. *In:* Keulartz, J. and Leistra, G. (eds) *Legitimacy in European nature conservation policy: case studies in multilevel governance.* The Netherlands: Springer.

Etherington, L. 2006. Digging beneath the surface: transposition, implementation and evaluation of European environmental law. *Journal of Environmental Policy and Planning*, 8, 107–133.

European Commission 1987. The Single Act: a new frontier for Europe. Communication from the Commission (COM(87) 100) to the Council. *Bulletin of the European Communities, Supplement 1/87.* Luxembourg: Office for Official Publications of the European Communities.

European Commission 2007. Guidance document on the strict protection of animal species of Community interest under the Habitats Directive 92/43/EEC. Final version, February 2007.

European Commission 2009. Beyond 2010 – Options for EU biodiversity policy. MEMO/09/177, 23 April 2009.

European Commission 2010a. Guidance document: wind energy developments and Natura 2000. October 2010.

European Commission 2010b. Options for an EU vision and target for biodiversity beyond 2010. 19 January 2010.

European Commission 2011. Our life insurance, our natural capital: an EU biodiversity strategy to 2020 (3 May 2011). COM(2011) 244 final.

European Commission 2013. Communication from the Commission to the European Parliament, the Council, the European Economic and Social Committee and the Committee of the Regions. Regulatory Fitness and Performance (REFIT): Results and Next Steps. Brussels, 2 October 2013. COM(2013) 685 final.

European Commission 2016a. Commission Staff Working Document. Fitness Check of the EU Nature Legislation (Birds and Habitats Directives). Brussels, 16 December 2016, SWD(2016) 472 final.

European Commission 2016b. *Statistics on environmental infringements* [Online]. Available: http://ec.europa.eu/environment/legal/law/statistics.htm [accessed 16 March 2017].
European Commission 2017. Natura 2000 nature and biodiversity newsletter. Number 41, February 2017.
European Council 2001. Presidency conclusions of the Göteborg European Council, 15–16 June 2001 (SN 200/1/01 REV 1).
European Parliament 2015. *Briefing: How the EU budget is spent: LIFE programme. February 2015* [Online]. Available: www.europarl.europa.eu/RegData/etudes/BRIE/2015/548992/EPRS_BRI(2015)548992_EN.pdf [accessed 28 March 2017].
European Parliament 2016. *Briefing: How the EU budget is spent: Common Agricultural Policy – Pillar I. July 2016* [Online]. Available: www.europarl.europa.eu/RegData/etudes/BRIE/2016/586622/EPRS_BRI(2016)586622_EN.pdf [accessed 28 March 2017].
Fairbrass, J. 2000. EU and British biodiversity policy: ambiguity and errors of judgement. *CSERGE* Working Paper GEC 2000-4: Centre for Social and Economic Research on the Global Environment, University of East Anglia and University College London.
Fairbrass, J. and Jordan, A. 2001. European Union environmental policy and the UK government: a passive observer or a strategic manager? *Environmental Politics*, 10, 1–21.
Ferranti, F., Beunen, R. and Speranza, M. 2010. Natura 2000 network: a comparison of the Italian and Dutch implementation experiences. *Journal of Environmental Policy and Planning*, 12, 293–314.
Gammell, A. 1999. The contribution of BirdLife International to the Birds Directive and the management of birds and their habitats in Europe. 20 years with the EC Birds Directive. Proceedings from a conference on the Council's Directive on the Conservation of Wild Birds, 18–19 November 1999 Elsinore, Denmark. The Danish National Forest and Nature Agency.
Habitats Directive 1992. Consolidated version of Council Directive 92/43/EEC of 21 May 1992 on the conservation of natural habitats and of wild fauna and flora. *OJ L 206*, 22 July 1992, p. 7. Consolidation date 1 July 2013.
Hiedanpää, J. 2002. European-wide conservation versus local well-being: the reception of the Natura 2000 reserve network in Karvia, SW-Finland. *Landscape and Urban Planning*, 61, 113–123.
Holder, J. 1997. Case C-44/95, R. v. Secretary of State for the Environment, ex parte Royal Society for the Protection of Birds, judgment of the Court of Justice 11 July 1996, [1996] ECR 1-3805. *Common Market Law Review*, 34, 1469–1480.
Jackson, A. L. R. 2011. Renewable energy vs. biodiversity: policy conflicts and the future of nature conservation. *Global Environmental Change*, 21, 1195–1208.
Jones, G. 2012a. The Bern Convention and the origins of the Habitats Directive. *In:* Jones, G. (ed.) *The Habitats Directive: a developer's obstacle course?* Oxford: Hart Publishing.
Jones, G. 2012b. George Osborne a 'bloody idiot' on wildlife protection. *The Guardian*, 18 June 2012.
Jordan, A. 1998. Step change or stasis? EC environmental policy after the Amsterdam treaty. *Environmental Politics*, 7, 227–236.
Jordan, A., Brouwer, R. and Noble, E. 1999. Innovative and responsive? A longitudinal analysis of the speed of EU environmental policy making, 1967–97. *Journal of European Public Policy*, 6, 376–398.
Juncker, J.-C. 2014. Mission letter dated 1 November 2014 from Jean-Claude Juncker, President of the European Commission, to Karmenu Vella, Commissioner for Environment, Maritime Affairs and Fisheries.

Kati, V., Hovardas, T., Dieterich, M., Ibisch, P. L., Mihok, B. and Selva, N. 2015. The challenge of implementing the European network of protected areas Natura 2000. *Conservation Biology*, 29, 260–270.

Krämer, L. 1993. The interdependency of Community and Member State activity on nature protection within the European Community. *Ecology Law Quarterly*, 20, 25–45.

Laffan, B. 2004. Birds, bears and bogs: EU nature conservation policy in six states. *OEUE Phase II: Occasional paper – 0.3 – 08.04*. Dublin European Institute, University College Dublin.

Laffan, B. and O'Mahony, J. 2008. 'Bringing politics back in'. Domestic conflict and the negotiated implementation of EU nature conservation legislation in Ireland. *Journal of Environmental Policy and Planning*, 10, 175–197.

Ledoux, L., Crooks, S., Jordan, A. and Kerry Turner, R. 2000. Implementing EU biodiversity policy: UK experiences. *Land Use Policy*, 17, 257–268.

Leibenath, M. 2008. Legitimacy of biodiversity policies in a multi-level setting: the case of Germany. *In:* Keulartz, J. and Leistra, G. (eds) *Legitimacy in European nature conservation policy: case studies in multilevel governance*. The Netherlands: Springer.

Mehtälä, J. and Vuorisalo, T. 2007. Conservation policy and the EU Habitats Directive: favourable conservation status as a measure of conservation success. *European Environment*, 17, 363–375.

Meyer, J. H. 2010. Saving migrants: a transnational network supporting supranational bird protection policy in the 1970s. *In:* Kaiser, W., Gehler, M. and Leucht, B. (eds) *Transnational networks in regional integration: informal governance in Europe 1945–83*. Basingstoke: Palgrave Macmillan.

Meyer, J. H. 2011. Green activism: the European Parliament's Environmental Committee promoting a European environmental policy in the 1970s. *Journal of European Integration History*, 17, 73–85.

Meyer, J. H. 2014. Getting started: agenda-setting in European environmental policy in the 1970s. *In:* Laursen, J. (ed.) *From crisis to new dynamics: the European Community 1974–83*. Brussels: Bruylant.

Moravcsik, A. 1998. *The choice for Europe: social purpose and state power from Messina to Maastricht*. Ithaca, NY: Cornell University Press.

Morris, R. K. A. 2011. The application of the Habitats Directive in the UK: compliance or gold plating? *Land Use Policy*, 28, 361–369.

Osborne, G. 2011. *Autumn forecast statement by the Chancellor of the Exchequer, Rt Hon George Osborne MP, 29 November 2011* [Online]. Available: www.gov.uk/government/speeches/autumn-forecast-statement-by-the-chancellor-of-the-exchequer-rt-hon-george-osborne-mp [accessed 28 March 2017].

Paavola, J. 2004. Protected areas governance and justice: theory and the European Union's Habitats Directive. *Environmental Sciences*, 1, 59–77.

Peterson, J. 2015. The Commission and the new intergovernmentalism: calm within the storm? *In:* Bickerton, C., Hodson, D. and Puetter, U. (eds) *The new intergovernmentalism: states and supranational actors in the post-Maastricht era*. Oxford: Oxford University Press.

Pierson, P. 1996. The path to European integration. *Comparative Political Studies*, 29, 123–163.

Pinton, F. 2001. Conservation of biodiversity as a European directive: the challenge for France. *Sociologia Ruralis*, 41, 329–342.

Pinton, F. 2008. Between European injunction and local consultation: analysing the territorialisation process for a public nature conservation initiative in France. *In:* Keulartz, J. and Leistra, G. (eds) *Legitimacy in European nature conservation policy: case studies in multilevel governance*. The Netherlands: Springer.

Rauschmayer, F., Van Den Hove, S. and Koetz, T. 2009. Participation in EU biodiversity governance: how far beyond rhetoric? *Environment and Planning C: Government and Policy*, 27, 42–58.

RGI 2011. European grid declaration on electricity network development and nature conservation in Europe. Opened for signature 10 November 2011. Renewables Grid Initiative.

Sandholtz, W. and Stone Sweet, A. 2012. Neo-functionalism and supranational governance. *In:* Jones, E., Menon, A. and Weatherill, S. (eds) *The Oxford handbook of the European Union*. Oxford: Oxford University Press.

SDC 2007. Turning the tide: Tidal power in the UK. Sustainable Development Commission, London, October 2007.

Shoard, M. 1990. Too precious, too Scottish to be left with the Scots. *The Times*, 30 June 1990.

Stoll-Kleemann, S. 2001. Opposition to the designation of protected areas in Germany. *Journal of Environmental Planning and Management*, 44, 109–128.

Stone Sweet, A. 2004. *The judicial construction of Europe*. Oxford: Oxford University Press.

Stuffmann, C. 1994. The European Community and EECONET. *In:* Bennett, G. (ed.) *Conserving Europe's natural heritage: towards a European ecological network*. London: Graham and Trotman.

UK Government 2012. Review of the balance of competences between the United Kingdom and the European Union. Presented to Parliament by the Secretary of State for Foreign and Commonwealth Affairs by command of Her Majesty. July 2012. London: The Stationery Office Limited.

UNCBD 2002. COP 6 Decision VI/26 (7–19 April 2002).

UNCBD 2010. Annex to COP 10 Decision X/2. Strategic plan for biodiversity 2011–2020 (18–29 October 2010).

Van Den Belt, H. 2008. The local implementation of nature policy: deliberative democracy or interest politics? *In:* Keulartz, J. and Leistra, G. (eds) *Legitimacy in European nature conservation policy: case studies in multilevel governance*. The Netherlands: Springer.

Van Hoorick, G. 1997. *Internationaal en Europees natuurbehoudsrecht: de verdragen, de Europese verordeningen en richtlijnen en hun tenuitvoerlegging in België en Nederland*. Antwerpen: Intersentia.

Winter, G. 2005. Conflicts between ownership rights and the implementation of the "Natura 2000" Network. *In:* Pâques, M. (ed.) *Le Droit de propriété et Natura 2000*. Bruxelles: Bruylant.

Witte, G. 2017. Trump tried and failed to build a wall in Ireland. That could mean big trouble for Europe. *The Washington Post*, 6 February 2017.

WWF 2006. Natura 2000 in Europe: an NGO assessment. Budapest.

2

WHY HISTORY MATTERS

> [W]hat one makes of the EC depends on whether one examines a photograph or a moving picture. Just as a film often reveals meanings that cannot be discerned from a single photograph, a view of Europe's development over time gives us a richer sense of the nature of the emerging European polity.
>
> *(Pierson, 1996, p. 128)*

Historical institutionalism: accounting for EU nature policy

Introduction

This book applies an historical institutionalist analysis to EU nature policy and law. Historical institutionalism is a research approach with deep roots,[1] whose leading exponent in the EU integration context is Paul Pierson.[2] When Pierson (1996) published "The path to European integration: a historical institutionalist analysis", the dominant paradigm in EU integration theory was Moravcsik's (1991, 1993) liberal intergovernmentalism. Pierson (1996, p. 123) set out to test the arguments of such intergovernmentalist accounts by constructing a framework for a long-term historical analysis, summarising his argument thus:

> Observers of the European Community have criticized "intergovernmentalist" accounts for exaggerating the extent of member-state control over European integration. This article grounds these criticisms in a historical institutionalist analysis, stressing the need to study European integration as a process that unfolds over time. Losses of control result not only from the autonomous actions of supranational organizations, but from member-state preoccupation with short-term concerns, the ubiquity of unintended consequences,

> and the instability of member-state policy preferences. Once gaps in control emerge, change-resistant decision rules and sunk costs associated with societal adaptations make it difficult for member states to reassert their authority.

Pierson (1996) contrasts historical institutionalism's long-term approach with that of liberal intergovernmentalists such as Moravcsik, who arguably provide a snapshot account focused on the grand bargains that punctuate EU history (i.e., treaty-making and treaty amendment), overlooking unintended consequences of such bargains and of everyday policymaking between grand bargains.

An historical institutionalist approach, Pierson (1996, p. 127) explains, "is *historical* because it recognizes that political development must be understood as a process that unfolds over time. It is *institutionalist* because it stresses that many of the contemporary implications of these temporal processes are embedded in institutions—whether these be formal rules, policy structures, or norms". Pierson's (1996, p. 126) aim, essentially, was to provide an account of Member State constraint at EU level, by focusing on "why gaps emerge in member-state control over the evolution of European institutions and public policies, why these gaps are difficult to close, and how these openings create room for actors other than member states to influence the process of European integration while constraining the room for maneuver of all political actors".

Factors leading to gaps in Member State control

In framing his argument, Pierson (1996) focused first on four factors that, he argued, were likely to lead to gaps in Member State control over EU policies: first, he highlighted the importance of *the autonomous actions of European institutional actors*, arguing that EU institutions such as the Commission, the CJEU and the European Parliament are "always looking for opportunities to enhance their powers . . . The Council, to be sure, continues to stand watch over proposed legislation and actively protects member-state interests. Yet the Commission, Parliament and Court possess considerable ability to advance their own interests" (Pierson, 1996, p. 133). Second, he emphasised *the restricted time horizons of political decision makers*, arguing, in essence, that given the nature of electoral politics, policy-makers are frequently most interested in the short-term consequences of their actions, meaning that long-term considerations weigh less heavily in decisions. Third, Pierson highlighted *the large potential for unintended consequences* of policy, meaning that, even if long-term factors were taken into consideration, "unintended consequences are likely to be widespread. Complex social processes involving a large number of actors always generate elaborate feedback loops and significant interaction effects that decision makers cannot hope to fully comprehend" (Pierson, 1996, p. 136). Finally, Pierson (1996, p. 140) noted the importance of *shifts in Chiefs of Government (COG) policy preferences* over time, owing, for example, to changes in government, etc.: "The result, over time, is that evolving arrangements will diverge from the intentions of original designers, while any newly arriving COG is likely to find institutional and policy arrangements considerably out of synch with its own preferences". Having identified these four sources of gaps in Member State control, Pierson (1996, p. 140) emphasises that most of these processes have a temporal quality: "The role of restricted time horizons, unintended consequences, and shifting member-state preferences will only be evident if we examine political processes over time".

Why gaps in Member State control prove hard to close

The final part of Pierson's (1996) analysis involved tackling an obvious counterpoint: where gaps in Member State control over EU policy or unintended consequences of such policy become clear, why don't the Member States simply take corrective action? The answer, Pierson (1996) argues, is that gaps, even once identified, can prove hard to close, for three broad reasons: first, *the resistance of EU institutional actors*, which are, Pierson argues, likely to resist Member States' attempts to exercise greater control over their activities. Second, there are *institutional obstacles to reform within the EU*; e.g., Member States will often share different views on important policy issues, and the requirement for unanimity or even qualified majority to make changes "takes time, resources and political capital" (Palmieri, 1997, p. 8). And third, there is the issue of *path dependence, and the sunk costs* associated with previous policy decisions; that is, initial policy decisions can become self-reinforcing: "When actors adapt to the new rules of the game by making extensive commitments based on the expectation that these rules will continue, previous decisions may lock in member states to policy options that they would not now choose to initiate" (Pierson, 1996, p. 144). If such barriers are sufficiently high, argues Pierson (1996, p. 142), "learning will not provide a sufficient basis for correction, and member-state control will be constrained".

Testing accounts of EU integration in the nature conservation field

Pierson (1996) thus focused his analysis on critiquing intergovernmentalist accounts of EU integration, such as Moravcsik's (1991, 1993) liberal intergovernmentalism, which, in its strong version, "denied the importance of supranational actors as independent actors in EU decision making and insisted that governments remained in control of the process of European integration" (Bache et al., 2011, p. 18). Moravcsik (1998), however, left behind the strong version of his theory, argue Sandholtz and Stone Sweet (2012b, p. 31), for a "far weaker" version, in order to address the issue of unintended consequences. Moravcsik (1998, p. 18) puts his own central argument thus: "European integration can best be explained as a series of rational choices made by national leaders. These choices responded to constraints and opportunities stemming from the economic interests of powerful domestic constituents, the relative power of each state in the international system, and the role of international institutions in bolstering the credibility of interstate commitments".

Moravcsik's "only way out" regarding the existence of unintended consequences, per Sandholtz and Stone Sweet (2012b, p. 31), "was to argue that the Member States sometimes delegate substantial independent authority to EU bodies (namely, the [CJEU]) so that the Court could enforce incomplete contracts, thereby enhancing the credibility of national commitments to common policies as well as the effectiveness of those policies", an argument criticised by Sandholtz and Stone Sweet (2012a) on the grounds that it is non-falsifiable, amongst other things.[3] For Moravcsik's part, Moravcsik and Schimmelfennig (2009, p. 75) argue that liberal intergovernmentalism assumes the existence of "unforeseen or initially undesired policies" and may be able to explain them insofar as such policy changes result from interstate decisions or tacit intergovernmental consent, even if other actors (e.g., the CJEU) are involved in the elaboration of such decisions. The key question, per Moravcsik and Schimmelfennig (2009, p. 75), is whether the uncertainty in the initial bargain "is so great as to divert integration fundamentally from its course, as [historical institutionalist] theory suggests". This is an empirical question, with the evidence tending to support the liberal intergovernmentalist account, claim Moravcsik and Schimmelfennig (2009).

Neo-functionalism and supranational governance

In his original paper on historical institutionalism and EU integration, Pierson (1996, p. 125) criticised neo-functionalism on the basis that it appears to attribute greater autonomy to supranational actors than can be sustained, in particular given the strong institutional position of Member States in the EU. However, Sandholtz and Stone Sweet (2012a, p. 21) have since refined neo-functionalist theory,[4] which, on their account, posits a basic feedback loop: "(a) increasing cross-border transactions activates (b) supranational governance (dispute resolution and rule-making), which facilitate (c) a subsequent expansion of cross-border transactions, which translates into greater social demand for new forms of supranational governance (spillover)". This concept of spillover explains how a theory that is based on cross-border transactions can potentially apply to a conceptually distant area such as nature conservation policy:

> With the removal of tariffs and quotas, for example, differences in national regulatory standards – for the environment, health and safety, technical compatibility, and so on – become more apparent as obstacles to exchange. Economic actors seeking to benefit from intra-Community exchange will then target these obstacles, both by attacking regulatory barriers through litigation and by pressuring EC legislative institutions to widen the jurisdiction of the EC into new domains. Transactors will always prefer, other things being equal, to live under (or adapt to) one set of rules rather than six, or twelve, or fifteen. As governments and EC institutions respond, spillover occurs.
>
> *(Stone Sweet and Sandholtz, 1998, p. 15)*

This makes an empirical claim about the mechanism of spillover, which may or may not hold up in the context of nature conservation (arguably not, in light of the history detailed in this book).

However, Stone Sweet and Sandholtz (1998) note that this is not the only mechanism driving increased supranational governance, citing additional explanatory factors such as globalisation, multilateral approaches to global problems, and pressure from transnational actors, all of which can exert pressure directly on EU institutions such as the European Commission. Moreover, the Commission and other EU institutions can act as policy entrepreneurs, with or without outside pressure: "The Commission regularly creates – and sometimes funds – roundtables, working groups, and committees composed of firms and other non-state actors. The working groups help the Commission design programs and draft legislation; these same groups then become advocates of the Commission's proposals vis-à-vis national governments and other actors" (Sandholtz and Stone Sweet, 2012a, p. 23). As this book shows, such explanatory factors are a much better fit than a transactions-based explanation in the context of nature conservation policy.

It is important to recognise that Stone Sweet and Sandholtz (1998, pp. 8–9) do not seek to characterise EU integration as uniformly supranational in nature. Instead, they argue that integration is uneven, meaning that different EU policy sectors (and the EU as a whole) may be placed at different places on a continuum from intergovernmental politics at one pole to supranational politics at the other. This continuum, they argue, measures movement from intergovernmental to supranational governance across three interrelated dimensions: EU rules; EU organisations ("those governmental structures, operating at the European level, that produce, execute, and interpret EC rules"); and transnational society ("those non-governmental actors who engage in intra-EC exchanges – social, economic, political – and thereby influence, directly or indirectly, policymaking processes and outcomes at the European level"). Movement along the continuum in respect of any given policy area will measure the "growing presence and intensity of each of these factors" (Stone Sweet and Sandholtz, 1998, p. 9).

Sandholtz and Stone Sweet's (2012a, p. 25) theory generates a number of testable hypotheses, including – of particular relevance to this book – the following: "EU organizations like the Commission and Court of Justice will routinely produce supranational outcomes that the Member States would not have produced on their own"; and "Supranational governance will routinely produce outcomes that conflict with the revealed preferences of the most powerful states".

A focus on testing these hypotheses is justified not least by Sandholtz and Stone Sweet's (2012a, p. 28) bold claim that neo-functionalism is "the most theoretically viable and empirically productive general theory of European integration".

New intergovernmentalism

New intergovernmentalism is not "a new grand theory of regional integration", (Bickerton et al., 2015a, p. 45) since it makes claims only regarding integration after the Maastricht Treaty in 1992. Since this time, claim Bickerton et al. (2015a), EU integration has been intergovernmental in character; and this is not a temporary deviation from supranationalism but rather is a distinct phase – a new normal – which has become entrenched.

In essence, the new intergovernmentalists argue that since Maastricht there has been a huge amount of EU integration but that this has happened in the absence of delegation to the 'old' supranational EU institutions of Commission and CJEU. "Under the traditional Community method", explain Bickerton et al. (2015b, p. vii),

> the European Commission is involved at key stages of policy initiation, implementation, and monitoring and it falls to the [CJEU] to uphold the rule of law. Since 1992, national governments have tended to sidestep such delegation in favour of deliberation in settings such as the European Council, the Eurogroup, and the Euro Summit. Delegation, where it has taken place, has tended to involve the creation and empowerment of *de novo* bodies [e.g., the European Central Bank, the European External Action Service (EEAS), Frontex, and the European Stability Mechanism].

Per Bickerton et al. (2015a), existing theoretical approaches to EU integration find it difficult to explain this "paradoxical tendency" towards integration without supranational policy-making post-1992. Thus, according to the new intergovernmentalists, the supranational governance school of Sandholtz and Stone Sweet (2012a) struggles to make sense of the post-Maastricht period because their approach is focused primarily on policy areas in which the EU has placed policy-making powers in the hands of the traditional supranational institutions (Commission, CJEU). Equally, the new intergovernmentalists criticise Moravcsik's liberal intergovernmentalism on the basis that its characterisation of the EU tending towards a 'stable institutional equilibrium' after Maastricht "is difficult to square with the significant increase in the scope of EU activity since 1992 and the shifts in the relative roles of core EU institutions within these processes" (Bickerton et al., 2015a, p. 17). More bluntly, Auer (2016, p. 477) argues that Moravcsik's accounts of the post-1992 period "are no longer credible".[5]

It is important to emphasise that the new intergovernmentalists do not view intergovernmentalism as a synonym for a rollback or cessation of integration. Quite the contrary, in fact. Instead, integration has continued apace, per the new intergovernmentalists, but without delegation to traditional supranational EU bodies, and in spite of some public scepticism.[6]

The new intergovernmentalists have put forward six hypotheses that seek to "elucidate, explain and understand" the implications of the "integration paradox" which characterises their theory (Bickerton et al., 2015a, p. 29):

> Hypothesis 1: Deliberation and consensus have become the guiding norms of day-to-day decision-making at all levels [in the EU].
>
> Hypothesis 2: Supranational institutions are not hard-wired to seek ever-closer union.

Hypothesis 3: Where delegation occurs, governments and traditional supranational actors support the creation and empowerment of de novo institutions [instead of delegating power to the Commission or the CJEU, for example].

Hypothesis 4: Problems in domestic preference formation have become standalone inputs into the European integration process.

As Bickerton et al. (2015a, pp. 33–34) explain,

> Preference formation in Europe in the post-Maastricht period has . . . become inseparable from a growing frustration with procedural issues and scepticism towards the idea of majoritarian-based representative democracy. How interests are identified and represented has become as much of an issue as has that of what those interests are . . . In short, the claim is that as the dynamics of preference formation change at the domestic level, so will those changes, to a considerable degree, shape the EU's policy-making dynamics.

Hypothesis 5: The differences between high and low politics have become blurred.

Hypothesis 6: The EU is in a state of disequilibrium.

The new intergovernmentalists' approach is not without its critics. Bulmer (2015, p. 294), for example, describes new intergovernmentalism as a "strong tendency rather than a blanket characterisation of EU integration, since there is conflicting evidence of post-Maastricht moves to greater supranationalism".[7] Schimmelfennig (2015) raises several criticisms, claiming that the features that Bickerton et al. (2015a) identify as intergovernmental cannot in fact be properly described as such;[8] that the post-Maastricht period is not predominantly intergovernmental; that the features of integration that the new intergovernmentalists identify as "new" in the post-Maastricht period are not new and, in fact, featured pre Maastricht; and that the theory is incomplete as a theory of EU integration, as "many hypotheses are unconditional, weak or hard to test". More generally, Schimmelfennig (2015) argues that new intergovernmentalism is not a general theory of EU integration because, he says, it focuses on a specific set of policies rather than 'the EU' as a whole.

In their rejoinder, Bickerton et al. (2015c) take issue with each of Schimmelfennig's (2015) arguments. First, on the question of novelty, Bickerton et al. (2015c) argue that the pattern of integration they identify in the post-Maastricht period has no *systematic* precedent, even if there were isolated occurrences pre-1992. Second, regarding whether the post-Maastricht period is characterised by increased intergovernmentalism or supranationalism, Bickerton et al. (2015c) note, for example, that an extension in the application of the EU's ordinary legislative procedure (in which qualified majority voting is the norm) does not necessarily point to increased supranationalism, because in practice the procedures may not end up being used (e.g., the Commission may not propose new EU legislation in the area). They argue, further, that their claim that deliberation and consensus are now the guiding norms of EU decision-making cannot be undone by Schimmelfennig's (2015) suggestion that these are attributes of supranationalism, because their claim is rather that deliberation and consensus have become uncoupled from supranational dynamics since Maastricht. Third, Bickerton et al. (2015c) dispute Schimmelfennig's (2015) characterisation of the so-called 'de novo' bodies to whom tasks have been delegated and seek to show that these bodies are quite different from the Commission, arguing, for example, that governments have greater direct input into these bodies (though citing in this regard only the European Stability Mechanism and the EEAS).[9] Most fundamentally, Bickerton et al. (2015c, p. 734) take issue with Schimmelfennig's (2015) failure to engage substantively with their hypotheses: or, as they put it, "his attempts to close down, rather than to promote, empirical debate".

In contrast, the present book seeks to engage empirically with Bickerton et al.'s (2015a) hypotheses for two principal reasons: first, as Bickerton et al. (2015a, p. 316) argue – and as highlighted by Bulmer (2015) – further research is needed to assess whether new intergovernmentalism as an analytical approach is "merely applicable to those new policy areas brought within the scope of EU activity since 1992, or does it also have something to say about those policy areas that have been long-standing components of the EU's action", such as EU nature policy. Second, Bulmer (2015, p. 303) calls specifically for historical institutionalist testing of the new intergovernmentalists' analysis: "just as the editors argue for a move away from micro analysis to the wider picture of integration, so I would welcome a similar shift from snapshot analysis to exploring the 'movie' of European integration". This book seeks to answer these two challenges, in the spirit of Bickerton et al.'s (2015a, p. 45) aim to "stimulate debate rather than to settle it".

Notes

1 As Steinmo (2008) notes, "Many of the most interesting and important studies of politics – from Karl Polanyi's classic *Great Transformations*, to Theda Skocpol's *States and Social Revolutions* and Philippe Schmitter's *Still a Century of Corporatism?* – would clearly be categorized as historical institutionalist were they written today".
2 Other authors have of course also made important contributions to the field – for an overview see Fioretos et al. (2016), and for an alternative take specifically in the EU integration context, see Büthe (2016).
3 That is, "[w]hen EU organizations carry out the preferences of the powerful member states, they supposedly confirm the theory that governments control EU development. But when EU organizations do not adhere to the preferences of member governments, they also supposedly confirm the theory (by carrying out member state desires for EU-level enforcement of incomplete contracts). Any possible outcome would "fit" the theory, rendering it immune to falsification" (Sandholtz and Stone Sweet, 2012a, p. 28).
4 See Niemann and Schmitter (2009).
5 For his part, Moravcsik (2016) gives Bickerton et al.'s (2015a) arguments only very brief analysis in a book review.
6 Thus "the new intergovernmentalism sees the tendency towards integration without traditional forms of delegation since Maastricht as a choice by national governments that remain committed to cooperative solutions, but deeply reluctant to cede further powers to the Commission and the Court of Justice even in domains that have traditionally been

seen as technocratic. Supranational institutions have been complicit in this shift rather than resistant to it, the new intergovernmentalism contends, but the significant increase in the scope of EU activity since 1992 rests uneasily with public opinion. National governments' reticence about empowering Community institutions can be seen as a concession in this context but it is a partial one at best given policy-makers' determination to press ahead with integration through other means" (Bickerton et al., 2015a, pp. 4–5).

7 On which see also Dehousse (2015).

8 E.g., on delegation to 'de novo' bodies: "It is true that governments have been reluctant to empower the Commission in the policy areas integrated since Maastricht. They delegated power to other supranational institutions, however, that are not generally less autonomous than the Commission".

9 Interestingly, Moravcsik (2016, p. 179) evidently sees this as a particular weakness of Bickerton et al.'s (2015a) argument, commenting that the plausibility of their analysis rests *inter alia* on "explaining away the powerful European Central Bank". In this vein, Bulmer (2015, pp. 299 and 294) argues that "the ECB is arguably a traditional supranational body . . . To see the ECB's creation as evidence of the new intergovernmentalism would seem problematic, since it could equally be considered as a paradigmatic case of the Community method".

References

Auer, S. 2016. The new intergovernmentalism: states and supranational actors in the post-Maastricht era. *International Affairs*, 92, 476–478.

Bache, I., George, S. and Bulmer, S. 2011. *Politics in the European Union*. Oxford: Oxford University Press.

Bickerton, C., Hodson, D. and Puetter, U. 2015a. The new intergovernmentalism and the study of European integration. *In:* Bickerton, C., Hodson, D. and Puetter, U. (eds) *The new intergovernmentalism: states and supranational actors in the post-Maastricht era*. Oxford: Oxford University Press.

Bickerton, C., Hodson, D. and Puetter, U. 2015b. *The new intergovernmentalism: states and supranational actors in the post-Maastricht era*. Oxford: Oxford University Press.

Bickerton, C., Hodson, D. and Puetter, U. 2015c. Something new: a rejoinder to Frank Schimmelfennig on the new intergovernmentalism. *Journal of Common Market Studies*, 53, 731–736.

Bulmer, S. 2015. Understanding the new intergovernmentalism: pre- and post Maastricht EU studies. *In:* Bickerton, C., Hodson, D. and Puetter, U. (eds) *The new intergovernmentalism: states and supranational actors in the post-Maastricht era*. Oxford: Oxford University Press.

Büthe, T. 2016. Supranationalism. *In:* Fioretos, O., Falleti, T. G. and Sheingate, A. (eds) *The Oxford handbook of historical institutionalism*. Oxford: Oxford University Press.

Dehousse, R. 2015. *The new supranationalism*. European Consortium for Political Research General Conference, Montreal.

Fioretos, O., Falleti, T. G. and Sheingate, A. (eds) 2016. *The Oxford handbook of historical institutionalism*. Oxford: Oxford University Press.

Moravcsik, A. 1991. Negotiating the Single European Act: national interests and conventional statecraft in the European Community. *International Organization*, 45, 19–56.

Moravcsik, A. 1993. Preferences and power in the European Community: a liberal intergovernmentalist approach. *Journal of Common Market Studies*, 31, 473–524.

Moravcsik, A. 1998. *The choice for Europe: social purpose and State power from Messina to Maastricht*. Ithaca, NY: Cornell University Press.

Moravcsik, A. 2016. The new intergovernmentalism: states and supranational actors in the post-Maastricht era. *Foreign Affairs*, 95, 179.

Moravcsik, A. and Schimmelfennig, F. 2009. Liberal intergovernmentalism. *In:* Wiener, A. and Diez, T. (eds) *European integration theory*, 2nd edn. Oxford: Oxford University Press.
Niemann, A. and Schmitter, P. 2009. Neofunctionalism. *In:* Wiener, A. and Diez, T. (eds) *European integration theory*, 2nd edn. Oxford: Oxford University Press.
Palmieri, D. R. 1997. Extending the reach of the EU Commission: Eurogroup subsidization and the implementation process. Biennial meeting of the European Community Studies Association, May 1997, Seattle, Washington.
Pierson, P. 1996. The path to European integration. *Comparative Political Studies*, 29, 123–163.
Sandholtz, W. and Stone Sweet, A. 2012a. Neo-functionalism and supranational governance. *In:* Jones, E., Menon, A. and Weatherill, S. (eds) *The Oxford handbook of the European Union*. Oxford: Oxford University Press.
Sandholtz, W. and Stone Sweet, A. 2012b. *Neo-functionalism and supranational governance. Unabridged version* [Online]. Available: https://works.bepress.com/alec_stone_sweet/ [accessed 14 June 2017].
Schimmelfennig, F. 2015. What's the news in 'new intergovernmentalism'? A critique of Bickerton, Hodson and Puetter. *Journal of Common Market Studies*, 53, 723–730.
Steinmo, S. 2008. What is historical institutionalism? *In:* Della Porta, D. and Keating, M. (eds) *Approaches and methodologies in the social sciences: a pluralist perspective*. Cambridge: Cambridge University Press.
Stone Sweet, A. and Sandholtz, W. 1998. Integration, supranational governance, and the institutionalization of the European polity. *In:* Sandholtz, W. and Stone Sweet, A. (eds) *European integration and supranational governance*. Oxford: Oxford University Press.

3

TOWARDS AN EU BIRDS DIRECTIVE

Three phases in the history of EU environmental policy

As Lenschow (2015, p. 320) records, many authors have identified three phases in the evolution of EU environmental policy. The present analysis draws on such earlier identifications (principally Knill and Liefferink (2013) and Wurzel (2008)) to describe the following three phases: 1958–1987 (foundational period); 1987–1992 ('golden era'); and 1992 to present (weakening environmental policy dynamics).[1] These phases are introduced in turn in advance of the relevant chapters of this book: the first phase is introduced immediately below, the second in Chapter 6, and the third in Chapter 10. The three phases dovetail with the key phases in the development of EU nature conservation policy: the Birds Directive (1979) was negotiated and adopted during the first phase, "against all odds" (Meyer, 2010b, p. 179); the Habitats Directive (1992) was negotiated and adopted during the second phase, when the policy and institutional dynamics were arguably at their strongest; and the Habitats and Birds Directives have both fallen to be implemented largely within the third phase, characterised by a weaker environmental policy landscape.

First phase (1958–1987): foundational period

At its foundation in 1958,[2] the European Economic Community (EEC) had no environmental policy mandate (Jordan and Adelle, 2013). Today the EU is responsible for a broad array of policy areas, going far beyond the original tasks of the EEC. The creation of an extensive EU environmental policy has been one part of this remarkable transformation. As Jordan (1998, p. 17) comments, "It is doubtful that anyone in the early 1970s could have envisaged the vast role that the EU would eventually play with respect to the environment". Yet already by 1992 it was virtually impossible to understand the environmental policy of any EU Member State without referring to the EU's own policy stance (Haigh, 1992): as such, EU environmental policy went "from silence to salience within thirty years", as Weale (1999, p. 40) puts it.

By the late 1960s a loose network had emerged advocating and supporting the adoption of a common EU environmental policy, including the European Parliament (Meyer, 2011b), some Member States, some NGOs, and even some companies, keen on achieving a level playing field (Wurzel, 2008). More generally, the 1960s and early 1970s had seen a rapid growth in environmental interest and awareness, kindled by landmark books such as Rachel Carson's *Silent Spring* in 1962 (Carson, 2002) and the Club of Rome's *The Limits to Growth* in 1972 (Meadows et al., 1972).[3] In addition, major environmental incidents such as the Torrey Canyon oil tanker disaster off Land's End in England in 1967 (McGurren, 1971) were seen by global audiences on television for the first time, and this no doubt contributed to a heightened sense of environmental awareness (Meyer, 2011a). Major international NGOs such as Friends of the Earth and Greenpeace were founded in this period (McCormick, 1993), and the first Earth Day event, on 22 April 1970 in the US, attracted an estimated 20 million Americans and helped to shift the political landscape of the time (Fried, 1998).

Preliminary work on establishing an EU environmental policy was carried out by the European Commission in the late 1960s and early 1970s, with particularly prominent roles at commissioner level played by the Italian federalist Altiero Spinelli – one of the founding fathers of the EU – and Dutch federalist Sicco Mansholt (Scichilone, 2009, 2010). Spinelli was a political strategist and former communist who had been imprisoned and confined for sixteen years (1927–1943) aged 20 for opposing the fascist regime in Italy (Pinder, 2007). In July 1970 he was made Commissioner for Industrial Affairs, Science and Technology (DG III) (Bussière, 2014a; Segal, 1973), but, according to Robert Toulemon, then Director General of DG III, insofar as the industrial portfolio was concerned, Spinelli was only really interested in environmental issues. He recalls Spinelli's refrain, "Industry is a tiger which needs to be tamed!" (Toulemon, 2003, p. 39).

For his part, Mansholt, who was architect of the common agricultural policy as Agriculture Commissioner from 1958–1972, was also very interested in environmental and development issues, and had been profoundly affected by the Club of Rome's ideas (Meadows et al., 1972) and by his friend Professor Jan Tinbergen, winner of the Nobel Prize for Economics in 1969 (Bussière, 2014b; Van der Harst, 2014). In this vein, Mansholt wrote a lengthy letter to Commission President Malfatti in February 1972 calling for an immediate U-turn in the EU's economic policy: an abandonment of economic growth maximisation, centralised economic planning at EU level, and the development of a "clean and recycling" economy, with GNP to be replaced by gross national happiness as a measure of progress (Mansholt, 1972a; Wurzel, 2012; Van der Harst, 2014; Bussière, 2014b).[4] To the surprise of many, Mansholt was soon after made interim Commission President in March 1972 (Van der Harst, 2014), a post he held until January 1973 (Blair, 2006), a period coinciding with the UN Conference on the Human Environment in Stockholm in June 1972. The Stockholm Conference gave Mansholt the opportunity to repeat his *Limits to Growth*-inspired ideas on the international stage (Scichilone, 2010; Mansholt, 1972b), albeit expressed somewhat less radically than the candid "negative growth" remarks he made after leaving the Commission (*The Ecologist*, 1974). His "remarkable [Stockholm] speech . . . was acclaimed by young demonstrators" (Van der Harst, 2014, p. 178). While Mansholt and Spinelli (and other commissioners of the time) did not see eye to eye regarding Mansholt's economic views (Scichilone, 2009; Van der Harst, 2014), in any event they together represented a powerful force in driving forward the establishment of an EU environmental policy.

By 1971 Michel Carpentier, a French head of unit in Spinelli's DG III, had been put in charge of a new 'Environmental Problems' unit within the DG (European Commission, 1971a), attached directly to the Director General (Bussière, 2014a).[5] This small unit worked hard on developing a nascent EU environmental policy, starting "practically from scratch" (Carpentier, 1999). A first Communication issued for comments in July 1971 (European Commission, 1971b), followed by a second Communication in March 1972 proposing an environmental action programme (European Commission, 1972). A few months after the UN's Stockholm conference in October 1972 the Heads of State or Government of the Member States of the enlarged EEC (which the UK, Ireland and Denmark attended, since they would join the EEC only three months later) issued a declaration following their Paris summit meeting stressing the value of an environmental policy "dans" (not "de la")[6] the Community, and inviting the Community institutions[7] to draw up an action programme. This nuance of language is important, reflecting the fact that some countries – particularly the UK, Denmark and France – remained doubtful of the idea of broadening the EU's competences in this direction (Carpentier, 1999).

Early in 1973, the Commission formally created what would later become DG XI and later still DG Environment, at its foundation the Environment and Consumer Protection Service (ECPS), an independent service reporting to the Italian Commissioner Scarascia-Mugnozza, then a Vice-President of the Commission with responsibility for environmental policy, amongst other things (European Commission, 1973).[8] The ECPS started with a small staff but grew quickly (Flattau, 1979; Schön-Quinlivan, 2013), and was headed by Michel Carpentier from 1973–1981. The ECPS had three divisions initially: Division 1: for the Prevention of Pollution and Nuisances (covering water, air, waste, chemicals and noise), headed by Stanley Johnson (father of British politician Boris) – a then 32-year old Englishman who played a "catalytic role" (Cashman, 2012), "initiated a lot of things in EU environment policy" (Henningsen, 2012), and would later play a crucial role in the history of the Habitats Directive; Division 2: Quality of Life and General Affairs, headed by a German, Dieter Hammer; and Division 3: Consumer Affairs, headed by another Englishman, John Braun (European Commission, 1973).

By all accounts Michel Carpentier was a formidable figure who was central to the rapid development of EU environmental policy in this period. He is variously described by colleagues as a "force of nature" (Brusasco-Mackenzie, 2011) and a "phenomenon" (Krämer, 2011). Johnson (2009, p. 262) adds,

> I liked and admired Michel Carpentier at first sight. A man of explosive energy and dogged determination, he had . . . a marvellous sense of humour and the instincts of a street fighter. He was exactly the right person to lead the fledgling environment and consumer affairs directorate. Other departments in the Commission would be only too keen to reduce it to impotence. It was Carpentier's job, by fair means or foul, to make sure that didn't happen.

In his new role, Carpentier (1973, p. 10) summarised the Commission's vision of a "reasoned, progressive and enlightened environmental policy" at a symposium in February 1973. Just five months later, on 19/20 July 1973, the Environment Council held its first ever meeting and reached political agreement regarding the First Environmental Action Programme proposed by the Commission after a difficult meeting (Johnson, 2011; CEC, 1973; Carpentier, 1999, 2004).[9] This was later formalised on 22 November 1973 by way of an unusual joint Declaration of the Council and representatives of the governments of the Member States meeting in the Council (First Environmental Action Programme, 1973).[10]

While the First Environmental Action Programme (1973) gave "environmental policy a sound political and intellectual platform from which to develop" (Jordan, 1998, p. 14), a firm legal platform remained absent: that is, there remained no explicit legal basis in the Treaty for the making of EU environmental legislation. The result was that, during this first phase in EU environmental policy, there were two key sources of new policy: first there were the non-binding Environmental Action Programmes – the First Environmental Action Programme (1973) was followed by the Second and Third Environmental Action Programmes in 1976 and 1983, respectively (Second Environmental Action Programme, 1977; Third Environmental Action Programme, 1983). Second, there was the Information and Standstill Agreement (1973) – self-described as a "gentleman's agreement" – which added environmental measures to an Information and Standstill Agreement of 1969, in view of France's decision to block Treaty changes to provide explicitly for the making of EU environmental legislation (Wurzel, 2012). The Information and Standstill Agreement (1973) committed the Member States to inform the Commission as soon as possible of, amongst other things, any draft national legislation concerning the protection or improvement of the environment that could directly affect the functioning of the common market, or which was relevant to the EU's programme for the reduction of pollution and nuisances and the protection of the natural environment, or which was of particular interest to the EU and the Member States from the point of view of the protection of public health or of the natural environment, particularly where the legislation could have repercussions for other Member States.[11] The standstill element of the agreement then gave the Commission two months to indicate its intention to propose EU legislation on the subject in place of the draft national legislation, and five months to make its proposal to the Council. By 1980, over 250 draft national laws had been notified in this way (European Commission, 1980) and, according to Rehbinder and Stewart (1985, p. 17), the Information and Standstill Agreement (1973) had the result that "Strongly environmentalist member states were more or less able to set the pace and direction for Community action by unilaterally proposing strong national environmental legislation, which forced the Community to react with measures of its own".

Whether deriving from an Environmental Action Programme or the Information and Standstill Agreement (1973), any EU environmental legislative proposal had to overcome an additional hurdle: such proposals needed by law to be based on a provision of the Treaty of Rome (1957), since the EU is of course only allowed to exercise powers that have been expressly allocated to it by the Treaty (Jordan and Adelle, 2013). Ultimately, until 1987 when the Treaty was amended to provide a specific environmental base, the two Treaty provisions most commonly relied upon by the Commission in seeking to make EU environmental legislation were (one or both of) Articles 100 and 235 of the Treaty of Rome (1957) (Knill and Liefferink, 2013). Article 100 allowed the Council to adopt EU directives to approximate differing national laws, provided that the differences "directly affect the establishment of the common market". Article 235, in turn, provided a legislative power where action by the EU should prove necessary to attain, in the course of the operation of the common market, one of the objectives of the EU, but the Treaty had not provided the necessary powers (i.e., the implied powers doctrine).

Despite this relatively weak legal base, the European Commission, which holds the sole right of legislative initiative in EU environmental matters (Schön-Quinlivan, 2013), set about implementing the First Environmental Action Programme (and its successors) with gusto during this first phase, leading to a "rapid and profound transformation" of EU environmental policy (Jordan, 1998, p. 17). However, until 1987, "the Commission had to go to great lengths to justify measures such as protection for bathing water and wild birds that frankly had little to do with the establishment of a common market" (Jordan and Jeppesen, 2000, p. 64).

Introduction to European bird conservation

Bird conservation has a long pedigree in Europe,[12] evidenced by the early establishment of NGOs such as the UK's (then all-female) Society for the Protection of Birds (precursor of the Royal Society for the Protection of Birds (RSPB)), founded in 1891 (RSPB, undated), Germany's *Deutscher Bund für Vogelschutz* (DBV), founded in 1899 (Chaney, 2008), and France's *Ligue de Protection des Oiseaux*, founded in 1912 (Meyer, 2010a).

By the time the Treaty of Rome (1957) was signed establishing the EEC, there were already several international bird conservation conventions in force with European countries as parties, including the Convention for the Preservation of Wild Animals, Birds and Fish in Africa 1900; the Convention for the Protection of Birds Useful to Agriculture 1902; and the Convention for the Protection of Birds 1950 (Bowman, 1999). Following these early international conservation initiatives, the 1960s saw a widespread public awakening to the issue of bird conservation, triggered by Rachel Carson's now classic book *Silent Spring*, which, when published in 1962, brought to public attention the impacts on birds of the widespread use of chemical pesticides, and helped to launch the modern environmental movement (Carson, 2002; Lytle, 2007). Against this backdrop, the hunting of migratory birds in Italy, Belgium and France (amongst other places) came to be an issue of major concern in Europe in the late 1960s, with NGOs joining forces across Europe to campaign regarding the issue (Meyer, 2010b).

The intergovernmental Council of Europe – an entirely separate organisation from the EU – was active during this period, passing two Resolutions on the subject of bird conservation in October 1967, and an additional Resolution in 1973 (Council of Europe, 1967a, 1967b, 1973). Interestingly, all three Resolutions focused on habitat protection as well as on controlling hunting and pesticide use. For example, one of the 1967 Resolutions recommended member governments (which included the six founding Member States of the EEC, as well as Denmark, Ireland and the UK, the three countries that would join the EEC in 1973): "To take any measures required for the greater and more effective protection of rare or disappearing species of birds in Europe, such measures to cover not only their legal protection but also their habitats and conditions of life" (Council of Europe, 1967a). However, a central problem with the Council of Europe's work was its lack of enforceability: as Hans Edgar Jahn, a German MEP who worked extensively on bird conservation issues (Meyer, 2010b), commented in 1973, "Experience has shown, that, despite all its good intentions, the Council of Europe in practice moves very slowly and must in any case confine itself to the formulation of recommendations" (European Parliament, 1973). Thus, as Meyer (2010b, p. 177) puts it, those who lobbied the EU institutions regarding the need for EU bird conservation legislation did so "because they deemed the Council of Europe ineffective".

Bird conservation and the First Environmental Action Programme (1973)

Introduction

While the European Commission's (1971b) first Communication about an EU environmental policy, published for comments in July 1971, contained a generic endorsement of conservation as a key task of the EU, the Communication contained no proposals for EU level action in the field of nature conservation generally or bird conservation specifically.[13] This reflected the position expressed by the Commission in March 1968 in response to a question posed in the European Parliament by the German MEP Hans Richart, who asked whether the Commission had any plans to propose harmonisation of bird protection legislation or, if it felt that the EU did not have legal competence, whether the Commission would facilitate an intergovernmental agreement. The Commission responded that it did not regard the EU as competent in the area, nor did it feel in a position to make any suggestions to the Member States regarding the issue (Meyer, 2014). And yet, when the EU came to adopt its First Environmental Action Programme (1973) just two years after the 1971 Communication, bird conservation was included as a subject for priority action, notwithstanding the subject's only tenuous link to the common market.

Activism by MEPs, NGOs and the public

There are several potential explanations for this. First, MEPs and bird conservation NGOs – in some cases working together – were active in the intervening period, calling for action to conserve migratory birds, particularly in light of the "mass extermination" of wild birds in Italy, Belgium and France (Jahn, 1975; Meyer, 2010b). To that end, for example, Luxembourger MEP Jean-Pierre Glesener posed a written parliamentary question to the Commission on 10 September 1971 regarding the "scandalous extermination" of migratory birds in Belgium and Italy, and asked the Commission whether it intended to propose common legislation to protect such birds (Glesener and European Commission, 1971). The Commission responded that, whilst it had not proposed such measures in its first Communication, it did not rule out the possibility of proposing such measures at some future date (Glesener and European Commission, 1971). In April 1972, at the "Industry and Society in the European Community" conference in Venice, then Commission President Mansholt spoke of the need to tackle the destruction of birds (Seefeld and European Commission, 1972). When questioned the following month by the European Economic and Social Committee[14] – an organisation intended to form a bridge between the EU and civil society – as to why the Commission had not included the issue of migratory birds in its 1971 Communication regarding an environmental action programme, the Commission replied that the EU "would surely be able to act in this area if the [Economic and Social Committee or the European Parliament]" so demanded, in effect inviting such a demand, as Meyer (2010b) argues.

Hans Edgar Jahn MEP followed up with a written question to the Commission on 15 February 1973, in which he highlighted the fact that, since the date of MEP Glesener's question in 1971, Belgium had banned bird-catching throughout its territory, but Italy had repealed a similar protective law in January 1970, and since then around 200 million migratory birds had been killed each year in Italy (Jahn and European Commission, 1973). MEP Jahn noted that this had triggered worldwide protests, including calls for boycotts of Italian products and services (Jahn and European Commission, 1973). In its reply, the Commission said that, "For a number of weeks now the Commission has been receiving letters of protest and petitions from individuals, animal protection societies and action committees"[15] and noted that "the Commission is at present considering the advisability of incorporating proposals on this subject in the [first] action programme on the environment, now under examination" (Jahn and European Commission, 1973). Jahn reinforced the point in a report on behalf of the Parliament's Committee on

Public Health and the Environment, published in June 1973 – just one month before the First Environmental Action Programme (1973) was adopted – in which he called on the Commission to include in the Action Programme "a proposal for the prohibition of bird destruction throughout the Community" rather than merely promoting joint action by the Member States in the Council of Europe and other international organisations (European Parliament, 1973).

A directive on birds to make Europe popular in the UK?

Such actions by MEPs, NGOs and individuals were clearly one trigger for the inclusion within the First Environmental Action Programme (1973) of wording raising the possibility of EU legislation on bird conservation (Jahn, 1975, p. 10). A second potential contributory factor is provided by Hubert David, the first Secretary General of the European Environmental Bureau, a Brussels-based NGO coalition founded in 1974, who recalls that "it has been said many times that the Birds Directive was proposed by the Commission to make Europe popular in the UK" (David, 2012). The UK, which has a longstanding and popular[16] tradition of bird conservation (RSPB, undated), also has a longstanding and popular tradition of Euroscepticism (Forster, 2002).

As noted above, the UK joined the EU in January 1973, six months before the First Environmental Action Programme (1973) was adopted, but the first decade of the UK's EU membership was turbulent (Dinan, 2007). Almost immediately on joining the EU, a Labour government under Harold Wilson came to power having promised in its manifesto to renegotiate the terms of the UK's accession and put the result to an in-out referendum (Dinan, 2007). The other EU Member States were "dismayed", and although continued EU membership was carried then by a large majority, the issue would raise its head again less than five years later, in 1979, when Margaret Thatcher came to power in the UK seeking (and ultimately winning) a large annual EU budget rebate for the UK (Dinan, 2007, p. 1127). It therefore seems perfectly plausible that one factor in favour of proposing an EU directive on birds may have been the issue's perceived popularity in the UK. Nevertheless, the UK's then Secretary of State for the Environment, Peter Shore of the Labour Party, who was an "arch Euro-sceptic" (Fairbrass and Jordan, 2001, p. 9), reportedly did not want the Birds Directive to be a popular success for the EU (Stuffmann, 1999). His wishes in that regard would ultimately be thwarted: in 1981, two years after the adoption of the Birds Directive, the *New Scientist* reported that "many environmentalists would like to see a greater [EU] emphasis on nature conservation and land management. Without doubt, the most popular directive so far has been the one to protect wild birds, which has no discernible connection with the EEC's *raison d'être* – trade" (Caufield, 1981, p. 417). Michel Carpentier agreed: "The Birds Directive . . . has been very popular. The House of Commons said that it was the best one we've ever done" (Caufield, 1981, p. 417).

Towards a directive on birds

Introduction

The First Environmental Action Programme (1973) records the official reasons used at the time to justify an EU level policy on bird conservation:

> Hundreds of millions of migratory birds and songbirds are captured and killed in Europe every year provoking worldwide protests against the countries which allow the trapping of birds.
>
> This massive destruction provokes a serious threat to the ecological balance in Europe, because it causes plant parasites to proliferate. As a result, the campaign against such parasites requires use on a bigger scale of insecticides which are sometimes harmful to man and to the natural environment.
>
> Policy for the protection of the environment should therefore include measures to prevent the large-scale destruction of birds, particularly songbirds and migratory birds, and more generally to protect the existence of certain animal species threatened with extinction.

To that end, the First Environmental Action Programme (1973) called for the promotion of joint action by the Member States in the Council of Europe and other international organisations, and mandated a "study with a view to possible harmonization of national regulations on the protection of animal species and migratory birds in particular", with a deadline of 31 December 1974.

Two points immediately stand out: first, the aim at the time the First Environmental Action Programme (1973) was produced seems principally to have been to address hunting of migratory birds – that is, species protection rather than habitat protection measures. Second, the more general stated aim of protecting the existence of certain animal species (other than birds) threatened with extinction was addressed only in a piecemeal manner at EU level until the Habitats Directive (1992) was adopted many years later.[17]

Between the date of the First Environmental Action Programme (1973) and the Birds Directive proposal in 1976 (European Commission, 1976b), the European Commission (1974) issued a Recommendation to Member States in December 1974 concerning the protection of birds and their habitats. This Recommendation highlighted the main conclusions of the ornithological study promised in the First Environmental Action Programme (1973), which was produced under the direction of Professor Dr Bernhard Grzimek. The study highlighted loss of habitats as a major issue for European birds in addition to hunting, and the Commission recommended that the Member States, if they had not already done so, should accede as quickly as possible to the recently concluded Ramsar Convention (1971) on the conservation of wetlands of international importance (which had not yet entered into force owing to too few ratifications), as well as the Convention for the Protection of Birds 1950, mentioned earlier in this chapter (European Commission, 1974).

The Grzimek study

As noted above, the ornithological study to inform Commission policy had been awarded to Grzimek, "one of the towering figures in African conservation" (Adams and McShane, 1996, p. xv), who, as the director of Frankfurt zoo, the host of a popular television programme, and German Chancellor Willy Brandt's special representative for the environment from 1970–1973, was "easily the most prominent figure in German environmentalism at the time" (Meyer, 2010a, p. 96). However, Grzimek was a veterinarian by training, not an ornithologist, despite being considered "an authority on all issues relating to the environment", which served as the "main motivation" behind his appointment to oversee the study, according to Claus Stuffmann, the Commission's senior negotiator in respect of the Birds Directive (Meyer, 2010b, p. 182; Meyer 2010a). Grzimek was in contact at the time with Gerhard Thielcke, head of the German section of the International Council for Bird Preservation (ICBP; now BirdLife International) and director of the ornithological station Radolfzell at Lake Constance in Germany (Meyer, 2010b). The ornithological work for the Commission study was delegated to two PhD students working in Radolfzell, who produced a report which, in the Commission's view, lacked the prestige and quality of an eminent ornithologist, relied heavily on data from Germany, and focused on hunting in Italy, Belgium and France, which "very undiplomatic approach [and the "'undiplomatic' language used" (Meyer, 2010a, p. 97)] made it very difficult for the Commission to use the report" (Meyer, 2010b, p. 186). The Commission therefore started looking for "more neutral expertise" (Meyer, 2010b, p. 186), having been advised internally that they couldn't do anything on the basis of the report (Temple Lang, 2012).

A petition – No.8/74 'Save the Migratory Birds' – had been submitted to the European Parliament, the European Commission, the Council and the Dutch Foreign Ministry on 26 August 1974 by the Dutch NGO *Stichting Mondiaal Alternatief*, supported by partner organisations from around the world, including national and international animal protection societies (Jahn, 1975; Meyer, 2011b, 2010a), and this gave MEP Hans Edgar Jahn, appointed rapporteur by the European Parliament's Environment Committee, another opportunity to emphasise the need for EU legislation. In the February 1975 parliamentary debate concerning the petition, Commissioner Guido Brunner conceded that the Commission would propose a directive if the Recommendation issued at the end of 1974 did not yield "satisfactory results" (Meyer, 2011b, p. 83). Jahn's (1975) report, and the European Parliament's (1975) resolution adopted in light of the report, called for urgent EU measures to be adopted, and Jahn then entered a period of working closely with environmental NGOs and activists (Meyer, 2011b). That such actions ultimately had an important impact was later confirmed by the Commission, which explained that its Birds Directive proposal, put forward in December 1976, was a response to the "wishes expressed repeatedly by the European Parliament in several of its Resolutions, debates and petitions (see especially the Resolution of 21 February 1975)[18] and to the requests of a broad section of public opinion indicated in more than 50,000 letters received by the Commission alone" (CEC, 1977, p. 2).

"Even if [Grzimek's] Frankfurt Zoological Society lacked thorough expertise in ornithology", comments Meyer (2010a, p. 96), "it was proficient in networking". Once its study for the Commission was out of the way in July 1975, Grzimek's assistant wrote to a newly-formed activist bird protection group in Germany to encourage them to lobby the Commission to propose a directive, and the Zoological Society itself called for a directive once its contractual relationship with the Commission had ended (Meyer, 2010a). In parallel with such lobbying efforts by NGOs and the European Parliament – which continued throughout the negotiating process for the directive, including coordinated NGO letter-writing campaigns as late as 1978 (Meyer, 2010a) – in 1974 the Commission had established a working group within the ECPS to discuss the idea of a legislative proposal on birds, and included within its ranks Commission officials with expertise and interest in bird conservation issues (Meyer, 2010b). As Stuffmann (1999, p. 16) explains, the Commission's formal proposal was "thoroughly prepared in three phases: a number of studies, a group of independent experts and a group of experts designated by the governments. The Commission was precautious to include the hunting and farming communities in these consultations".

Key influential individuals in the development of the Birds Directive

One official involved in the ECPS's working group was an Irishman, John Temple Lang, then a lawyer with the Commission's Legal Service, who was heavily involved in bird conservation issues in Ireland and internationally, including with the ICBP (IELA, undated; Meyer, 2010b).[19] Temple Lang was very critical of the earlier Grzimek-directed study, and recommended that the Commission should look to Stanley Cramp, a leading British ornithologist (Simmons, 1989; Temple Lang, 2012), for their desired additional study (Meyer, 2010b; Temple Lang, 2012).[20] This the Commission did, and Cramp produced a slim, straightforward volume,[21] containing clear priorities for action in a concluding chapter entitled "What should be done?":

> Firstly, habitats must be preserved, for without a variety of habitats there can be no variety in any forms of wildlife. The most threatened habitats are clearly wetlands of all kinds; The second main line of progress must be to improve protective legislation; The third aim must be to reduce environmental pollution; A fourth need is for increased research into bird populations and their trends.
>
> *(Cramp, 1977, pp. 49–52)*[22]

Another important contact of John Temple Lang's was Alistair Gammell, then working for the RSPB (Meyer, 2010b). As Gammell (2012) highlights, when an EU legislative instrument on birds was first mooted the EU's significance as an environmental regulator remained unclear:

> we [the RSPB] heard about the idea of this Birds Directive being discussed in Brussels, and we took a view that either . . . the European Union was going to be important, or not, and we weren't quite sure which, but on balance we thought that it probably would be important, and therefore it would probably be important that we engaged in this new legislation, and as a result I went to Brussels to start to lobby . . . Commission officials were very happy to take what expertise I could offer . . . so I had, more or less, very easy access right into the Commission building and worked with Commission officials at that time.

The role of hunting organisations

More generally, the consultation process regarding the directive "was informal and very open to groups which could demonstrate an interest or contribute expertise [and] included bird protection organizations, environmentalists, hunting organizations and national governments" (Meyer, 2010b, p. 187). Although hunting organisations clearly played an important role in the history of the Birds and Habitats Directives, Meyer (2010b, p. 190) emphasises "[t]he relative weakness and the relatively late intervention of the hunting groups at the European level" in respect of the Birds Directive. The Paris-based International Council for Game and Wildlife Conservation (CIC), which traditionally represented hunting interests at international level, "did not prove to be very influential in the EC in the context of the birds directive", according to Meyer (2010a, p. 92), and the EU hunting coalition FACE was not itself established until March 1977, several months after the Birds Directive had been proposed (FACE, 2003).

As Meyer (2010b, p. 191) explains,

> What the [ECPS] valued highly was scientifically sound, thus credible and apparently even-handed, expertise. While a broad range of groups was consulted to produce a legislative proposal acceptable to the member states, this gave an advantage to the traditional bird protection organizations, who, like the RSPB, had volunteered to cooperate early onwards. Conversely, hunting organizations' high-handed claim to being the real experts in the field appeared less convincing to the Commission officials.

"Hunters' dilatory stance concerning EC legislation", adds Meyer (2010b, p. 191), citing Stuffmann, "excluded them as a potential ally, as soon as the Commission officials had earmarked the issue as a promising area for advancing European integration". Nevertheless, certain hunters worked very hard to ensure hunting interests were fully aired in respect of the Birds Directive proposal, including a

French hunting representative, Raymond Pouget,[23] who was "fighting like hell", according to Joost van der Ven, a Dutch nature conservation NGO representative involved with the ICBP at the time (van der Ven, 2012). As Mireille Jardin, a member of the French delegation for the negotiation of the Birds Directive, comments, "in France the hunting lobbying was very powerful and . . . opposition to the adoption of the directive was very strong. I remember in the south of France [huge] inscriptions on walls everywhere against it!" (Jardin, 2013).

Networking by NGOs

On the bird protection side, Alistair Gammell of the RSPB became involved early in various networks. In addition to participating in discussions with the Commission, Gammell was appointed expert to assist in the drawing up of the Economic and Social Committee's opinion on the proposed Directive (ESC, 1977; Meyer, 2010b). He also produced a statement regarding the Birds Directive proposal for the EEB, the Brussels-based NGO coalition (Meyer, 2010b). At the initiative of the RSPB and its German equivalent the DBV (now NABU), representatives of various European bird protection groups met in 1976 in Germany to discuss cooperation, and in 1977 the groups decided to create a Working Group of European Bird Protection Societies (WEBS) (Meyer, 2010b; BirdLife International, 2012). WEBS marked the beginning of a new way for the ICBP to work together in Europe, with Gammell as its first secretary-general, working closely with Jean-Patrick Leduc of what is now France Nature Environnement (David, 2012) to collect and distribute information to allow national groups to lobby more effectively (Meyer, 2010b; BirdLife International, 2012).

As Gammell (1999, p. 9) would later reflect, "it is clear to me that the Birds Directive had the effect of joining together the Bird Protection Societies in common cause and thus was instrumental in forming what has now become BirdLife. Thus, the Birds Directive is not only our baby, but we too are its baby!" The same was true of the EU hunting coalition FACE, and to a certain extent also the EEB, founded in December 1974 and now the largest federation of environmental organisations in Europe, which was "strengthened by the precedent set by the birds directive in the area of nature protection", and pushed for further EU competence in the field, a large number of conservation bodies having joined the EEB at the end of 1979 (Meyer, 2010a, p. 91). Nevertheless, according to the EEB's first Secretary General, Hubert David, the conservation organisations were distinguished from other interest groups by the intelligence advantage they gained through working together, which contributed "enormously" to the successful adoption of the Birds Directive (David, 2012).

The crucial roles of Claus Stuffmann and John Temple Lang

In 1976,[24] Claus Stuffmann – a German – took over as the head of Division 3 (General planning and environmental improvement) at the ECPS, having spent the previous years as a member of Commissioner Spinelli's *cabinet* with responsibility for environmental matters (European Commission, 1976a; Audland, 1976; HAEU, 1970–1972). Stuffmann was therefore closely involved with Spinelli and Carpentier's team in helping to develop and launch EU environmental policy as a whole (Van de Velde, 2014), as well as later playing a central role in the detailed creation of EU nature conservation and other environmental policies with the ECPS and successor institutions. Stuffmann's personal commitment to European integration is said to have been forged by his experiences growing up in Germany during the Second World War (Devillers, 2013). He is reported to have commented during his working career that he had two passions in life: the project of European unity and nature conservation, and "he certainly acted that way" (Devillers, 2013). On joining the ECPS in 1976 Stuffmann took charge of work relating to the Birds Directive, and "masterminded the whole thing", according to Temple Lang (2012).[25]

In addition to the work of Stuffmann's division, John Temple Lang himself – whom Alistair Gammell describes as "very influential" and "the key guy" (Gammell, 2012) – had a crucial role to play:

> although he was working for the Legal Service of the Commission at the time . . . he was very much both a mentor to me, and to some extent to other Commission officials, because he was both a lawyer and a keen bird watcher, so he uniquely, I think he had probably much more vision as to what [the directive] could be than I did, and probably Commission officials did at the time. Because he combined those two things: a senior Commission lawyer with his knowledge as a bird watcher.

Drawing on this knowledge and expertise, Temple Lang – who had received a copy of a draft directive produced by other Commission officials, which covered the killing of song birds, etc. – recommended that the draft ought absolutely to include habitat conservation measures, and personally drafted Articles 3 and 4, the habitat conservation provisions of the Birds Directive proposal (Meyer, 2010b; Temple Lang, 2012; European Commission, 1976b), which provisions represented a huge leap forward for bird conservation in the EU, and laid the foundations for the Natura 2000 network. Temple Lang (2013) describes his idea as "extremely simple": namely, "to extend the [criteria for identifying areas of international importance][26] to all migratory species, and then in parallel to produce a comprehensive list of every place in Europe that met those criteria, and argue that there was a legal obligation to protect them all. This idea succeeded, even better than I had hoped".[27]

Notes

1 Note Wurzel's (2008, 2012) sub-division of the first phase identified here into an "infant phase" (1958–72) and an "adolescent phase" (1972–87), and his sub-division of the third phase into a "sedate phase" (1992–2005) and "selective activism" (since 2005). Note also that Knill and Liefferink (2013) begin the first phase at 1972.

2 The Treaty of Rome (1957) was signed in 1957 and came into force in 1958.

3 *The Limits to Growth* was commissioned by the Club of Rome and produced by an MIT research team.
4 *The Limits to Growth* (Meadows et al., 1972) was published in March 1972 (Golub and Townsend, 1977), whilst Mansholt's letter to the Commission President was dated a month earlier, February 1972 (Mansholt, 1972a). However, one of the authors of *The Limits to Growth* notes that by the summer of 1971 a draft of some of the key findings was circulating amongst members of the Club of Rome (Meadows and Meadows, 2007), and it seems Mansholt had sight of this document: his February 1972 letter refers to a July 1971 report of the System Dynamics Group at MIT, the group that produced *The Limits to Growth*. Club of Rome co-founder Alexander King therefore appears to have been in error when he commented of Mansholt's letter, "Mansholt had never read anything from the Club of Rome, not even *Limits to Growth*" (Murawiec and di Paoli, 1981, p. 22).
5 Carpentier (1999) states that the 'Environmental Problems' unit was working on developing an EU environmental policy in 1970 and 1971. The Commission directory shows him still in charge of his former unit (industrial policy in the nuclear and advanced technology sectors) as at September 1970; by September 1971 he is shown in charge of the 'Environmental Problems' unit (European Commission, 1970, 1971a).
6 This important nuance (Carpentier, 1999) is evident in the French original but is not captured by the English translation of the declaration.
7 Referring to the Commission, the European Parliament and the Economic and Social Committee (Paris Summit, 1972).
8 Spinelli could not hang on to all his responsibilities following the 1973 enlargement; having thus 'fathered' EU environmental policy, he handed over responsibility for the policy at this time (Van de Velde, 2014). Nevertheless, the ECPS remained in daily contact with Spinelli's *cabinet* (Pleinevaux, 2010).
9 Carpentier (1999, p. 293) gives the date of this key meeting as 22/23 November 1972 (with 1972 clearly a typo for 1973) and Carpentier (2004) gives the date as November 1973. However, these are misrecollections. November 1973 marked the formal making of the declaration; the difficult meeting at which political agreement was reached was 19/20 July 1973 (CEC, 1973).
10 At the Environment Council's meeting of 19/20 July 1973, France insisted upon this joint formulation on the basis that certain matters covered in the programme were for the Member States and not for the Community (CEC, 1973). This tricky issue required the intervention of Commission President Ortoli before political agreement was reached (CEC, 1973; Carpentier, 1999).
11 The Information and Standstill Agreement (1973) was amended in 1974 to clarify the application of the Agreement to measures falling within the scope of the First Environmental Action Programme (1973): see Amendment to the Information and Standstill Agreement (1974).
12 For overviews of the history of bird conservation, see Boardman (2006), Bowman et al. (2010), and Bowman (1999).
13 The closest the Communication comes is a section on the "Development and improvement of open spaces and the natural environment", which related to agricultural policy (European Commission, 1971b, p. 19).
14 The question was posed by "the former Commission official and head of the Luxembourg Chamber of Commerce Carlo Hemmer" (Meyer, 2011b, p. 81).
15 According to Fairbrass and Jordan (2001, p. 9), citing Alistair Gammell, then of the RSPB, a "sustained campaign by Dutch and German citizens directed at the Commission and European Parliament proved to be particularly influential".
16 Toke and Marsh (2003, p. 241) record that the "RSPB is, after the National Trust, the largest NGO in the UK. It has over a million members, a figure that is significantly larger than the combined memberships of all UK political parties". When the Birds Directive was proposed in 1976 the RSPB already had some 200,000 adult members and a junior membership of 76,000 (Hillaby, 1976). Note also that in 1976 the UK was the first of the then

nine EU Member States to accede to the Ramsar Convention on wetlands of international importance, a convention with particular significance for birds (Ramsar, 2013).

17 Several EU nature conservation and animal welfare laws were passed in the interim period, including: the Whales Regulation (1981) on common rules for imports of whale products and other cetacean products; the first CITES Regulation (1982), on international trade in endangered species; the Seal Pups Directive (1983); and the Leghold Traps Regulation (1991).

18 See European Parliament (1975). Claus Stuffmann later stated that preparatory activities regarding the Birds Directive were "speeded up by a strong resolution voted in ... 1975 by the European Parliament urging the Commission to come forward with concrete proposals as announced in its Action programme for the end of 1974" (Stuffmann, 1999, p. 16).

19 Such involvement continued after Temple Lang began working for the Commission in 1974 – he was Chair of the European Continental Section of the International Council for Bird Preservation from 1980–1984 and Co-secretary, Standing Committee on Applied Ornithology, International Ornithological Congress 1986–1994 (IELA, undated). Temple Lang (2012) reports that he was contacted by the ECPS, which had somehow heard about his involvement in the production of what ultimately became Ireland's Wildlife Act 1976, and that he was closer to being an expert on bird conservation matters than anyone else in the Commission at the time.

20 In addition to Cramp, Alistair Gammell of the RSPB states that the Dutch Professor M. F. Mörzer Bruijns helped the Commission in their early thinking regarding the Birds Directive (Gammell, 1999).

21 In contrast to the "lengthy report" produced by the Grzimek team (Jahn, 1975, p. 12).

22 While Cramp's study was formally published in 1977, presumably a version of the report was available to the Commission in advance of the Birds Directive proposal being sent to the Council in December 1976 (European Commission, 1976b). This would be consistent with the Grzimek study, which was formally published in 1976 but whose main conclusions were referred to in the Commission's December 1974 Recommendation (European Commission, 1974; Poltz and Conrad, 1976). The Explanatory Memorandum for the directive proposal refers to "studies" (plural), almost certainly meaning the Grzimek study and the Cramp study (European Commission, 1976b).

23 Henri Jaffeux, who worked on nature conservation for many years for the French government, explains that Raymond Pouget did not represent all hunters or all types of hunting; he was the president of the National Association of Waterfowl Hunters (ANCGE) (Jaffeux, 2013).

24 Presumably shortly after Spinelli left the Commission on 4 July 1976; Stuffmann is pictured with Spinelli and the rest of the *cabinet* on the eve of Spinelli's departure (Bussière et al., 2014, pp. 86 and 475).

25 In this capacity Stuffmann and his colleague Richard Geiser visited Italy in July 1977 to appraise themselves of the bird hunting situation on the ground (Bana, 2003).

26 Criteria for identifying areas of international importance for waterfowl and wetlands were well established by the time the Birds Directive was adopted in 1979 (Smart, 1976; Temple Lang, 1982, 1983), and updated criteria were agreed at the first meeting of the Conference of the Contracting Parties to the Ramsar Convention in Cagliari in 1980 (De Klemm and Créteaux, 1995).

27 "The apparent novelty", he adds, "lay in applying the criteria to roosts, bottlenecks, moulting areas, and not just to breeding and wintering areas" (Temple Lang, 2013).

References

Adams, J. S. and McShane, T. O. 1996. *The myth of wild Africa: conservation without illusion.* Berkeley and Los Angeles: University of California Press.

Amendment to the Information and Standstill Agreement 1974. Agreement of the representatives of the Governments of the Member States of the European Communities, meeting in Council of 15 July 1974 supplementing the Agreement of 5 March 1973 on information for the Commission and for the Member States with a view to possible harmonisation throughout the Communities of urgent measures concerning the protection of the environment. *OJ C 86*, 20 July 1974, p. 2.

Audland, C. J. 1976. Note regarding responsibilities of members of cabinets from Deputy Secretary General to Chefs de Cabinet and others. SEC(76) 1453, 8 April 1976.

Bana, G. 2003. Una signora sempre giovane: venticinque anni di impressioni. *In:* L'Ue, F. D. A. D. C. D. (ed.) *FACE: 25, 1977–2002. Already a quarter of a century: a commemorative book*. Brussels: FACE.

Birdlife International 2012. Ninety years spreading our wings: many partners, one BirdLife. *World Birdwatch* (December, Vol. 34(4)). London: Solo Publishing Ltd.

Birds Directive 1979. Council Directive 79/409/EEC of 2 April 1979 on the conservation of wild birds, *OJ L 103*, 25 April 1979, p. 1, since codified as Directive 2009/147/EC of the European Parliament and of the Council of 30 November 2009 on the conservation of wild birds. *OJ L 20*, 26 January 2010, p. 7.

Blair, A. 2006. *Companion to the European Union*. Abingdon: Routledge.

Boardman, R. 2006. *The international politics of bird conservation: biodiversity, regionalism and global governance*. Cheltenham: Edward Elgar.

Bowman, M. 1999. International treaties and the global protection of birds: part I. *Journal of Environmental Law*, 11, 87–120.

Bowman, M., Davies, P. and Redgwell, C. 2010. *Lyster's international wildlife law*. Cambridge: Cambridge University Press.

Brusasco-Mackenzie, M. 2011. Interview (in French) with Margaret Brusasco-Mackenzie conducted by Christian van de Velde in on 8 August 2011. Ref: INT114. HISTCOM.2: Histoire interne de la Commission européenne 1973–1986. Project coordinator: Université catholique de Louvain.

Bussière, É. 2014a. An improbable industrial policy. *In:* Dumoulin, M. (ed.) *The European Commission, 1958–1972: history and memories* (Deluxe edition), updated 2014. Luxembourg: Office for Official Publications of the European Communities.

Bussière, É. 2014b. Moves towards an economic and monetary policy. *In:* Dumoulin, M. (ed.) *The European Commission, 1958–1972: history and memories* (Deluxe edition), updated 2014. Luxembourg: Office for Official Publications of the European Communities.

Bussière, E., Dujardin, V., Dumoulin, M., Ludlow, P., Brouwer, J. W. and Tilly, P. (eds) 2014. *The European Commission, 1973–86 – history and memories of an institution*. Luxembourg: Publications Office of the European Union.

Carpentier, M. 1973. Introductory address by Mr Michel Carpentier to the Third European Symposium on Management, Davos, February 1973, III/2537/72 E.

Carpentier, M. 1999. La naissance de la politique de l'environnement. *Revue des Affaires Européennes*, 3–4, 284–297.

Carpentier, M. 2004. Interview with Michel Carpentier conducted by Éric Bussière and Arthe Van Laer on 5 January 2004. Ref: INT756. CONSHIST.COM Histoire interne de la Commission européenne 1958–1973. Project coordinator: Université catholique de Louvain.

Carson, R. 2002. *Silent spring: fortieth anniversary edition*. New York: Mariner Books.

Cashman, L. 2012. Interview with Liam Cashman, DG Environment lawyer, conducted by Andrew L. R. Jackson on 17 August 2012.

Caufield, C. 1981. Environment: Britain lags, Europe leads. *New Scientist* (13 August), 416–417.

CEC 1973. Minutes of the 251st meeting of the Council of the European Communities held in Brussels on Thursday 19 and Friday 20 July 1973. 1586/73 (PV/CONS 23).

CEC 1977. Note: Proposal for a Council Directive on bird conservation [R/3265/76 (ENV 145)] – Outcome of the meeting of the Working Party on the Environment on Friday 11 March 1977. Council of the European Communities. Brussels, 22 March 1977 (29.03). R/633/77 (ENV 46) (AGRI 174).

Chaney, S. 2008. *Nature of the miracle years: conservation in West Germany, 1945–1975.* Oxford: Berghahn Books.

CITES Regulation 1982. Council Regulation (EEC) No 3626/82 of 3 December 1982 on the implementation in the Community of the Convention on international trade in endangered species of wild fauna and flora. *OJ L 384*, 31 December 1982, p. 1.

Council of Europe 1967a. Resolution (67) 24 – Birds in need of special protection (adopted by the Ministers' Deputies on 27 October 1967).

Council of Europe 1967b. Resolution (67) 25 – Various causes of the disappearance of wild life (adopted by the Ministers' Deputies on 27 October 1967).

Council of Europe 1973. Resolution (73) 31 – on birds in need of special protection in Europe (adopted by the Committee of Ministers on 26 October 1973).

Cramp, S. 1977. *Bird conservation in Europe: a report. Prepared for the Environment and Consumer Protection Service, Commission of the European Communities.* London: HMSO.

David, H. 2012. Interview with Hubert David, founder and first Secretary General of the European Environmental Bureau, conducted by Andrew L. R. Jackson on 3 August 2012.

De Klemm, C. and Créteaux, I. 1995. *The legal development of the Ramsar Convention.* Gland, Ramsar Convention Bureau.

Devillers, P. 2013. Interview with Pierre Devillers of Belgium's IRSNB conducted by Andrew L. R. Jackson on 25 April 2013.

Dinan, D. 2007. Fifty years of European integration: a remarkable achievement. *Fordham International Law Journal*, 31, 1118–1142.

ESC 1977. Economic and Social Committee Opinion on the proposal for a Council Directive on bird conservation. *OJ C 152*, 29 June 1977, p. 3.

European Commission 1970. Directory of the Commission of the European Communities (September 1970).

European Commission 1971a. Directory of the Commission of the European Communities (September 1971).

European Commission 1971b. First Communication of the Commission about the Community's policy on the environment. SEC(71) 2616 final, Brussels, 22 July 1971.

European Commission 1972. Communication from the Commission to the Council on a European Communities' programme concerning the environment (presented 24 March 1972) (SEC(72) 666, 22 March 1972). *OJ C 52*, 26 May 1972, p. 1.

European Commission 1973. Directory of the Commission of the European Communities (October 1973).

European Commission 1974. Commission Recommendation of 20 December 1974 to Member States concerning the protection of birds and their habitats (75/66/EEC). *OJ L 21*, 28 January 1975, p. 24.

European Commission 1976a. Directory of the Commission of the European Communities (September 1976).

European Commission 1976b. Proposal for a Council directive on bird conservation (COM(76) 676 final). Brussels, 23 December 1976, R/3265/76 (ENV. 145), attaching a letter from the Commission to the Council dated 20 December 1976 with COM(76) 676 final as its enclosure.

European Commission 1980. Progress made in connection with the environment action programme and assessment of the work done to implement it. Communication from the Commission to the Council. COM(80) 222 final, Brussels, 7 May 1980.

European Parliament 1973. Report on behalf of the Committee on Public Health and the Environment on the proposals from the Commission of the European Communities to the Council (Doc. 62/73) on the programme of environmental action of the European Communities together with proposed measures to be taken in this field. Document 106/73, 27 June 1973, Rapporteur: Mr Hans Edgar Jahn.

European Parliament 1975. Resolution of 21 February 1975 on Petition No. 8/74: 'Save the migratory birds'. *OJ C 60*, 13 March 1975, p. 51.

FACE 2003. FACE: 25, 1977–2002. *Already a quarter of a century: a commemorative book. Fédération des associations de chasse de l'UE*. Brussels, FACE.

Fairbrass, J. and Jordan, A. 2001. European Union environmental policy and the UK government: a passive observer or a strategic manager? *Environmental Politics*, 10, 1–21.

First Environmental Action Programme 1973. Declaration of the Council and of the European Communities and of the representatives of the Governments of the Member States meeting in the Council of 22 November 1973 on the programme of action of the European Communities on the environment. *OJ C 112*, 20 December 1973, p. 1.

Flattau, E. 1979. Environment no longer ignored: Carpentier has gained Europe's ear. *Sarasota Journal*, 23 October 1979.

Forster, A. 2002. Anti-Europeans, anti-marketeers and Eurosceptics: the evolution and influence of labour and conservative opposition to Europe. *The Political Quarterly*, 73, 299–308.

Fried, A. 1998. US Environmental interest groups and the promotion of environmental values: The resounding success and failure of Earth Day. *Environmental Politics*, 7, 1–22.

Gammell, A. 1999. The contribution of BirdLife International to the Birds Directive and the management of birds and their habitats in Europe. 20 years with the EC Birds Directive. Proceedings from a conference on the Council's Directive on the Conservation of Wild Birds, 18-19 November 1999 Elsinore, Denmark. The Danish National Forest and Nature Agency.

Gammell, A. 2012. Interview with Alistair Gammell, formerly of the RSPB, conducted by Andrew L. R. Jackson on 31 July 2012.

Glesener, J.-P. and European Commission 1971. Written question no. 285/71, of 10 September 1971, by Mr Glesener to the Commission of the European Communities; and response from the Commission dated 12 November 1971. *OJ C 119*, 26 November 1971, p. 3.

Golub, R. and Townsend, J. 1977. Malthus, multinationals and the Club of Rome. *Social Studies of Science*, 7, 201–222.

Habitats Directive 1992. Consolidated version of Council Directive 92/43/EEC of 21 May 1992 on the conservation of natural habitats and of wild fauna and flora. *OJ L 206*, 22 July 1992, p. 7. Consolidation date 1 July 2013.

HAEU 1970–1972. Collaboration d'A. Spinelli aux activités de l'Europa Union Deutschland (Code: AS-271; Dossier: no. 000271). Historical Archives of the European Union.

Haigh, N. 1992. *Manual of environmental policy*. Harlow, UK: Longman.

Henningsen, J. 2012. Interview with Jørgen Henningsen, former Director at DG Environment, conducted by Andrew L. R. Jackson on 24 August 2012.

Hillaby, J. 1976. Out and about: playing at God. *New Scientist* (26 February), 460–461.

IELA undated. *Dr. John Temple Lang, Patron of the Irish Environmental Law Association* [Online]. Available: www.iela.ie/council-members/dr-john-temple-lang/ [accessed 30 June 2017].

Information and Standstill Agreement 1973. Agreement of the representatives of the Governments of the Member States meeting in Council of 5 March 1973 on information for the Commission and for the Member States with a view to possible harmonisation throughout the Communities of urgent measures concerning the protection of the environment. *OJ C 9*, 15 March 1973, p. 1.

Jaffeux, H. 2013. Personal communication: e-mail dated 18 April 2013 from Henri Jaffeux, ex French government official who worked on implementing the Habitats Directive, to Andrew L. R. Jackson.

Jahn, H. E. 1975. Document 449/74, Report drawn up on behalf of the Committee on Public Health and the Environment on Petition No. 8/74 'Save the migratory birds' (Rapporteur: Mr H. E. Jahn), 7 February 1975.

Jahn, H. E. and European Commission 1973. Written question no. 620/72, of 15 February 1973, by Mr Jahn to the Commission of the European Communities; and response from the Commission dated 10 April 1973. *OJ C 39*, 7 June 1973, p. 12.

Jardin, M. 2013. Personal communication: e-mail dated 12 April 2013 from Mireille Jardin, member of French government delegation that negotiated the Birds Directive, to Andrew L. R. Jackson.

Johnson, S. 2009. *Stanley I presume*. London: Fourth Estate.

Johnson, S. 2011. Interview with Stanley Johnson conducted by Christian van de Velde on 18 October 2011. Ref: INT185. HISTCOM.2: Histoire interne de la Commission Européenne 1973–1986. Project coordinator: Université catholique de Louvain.

Jordan, A. 1998. EU environmental policy at 25: the politics of multinational governance. *Environment*, 40, 14–20.

Jordan, A. and Adelle, C. 2013. EU environmental policy: Contexts, actors and policy dynamics. *In:* Jordan, A. and Adelle, C. (eds) *Environmental policy in the EU: actors, institutions and processes*, 3rd edn. Abingdon: Routledge.

Jordan, A. and Jeppesen, T. 2000. EU environmental policy: adapting to the principle of subsidiarity? *European Environment*, 10, 64–74.

Knill, C. and Liefferink, D. 2013. The establishment of EU environmental policy. *In:* Jordan, A. and Adelle, C. (eds) *Environmental policy in the EU: actors, institutions and processes*, 3rd edn. Abingdon: Routledge.

Krämer, L. 2011. Interview (in French) with Ludwig Krämer conducted by Christian van de Velde on 7 October 2011. Ref: INT194. HISTCOM.2: Histoire interne de la Commission Européenne 1973–1986. Project coordinator: Université catholique de Louvain.

Leghold Traps Regulation 1991. Council Regulation (EEC) No 3254/91 of 4 November 1991 prohibiting the use of leghold traps in the Community and the introduction into the Community of pelts and manufactured goods of certain wild animal species originating in countries which catch them by means of leghold traps or trapping methods which do not meet international humane trapping standards. *OJ L 308*, 9 November 1991, p. 1.

Lenschow, A. 2015. Environmental policy: contending dynamics of policy change. *In:* Wallace, H., Pollack, M. A. and Young, A. R. (eds) *Policy-making in the European Union*, 7th edn. Oxford: Oxford University Press.

Lytle, M. H. 2007. *The gentle subversive: Rachel Carson, Silent Spring, and the rise of the environmental movement*. Oxford: Oxford University Press.

Mansholt, S. L. 1972a. *Letter from Sicco Mansholt to Franco Maria Malfatti, President of the European Commission* [Online]. Luxembourg: CVCE Centre for European Studies. Available: www.cvce.eu/viewer/-/content/51303966-0532-46bc-89c7-271ef294eb13/en [accessed 30 June 2017].

Mansholt, S. L. 1972b. Press release: Address by Mr Sicco L. Mansholt, President of the Commission of the European Communities, to the Stockholm Conference (8 June 1972, PP/500/72-E).

McCormick, J. 1993. International nongovernmental organizations: prospects for a global environmental movement. *In:* Kamieniecki, S. (ed.) *Environmental politics in the international arena: movements, parties, organizations, and policy*. Albany: State University of New York Press.

McGurren, H. J. 1971. The externalities of a Torrey Canyon situation: an impetus for change in legislation. *Natural Resources Journal*, 11, 349.

Meadows, D. H. and Meadows, D. 2007. The history and conclusions of *The Limits to Growth*. *System Dynamics Review*, 23, 191–197.

Meadows, D. H., Meadows, D. L., Randers, J. and Behrens III, W. W. 1972. *The limits to growth: a report for The Club of Rome's project on the predicament of mankind*. New York: Universe Books.

Meyer, J. H. 2010a. Greening Europe? Environmental interest groups and the Europeanization of a new policy field. *In:* Kaiser, W. and Meyer, J. H. (eds) *Non-State actors in European integration in the 1970s: towards a polity of transnational contestation*. Leipzig: Leipziger Univ.-Verl.

Meyer, J. H. 2010b. Saving migrants: a transnational network supporting supranational bird protection policy in the 1970s. *In:* Kaiser, W., Gehler, M. and Leucht, B. (eds) *Transnational networks in regional integration: informal governance in Europe 1945–83*. Basingstoke: Palgrave Macmillan.

Meyer, J. H. 2011a. Appropriating the environment: How the European institutions received the novel idea of the environment and made it their own. Kolleg-Forschergruppe "The Transformative Power of Europe", KFG Working Paper Series.

Meyer, J. H. 2011b. Green activism: the European Parliament's Environmental Committee promoting a European environmental policy in the 1970s. *Journal of European Integration History*, 17, 73–85.

Meyer, J. H. 2014. Getting started: agenda-setting in European environmental policy in the 1970s. *In:* Laursen, J. (ed.) *From crisis to new dynamics: the European Community 1974–83*. Brussels: Bruylant.

Murawiec, L. and Di Paoli, D. 1981. Club of Rome founder Alexander King discusses his goals and operations. *Executive Intelligence Review* (23 June), 8.

Paris Summit 1972. Declaration of the first summit conference of the enlarged Community: Meetings of the Heads of State or Government, Paris, 19–21 October 1972. French (VO) and English versions. *Bulletin of the European Communities*, No. 10, 1972.

Pinder, J. 2007. Altiero Spinelli's European federal odyssey. *The International Spectator*, 42, 571–588.

Pleinevaux, C. 2010. Interview (in French) with Claude Pleinevaux conducted by Christian van de Velde on 8 October 2010. Ref: INT238. HISTCOM.2: Histoire interne de la Commission Européenne 1973–1986. Project coordinator: Université catholique de Louvain.

Poltz, W. and Conrad, B. 1976. *Vogelschutz in Europa: ein situationsbericht über den vogelschutz in den Staaten der Europäischen Gemeinschaft*. Greven: Kilda-Verlag.

Ramsar 2013. *Contracting Parties in order of their accession* [Online]. Available: http://archive.ramsar.org/cda/en/ramsar-about-parties-parties-in-order/main/ramsar/1-36-123%5E20715_4000_0__ [accessed 30 June 2017].

Ramsar Convention 1971. Convention on wetlands of international importance especially as waterfowl habitat (UN Treaty Series I-14583), amended in 1982 and 1987.

Rehbinder, E. and Stewart, R. 1985. *Environmental protection policy*. Berlin: Walter de Gruyter.

RSPB undated. *Our history* [Online]. Available: ww2.rspb.org.uk/about-the-rspb/about-us/our-history/ [accessed 30 June 2017].

Schön-Quinlivan, E. 2013. The European Commission. *In:* Jordan, A. and Adelle, C. (eds) *Environmental policy in the EU: actors, institutions and processes*, 3rd edn. Abingdon: Routledge.

Scichilone, L. 2009. The origins of the common environmental policy. The contributions of Spinelli and Mansholt in the ad hoc group of the European Commission, 1969–1972. *In:* Rasmussen, M. and Knudsen, A.-C. L. (eds) *The road to a united Europe: interpretations of the process of European integration*. Brussels: P.I.E. Peter Lang.

Scichilone, L. 2010. The European Commission and the beginnings of the EEC's environmental policy (1969–1973). *In:* Preda, D. and Pasquinucci, D. (eds) *The road Europe travelled along: the evolution of the EEC/EU institutions and policies*. Brussels: P.I.E. Peter Lang.

Seal Pups Directive 1983. Council Directive 83/129/EEC of 28 March 1983 concerning the importation into Member States of skins of certain seal pups and products derived therefrom. *OJ L 91*, 9 April 1983, p. 30.

Second Environmental Action Programme 1977. Resolution of the Council of the European Communities and of the representatives of the governments of the Member States meeting within the Council of 17 May 1977 on the continuation and implementation of a European Community policy and action programme on the environment. *OJ C 139*, 13 June 1977, p. 1.

Seefeld, H. and European Commission 1972. Written question no. 259/72, of 14 August 1972, by Mr Seefeld to the Commission of the European Communities; and response from the Commission dated 8 November 1972. *OJ C 122*, 24 November 1972, p. 11.

Segal, G. 1973. Europe's R&D men. *New Scientist* (1 February), 244–246.

Simmons, K. E. L. 1989. Obituaries: Stanley Cramp (1913–1987). *Ibis*, 131, 612–614.

Smart, M. 1976. Proceedings of the international conference on conservation of wetlands and waterfowl, Heiligenhafen, December 1974. Annex II, Recommendations for criteria to be used in identifying wetlands of international importance. Slimbridge, International Waterfowl Research Bureau.

Stuffmann, C. J. 1999. The upcoming of the Birds Directive. 20 years with the EC Birds Directive. Proceedings from a conference on the Council's Directive on the Conservation of Wild Birds, 18–19 November, Elsinore, Denmark. The Danish National Forest and Nature Agency.

Temple Lang, J. 1982. The European Community Directive on bird conservation. *Biological Conservation*, 22, 11–25.

Temple Lang, J. 1983. The Wildlife Act and EEC bird conservation measures. *In:* Blackwell, J. and Convery, F. J. (eds) *Promise and performance: Irish environmental policies analysed*. Dublin: The Resource and Environmental Policy Centre.

Temple Lang, J. 2012. Interview with John Temple Lang, ex-official with the European Commission's Legal Service, conducted by Andrew L. R. Jackson on 17 August 2012.

Temple Lang, J. 2013. Personal communication: e-mail dated 17 April 2013 from John Temple Lang, ex-official with the European Commission's Legal Service, to Andrew L. R. Jackson.

The Ecologist 1974. Interview with Sicco Mansholt. *The Ecologist*, 4 (August/September), 259–260.

Third Environmental Action Programme 1983. Resolution of the Council of the European Communities and of the representatives of the Governments of the Member States, meeting within the Council, of 7 February 1983 on the continuation and implementation of a European Community policy and action programme on the environment (1982 to 1986). *OJ C 46*, 17 February 1983, p. 1.

Toke, D. and Marsh, D. 2003. Policy networks and the GM crops issue: assessing the utility of a dialectical model of policy networks. *Public Administration*, 81, 229–251.

Toulemon, R. 2003. Interview (in French) with Robert Toulemon conducted by Gérard Bossuat, Éric Bussière and Anaïs Legendre on 17 December 2003. Ref: INT721. CONSHIST.COM Histoire interne de la Commission Européenne 1958–1973. Project coordinator: Université catholique de Louvain.

Treaty Of Rome 1957. Treaty establishing the European Economic Community. United Nations Treaty Series.

Van de Velde, C. 2014. Environmental and consumer protection. *In:* Bussière, E., Dujardin, V., Dumoulin, M., Ludlow, P., Brouwer, J. W. and Tilly, P. (eds) *The European Commission, 1973–86 – history and memories of an institution*. Luxembourg: Publications Office of the European Union.

Van der Harst, J. 2014. Sicco Mansholt: courage and conviction. *In:* Dumoulin, M. (ed.) *The European Commission, 1958–1972: history and memories* (Deluxe edition), updated 2014. Luxembourg: Office for Official Publications of the European Communities.

Van der Ven, J. 2012. Interview with Joost van der Ven, Dutch nature conservation NGO representative at the time of the Birds Directive negotiations, conducted by Andrew L. R. Jackson on 24 August 2012.

Weale, A. 1999. European environmental policy by stealth: the dysfunctionality of functionalism? *Environment and Planning C: Government and Policy*, 17, 37–51.

Whales Regulation 1981. Council Regulation (EEC) No 348/81 of 20 January 1981 on common rules for imports of whales or other cetacean products. *OJ L 39*, 12 February 1981, p. 1.

Wurzel, R. 2008. Environmental policy: EU actors, leader and laggard States. *In:* Hayward, J. (ed.) *Leaderless Europe*. Oxford: Oxford University Press.

Wurzel, R. 2012. From environmental disunion towards environmental union? *In:* Hayward, J. and Wurzel, R. (eds) *European disunion: between sovereignty and solidarity*. Basingstoke: Palgrave Macmillan.

4

NEGOTIATION OF THE BIRDS DIRECTIVE (1976–1979) AND ITS EARLY IMPLEMENTATION

Habitat protection in the Birds Directive

Introduction

While the hunting of migratory birds had been the trigger that led many to campaign for EU legislation on birds (Meyer, 2010), by the time the Birds Directive proposal emerged in December 1976 – having been adopted by the Commission "without great internal difficulties" (Stuffmann, 1999, p. 16) – the policy aim had shifted in an important respect, with habitat protection now firmly rooted in the proposal (European Commission, 1976), due in no small part to John Temple Lang's work. "It is clear", the Explanatory Memorandum to the Commission's proposal records, "that the situation of some 400 species of birds observed regularly in the territory of the Member States is giving rise to serious problems for several reasons . . . [S]ome sixty species are currently in danger of extinction, whilst not even one third of the species show normal reproduction rates" (European Commission, 1976). As with Cramp's (1977) report, the first cited reason for this state of affairs was habitat change and habitat loss: "Wildlife cannot be preserved without adequate space for the survival and reproduction of the species", records the Explanatory Memorandum for the proposal (European Commission, 1976).

It is a sign of how far nature conservation policy has come in the EU that in 1976 only seven of the then nine EU Member States had national bird protection laws and hunting control laws, France and Italy having no such provisions (European Commission, 1976). The aim of the proposed directive was to introduce a general scheme of protection to maintain the numbers of all species of naturally occurring birds in the wild state[1] in the European territory of the Member States at a level compatible with certain requirements – ecological, economic, recreational and scientific (European Commission, 1976). As Temple Lang (1982, p. 12) commented shortly after the directive's adoption, "The directive is comprehensive. It applies to all species of birds occurring naturally in the wild state in all the European territories of Member States, but not to Greenland or to the French overseas departments and territories. It applies to the habitats of all these species, as well as to their nests and eggs". Here we will focus on the habitat protection aspects of the proposed directive, since this triggered the creation of a network of EU protected areas for certain bird species, which network was later incorporated within Natura 2000.

In the words of the Commission's proposal,

> Because of the special situation of certain species, Article 4 provides for additional measures to protect the habitats. These measures relate to rare species or those in danger of extinction, on the one hand, and to migratory birds, on the other. Annex I lists the species regarded as rare or in danger of extinction for the purposes of this Directive. The Annex was drawn up in the light of the need for special protection measures at Community level to cover the European territory of the Member States. Each Member State will designate and manage, as special protection areas, those areas most suitable for the survival and reproduction of those species, taking into account the special requirements of migratory birds. There must be a coherent Community network of such areas, which is why the Commission is to be consulted. Finally, it is stipulated that any pollution, deterioration of the habitat or disturbances of birds which has a significant effect in respect of the objectives of these special protection measures must be prevented. This provision is a logical and necessary supplement ensuring that the special measures to protect the habitat make a fully effective contribution.
>
> *(European Commission, 1976)*

Member States' general reactions to the proposed directive

The Directive was proposed by the Commission on 20 December 1976, and fell to be negotiated during, successively, British, Belgian, Danish, German and French Presidencies (Stuffmann, 1999). There is nothing in the Member States' initial reactions to the draft Directive in general and to Article 4 in particular that indicates any recognition that something new and extraordinary had been proposed – that is, a coherent system of protected EU nature conservation areas that would ultimately, following the addition of sites under the Habitats Directive, come to cover almost one-fifth of the EU's land area as well as large marine areas. As Alistair Gammell of the RSPB comments,

> at the time of the Birds Directive being passed, I really don't think that it was in the minds of the British or any other government that this was primarily about habitat protection. It was primarily about not killing migratory birds, and controlling hunting . . . I'm not suggesting they didn't read and understand those articles concerning habitat protection and the need

> for SPAs, etc., but what I'm saying is that the emphasis of their thinking of what this was about wasn't in that direction at all, and that's what has changed, really, in the last 20 or 30 years, is that these things have become far more to do with habitat protection and less to do with hunting, although hunting is still an issue and still regularly crops up.
>
> *(Gammell, 2012)*

Hunting was a hot topic at the time, however, and the proposal did not sail through the Council, the rhythm of meetings to discuss the draft proposal being "unexpectedly dense", according to Claus Stuffmann.[2] During the Member States' first discussion of the Commission's formal proposal, at a meeting of the Working Party on the Environment on 11 March 1977, most Member States broadly welcomed the proposal, albeit that Germany and the Netherlands – widely agreed to be two of the initial environmental leader states (Wurzel, 2013, p. 90) – made particularly positive contributions, and France and Belgium, with powerful pro-hunting lobbies, expressed some reservations, which would later be voiced rather more forcefully.[3]

Denmark's reluctance to allow the EU to regulate nature conservation

Denmark's position is worthy of particular note: whilst characterised as a leader in EU environmental policy, historically Denmark has a reputation as a "forerunner" state, aiming to maximise its freedom to develop its own ambitious domestic environmental policy (Wurzel, 2013, p. 89), Denmark also has a reputation for Euroscepticism (Archer, 2000). Cichowski (2007, p. 164) highlights the fact that "public debate in Denmark over the last few decades has criticized EU regulations as presenting a threat to the potentially high-level of protection embodied in national legislation". As Pagh (1999, pp. 302 and 317) put it in 1999: "for almost two decades the political left-wing and the Greens in Denmark regarded EU and EU environmental law as the target rather than an ally in their attempts to ensure greater environmental protection . . . [P]ublic debate in Denmark on EU environmental law has been . . . to what extent does EU law prevent Denmark from having a high level of protection".[4]

Thus, while Denmark is regarded as one of the "green trio" of initial environmental leader states (together with Germany and the Netherlands) (Wurzel, 2013, p. 90), Denmark was often sceptical of EU environmental proposals, and of attempts to expand the EU's competences into new areas, particularly areas, such as nature conservation and land use planning, which had been expressed in Denmark to be off-limits for EU regulation during the Danish debate over whether or not to join the EU (Koester, 2012). In respect of the Birds Directive proposal, for example, of the then Member States, Denmark was the "most reluctant", according to Stuffmann (1999, p. 16).[5] Veit Koester, Denmark's negotiator at the time, explains:

> When the Birds Directive was negotiated, our position within the EU, the Danish official position within the EU, was that nature conservation and land use, land planning [were] not matters that should be built within the EEC . . . that they were, those areas were under national sovereignty. That was the official policy. But when the Birds Directive, this idea came up, we were sort of being split between arguing that nature conservation should be reserved for the members themselves to decide, and then on the other side, we had a situation where hunting and therefore protection of birds was an area which did not belong to the Ministry of Environment at that time, it was under the Ministry of Agriculture, so we had at that time constant battles with the Ministry of Agriculture, which was promoting hunting and then our Ministry, which promoted protection. So we saw probably some advantages in having this directive because this could be a means to sort of squeezing the Ministry of Agriculture . . . and then we had to give up, to a certain extent, but only to a limited extent, this idea of having land use and nature conservation as domains reserved for the Member States themselves, but it was not that difficult to give it up because it was sort of being limited to certain areas which were to some extent already protected by legislation, so we did not feel that we give in very much by having the habitat protection provisions in the [Birds] Directive.
>
> *(Koester, 2012)*[6]

The UK – widely perceived as "the dirty man of Europe" for much of the 1970s and 1980s (Wurzel, 2013, p. 90) – had a reputation, alongside Denmark, as a minimalist, traditionally fighting "to preserve national autonomy in many areas and [preferring] limiting integration to a much lower level of political cooperation", particularly when compared with maximalists such as the Commission, the European Parliament and (often) the six founding Member States, which have tended to favour closer cooperation (Jordan, 1998, p. 16). In addition to not wanting the Birds Directive to be a popular success for the EU, the UK's then Secretary of State for the Environment Peter Shore wrote in May 1978 that he was opposed in principle to the EU's becoming involved in the regulation of issues such as bird protection.[7] Shore evidently did not object strongly enough or hold sufficient sway to prevent the directive's (unanimous) adoption. For its part, Germany argued that given its federal structure the Länder would have difficulties with anything that would reduce their competence; however, it was made clear, according to Stuffmann (1999, p. 17), that the Bundesrat "would most probably not veto a directive containing very strict restrictions mainly of unselective hunting methods".

General comments on the draft Birds Directive

In terms of general comments on the draft Directive (all recorded in CEC (1977a, pp. 2–7)), the Netherlands expressed itself to be "very much in favour" of the proposal; Italy was "very favourably disposed", and said it would be "making a number of suggestions, aimed for example at improving the protection of habitats"; Ireland was "very constructive" and said that the proposed directive "was based on exactly the same principles as Irish legislation";[8] France was "very much in favour of the aims of the Directive", but noted the "not inconsiderable political interest" and argued that hunting – which was "essentially a recreational or leisure activity" – should be regulated at regional rather than EU level; in contrast, Germany said it was "very pleased with the form of the Directive, which it said was very important"; Denmark indicated its agreement with the aims and principles of the proposal, whilst noting that the Directive provided for very considerable harmonisation of national laws, and querying the criteria used by the Commission in drawing up the Annexes; Belgium approved of the aims and principles of the proposed Directive, but thought that considerable improvements could be made to the form and substance; echoing France, Belgium also expressed concerns about the balance that had been struck between protection and hunting, arguing that "hunting should not simply be presented as an exception that was tolerated, but as a specific form of recreational activity"; finally, the UK expressed approval for the form of the Directive, and noted that the problem was one that could not be satisfactorily solved at the national level; as a general comment, the UK "did not think the Directive posed any major problems".[9] How wrong this would later prove.[10] Nevertheless, the generally positive reception given to the Commission's proposal, and the detailed negotiations which followed, contradict to some degree Meyer's (2011, p. 84) claim of a "recalcitrant Council", or at least reveal this as an oversimplification of a complex institution.

Legal basis for the proposal

At the Council's second formal discussion of the Birds Directive proposal, at a meeting of the Working Party on the Environment on 23 and 24 June 1977, the issue of the legal basis for the proposed Directive came to the fore (CEC, 1977b). Clearly aware of potential competence problems – the EU not yet having specific legal powers to make environmental policy – the Commission had been keen to emphasise the common market credentials of its proposal, which came in the form, the Commission argued, of regulating the trade in dead game birds and in live birds of certain species. However, the link between this and the proposed directive's wider conservation measures for wild birds was not as evident as the Commission tried to make out in its Explanatory Memorandum for the proposal (emphasis added):

> The differing laws of the Member States in this field [i.e., trade in birds] are likely to affect trade between Member States and therefore have a direct effect on the functioning of the Common Market. *This is why* any harmonisation of legislation *must be accompanied* by Community action for the

> protection, conservation and management of all species of wild bird regularly observed in the territory of the Member States; such action would fall within the scope of Article 235 of the EEC Treaty.
>
> *(European Commission, 1976, paragraph II(3))*

Certain (unnamed) Member States expressed doubts about EU competence in the matter (Council Legal Service, 1977),[11] and the Working Party on the Environment therefore asked the Council Legal Service to draw up an opinion on the legal basis for the proposed directive, and the Commission Legal Service committed to doing the same (CEC, 1977b). Pending these opinions, Germany and Denmark entered general reservations about the entire proposal (CEC, 1977b). As Stuffmann (1999, p. 17) explains, "The fundamental political problem was in Denmark where Parliament had tied the hands of the government because it did not want to accept any Community competence in the area of nature conservation and therefore strongly questioned the validity of the legal basis".

On 25 July 1977, the Council Legal Service (1977) delivered its opinion. The Commission had based the proposed Directive on two powers – Articles 100 and 235 of the Treaty of Rome (1957) – and was considering adding a third power, Article 43, which provided for the making of legislation in respect of the common agricultural policy (Council Legal Service, 1977), the European Parliament having requested the addition of this base in its opinion on the proposed Directive of 14 June 1977 (European Parliament, 1977). Article 100 of the Treaty allowed the Council, acting unanimously on a proposal from the Commission, to issue directives to approximate differing national laws, provided that the differences "directly affect the establishment of the common market". However, the Council Legal Service was not convinced that "any provision of the draft directive aims to harmonise trade in birds between Member States" (Council Legal Service, 1977, paragraph 7). Article 6, the Legal Service argued, "is a prohibition on trade in birds, with the exception of those mentioned, which may be sold, and seems to have as its object the protection of birds, rather than any harmonisation of trade . . . No evidence has otherwise been offered by the Commission to show that national laws regarding trade in birds, or any other aspect of national legislation relating to them, affect the establishment or functioning of the Common Market" (Council Legal Service, 1977, paragraphs 7 and 8). So much, then, for Article 100 as a basis for the Birds Directive.

The focus then turned to Article 235 of the Treaty, which allowed the Council, acting unanimously on a proposal from the Commission, and after consulting the European Parliament, to legislate where action by the EU should prove necessary to attain, in the course of the operation of the common market, one of the objectives of the Community, but the Treaty had not provided the necessary powers. The Council Legal Service looked to Article 2 of the Treaty for the objectives of the Community: "to promote throughout the Community a harmonious development of economic activities, a continuous and balanced expansion, an increase in stability, an accelerated raising of the standard of living and closer relations between the States belonging to it". The question then was whether the protection of birds could be said to fall within these objectives; to this question, the Council Legal Service (1977) said, the Council had already given its own answer, pointing in this regard to the First Environmental Action Programme (1973) and Second Environmental Action Programme (1977), which stated that "improvement in the quality of life and the protection of the natural environment are among the fundamental tasks of the Community". On this basis, according to the Council Legal Service, Article 235 provided a legal power for making the Birds Directive. Finally, the Council Legal Service opined that Article 43 of the Treaty would not be an appropriate legal base for the proposed directive because furtherance of the common agricultural policy was "evidently not the object of the proposal" (Council Legal Service, 1977, paragraph 14).[12]

Notwithstanding the fact that the Birds Directive was ultimately adopted solely on the basis of Article 235, it is clear that certain delegations – amongst them Germany and Denmark – remained uneasy about the EU's move into nature conservation regulation. Germany prepared a draft statement providing that: "The German delegation points out that its agreement to this Directive cannot be interpreted as constituting a precedent for any further action by the European Communities in the field of nature conservancy" (CEC, 1977f, p. 30). This position later softened – or at least was explained in more detail – as follows:

> The German delegation is doubtful as to the adequacy of Article 235 of the EEC Treaty as the sole legal basis for the Directive on bird conservation. In its opinion Article 100 should also have been cited. [Germany] points to the fact that its agreement on this Directive is not to be interpreted as pre-judging any other Community activities based solely on Article 235 of the EEC Treaty, particularly in the field of nature conservation. Moreover the German delegation wishes to point out that, with respect to other Community activities, specifically in the fields of nature conservation and protection of the countryside, a thorough examination will have to be made of Community powers.
>
> *(CEC, 1978, p. 47)*[13]

Interestingly, we are faced here with an apparent loss of institutional memory: Germany's argument could have been countered by highlighting that (emphasis added) "the Paris Summit Conference [of Heads of State or Government of October 1972] had requested the Community institutions to set up an environmental policy by making *the widest possible use* of Article 235 of the Treaty", per Laurens Jan Brinkhorst of the Netherlands, who would later play a key role in respect of the Habitats Directive (CEC, 1973, p. 23) (see Chapters 6–9).[14]

Denmark continued to query the EU's competence to legislate for bird conservation right to the end, and in its own draft statement doubted

> whether the provisions of this [Birds] Directive correspond to the objectives defined in Article 2 of the EEC Treaty and whether Article 235 is applicable[15] . . . While the Danish Government has not wished to refute the arguments put forward in this matter by the other eight Member States and the legal departments [i.e., the Council and Commission Legal Services], it wishes to point out that it does not consider that the adoption of this Directive can create a precedent for the adoption of further Community acts of a similar nature.
>
> *(CEC, 1977f, p. 31, 1978)*

Reflecting in 1980 on seven years of EU environmental policy, the Commission commented:

> Recourse to Article 235, either on its own or in conjunction with another Article of the Treaty, has not caused any particular difficulties, apart from a few cases relating to the protection of wildlife. The Commission hopes that reservations in this sphere will be withdrawn since the international character of the matters involved, their effects on various policies (commercial policy, agricultural policy and fisheries policy) and public sensitivity on these matters are in themselves sufficient to justify the action undertaken by the Community in this respect.
>
> *(European Commission, 1980a, p. 6)*

The Commission's hopes were not readily fulfilled – it would take another eight years, as well as a Treaty change, before the Habitats Directive would be proposed.

Site selection: to consult the Commission or not?

At their first discussion of the proposed Birds Directive, on 11 March 1977, several Member States expressed their opposition to the proposed requirement to consult the Commission before designating special protection areas under Article 4 (all quotes in this paragraph taken from CEC (1977a, pp. 12–13)). France requested that "for practical reasons and for reasons of principle an *a posteriori* notification be introduced instead". The German and Danish delegations thought that "any duplication of the activities of other international organisations working in the same field should be avoided"[16] and that "no attempt should be made to introduce new and cumbersome procedures". Germany and Denmark argued that it would be enough to encourage bilateral cooperation and for the Member States to confine themselves to the biannual reporting requirements set out in Article 11 of the Commission's proposal (European Commission, 1976). The UK expressed some support for this view, suggesting that Article 4 should adopt the expression "if necessary in consultation with other Member States". Striking a discordant note were the Netherlands and Belgium, which argued that "it was particularly in this field of the protection of birds, which knew no frontiers, that prior consultation of the Commission and the other Member States was necessary to ensure the coherent approach desired and to strengthen co-ordination of the measures to be taken". The Commission responded that its sole purpose in proposing the consultation procedure was "to ensure a degree of coherence and rationality that did not exist at present in the framework of existing international organisations, which were consulted on a purely voluntary basis". The Commission also emphasised that "in any case the Member States would remain completely free to take decisions, as the Commission's opinion was only consultative".

At the second formal discussion of the proposal, at a meeting of the Working Party on the Environment on 23 and 24 June 1977, the UK – then holding the Presidency – had softened its earlier approach somewhat, now arguing that the relevant wording should read "in consultation with the Commission" (and not, per the Commission's proposal, "after" consulting the Commission (CEC, 1977b, European Commission, 1976) or the UK's earlier proposal "if necessary in consultation with other Member States" (CEC, 1977a)). Nevertheless, the final version of the Birds Directive (1979, Article 4(3)) merely required Member States to "send the Commission all relevant information so that it may take appropriate initiatives with a view to the coordination necessary to ensure that the [special protection areas] form a coherent whole which meets the protection requirements of these species in the geographical sea and land area" where the Directive applies. A Resolution adopted in parallel to the Birds Directive (1979) contained additional detail regarding this process, committing the Member States to notify the Commission within two years of the SPAs that they had classified (CEC, 1979).[17]

Selecting special protection areas

At a Working Group meeting on 28 and 29 July 1977, Italy asked that an indication be provided of the number and total area of zones to be classified as special protection areas (CEC, 1977b). France developed this idea at a Working Group meeting on 22 and 23 September 1977, suggesting that provision should be made in the context of Article 4 for the drawing up of a list in an annex specifying a number of special protected areas on a geographical basis, and suggested the Ramsar Convention's (1971) list of wetlands might be adopted for this purpose (CEC, 1977g). Italy suggested drawing up a list of areas "determined on the basis of the criteria of international conventions", and the Member States "would then indicate the minimum number (50 or 60%) from this list of areas which they particularly wished to protect" (CEC, 1977g, p. 4). Both the French and Italian approaches would of course have limited the conservation impact of Article 4: the French proposal appears to have been aimed at limiting Article 4 to wetlands, when in fact the proposed (and finally adopted) provision extended much further than this; and the Italian approach envisaged designating perhaps as few as 50% of the sites indicated as requiring protection pursuant to international conventions. However, the other Member States and the Commission argued that it did not seem advisable to determine geographically a certain number of special protection areas as part of the Birds Directive proposal, since this would delay the directive's adoption (CEC, 1977g).

In a summary report provided to the Council on 4 December 1977, the Committee of Permanent Representatives (COREPER) outlined the main outstanding issues, recording the fact that France and Italy regretted the lack of detail regarding the habitat protection parts of the Directive (e.g., the absence of criteria for selecting sites), and that those countries had requested the adoption of a Resolution inviting: (a) the Member States to announce the areas chosen for the application of Article 4, and (b) the Commission to make proposals concerning, among other things, the coordination of these areas, their status and characteristics (CEC, 1977f). Denmark and France separately proposed text for such a Resolution, which texts give a fascinating insight into the then potential future direction of EU nature conservation policy. The French draft called "upon the Commission to take the necessary co-ordinating steps to ensure that the network which [the Council] establishes can be integrated into a larger network, should the need arise" (CEC, 1977f, p. 24). Similarly, the Danish draft provided:

> The Council calls upon the Commission to recommend to Member States to, if needed, take protective measures in further areas in order to ensure that areas protected pursuant to Article 4 form a coherent system that can be integrated into future Community action in the field, cf. Point 5 below . . . [Point 5] The Council renews in this connection its intentions to consider suggestions from the Commission based on paragraphs 89–92, 108–112, 115–118, 130–133, 137 and 154–159 of the [Second Environmental Action Programme (1977)], the paragraphs being of special importance for the protection of the wild fauna of Europe.
>
> *(CEC, 1977f, p. 27)*

While one could argue that the French draft may simply have had in mind a larger network of special protection areas for birds under the Birds Directive (e.g., on the accession of new EU Member States such as Greece, Spain and Portugal, which had applied for EEC membership in 1975 (Greece) and 1977 (Spain and Portugal), respectively), the Danish draft was explicit: special protection areas under the Birds Directive may need to be integrated into "future Community action in the field", relating, for example, to "the protection of the wild fauna of Europe". It is important to note in this regard that the EU's Second Environmental Action Programme (1977) had been adopted on 17 May 1977, between the Council's first and second formal discussions of the Birds Directive proposal, and hence by the time of Denmark's and France's proposed Resolutions regarding site selection under the Birds Directive. The Second Environmental Action Programme (1977) noted the ongoing discussions regarding the proposed Birds Directive, and effectively sowed the first seeds at EU level of the idea of a broader EU nature conservation instrument applying to flora and fauna in addition to birds, as adverted to by Denmark's draft

Resolution. Amongst other things, the Second Environmental Action Programme (1977) set in motion the development of an ecological mapping system, which would later prove important in the context of the Habitats Directive (1992), and highlighted the Council of Europe's ongoing examination of the possibility of creating an international instrument to protect wildlife and to preserve biotopes (which ultimately became the Bern Convention (1979) on the conservation of European wildlife and natural habitats),[18] adding: "in so far as action is necessary at Community level in order to ensure that the instrument is satisfactorily applied, the Commission will submit appropriate proposals to the Council" (Second Environmental Action Programme, 1977, p. 26).

We are therefore presented with something of a conundrum: the wording of Denmark's proposed Resolution in respect of the Birds Directive clearly recognised the possibility of a larger network of EU protected areas in the future, covering wild fauna other than birds, for example; yet, at the same time, as we have seen, Denmark remained sceptical of the EU's competence to legislate in the area, and of extending the Commission's powers in this direction. One might perhaps view Denmark's draft Resolution as an example of environmental pragmatism: Denmark perhaps viewed the development of additional EU nature conservation policy as likely – perhaps even inevitable – given the text of the Second Environmental Action Programme (1977), and had the foresight to seek to secure the best environmental outcome (a coherent network), irrespective of whether Denmark fully supported at that time the idea of additional EU nature conservation measures. In any event, the final version of the Resolution followed France's proposed wording, yet clearly left the door open to the idea of a network of protected areas covering more than merely birds and their habitats: "The Council calls upon the Commission to take the necessary coordinating steps to see that the network thereby established fulfils the objectives of the [Birds] Directive and can be integrated into a larger network, should the need arise" (CEC, 1979).

Site protection

The proposed text of Article 4 in the European Commission's (1976) original proposal read:

> Article 4
>
> 1. The species listed in Annex I shall be the subject of more stringent protection measures aimed at the conservation of habitats in a sufficient number of areas to ensure the survival and reproduction of these species throughout the territory of the Member States. To this end, after consulting the Commission, the Member States shall designate and classify the most appropriate areas as special protected areas.
> 2. Each in respect of its own territory the Member States shall, after consulting the Commission, take similar measures concerning migratory

species in order to establish protected areas which are adequate in number and in area for the reproduction, moulting and wintering of these species and in order to provide staging points within their zones of migration. The Member States shall take the requisite measures to preserve recognized wetlands of international importance.

3. The Member States shall take appropriate measures to prevent, in the special protected areas referred to in paragraphs 1 and 2 of this Article, the pollution and deterioration of habitats and the disturbance of birds, should these factors have a significant effect having regard to the objectives of this Article.

Insofar as site protection was concerned, at the first discussion of the proposed Birds Directive, on 11 March 1977, France and Ireland struck the most negative tone, arguing for the entire first sentence of Article 4(1) to be deleted (CEC, 1977a, p. 12). The UK and Italy sought basic clarifications,[19] including querying what exactly was meant by special protection area, "which might be synonymous with a special nature reserve for a particular species" (CEC, 1977a, p. 12). At the second discussion of the proposal, in June 1977, Germany – ever a green leader – suggested (ultimately unsuccessfully – see CEC, 1977b) introducing a new Article 4(4), to read: "The Member States shall take appropriate measures in respect of the species listed in Annex I to reintroduce species or populations threatened with extinction or already extinct into biotopes within their natural habitat" (CEC, 1977b, p. 3). Ireland sought to weaken Article 4(3) – then the key site protection provision – at the same meeting by suggesting that the words "shall take appropriate measures to prevent, [within special protection areas,] the pollution and deterioration of habitats and the disturbance of birds", etc., be replaced by "shall take all practical steps to prevent or minimise", etc. This attempt was not ultimately successful, however, with the "appropriate measures" language being retained in the adopted version of the directive (see Birds Directive, 1979, Article 4(4)).

On 26 July 1977 – during a Belgian Presidency – the Commission produced a revised draft of certain provisions of the proposed Birds Directive, including Article 4 (European Commission, 1977a). The revised Article 4 provided:

Article 4

1. The rare or endangered species listed in Annex I shall be the subject of special conservation measures in respect of their habitat.

   To this end, and keeping in mind the necessity to protect the above species over the territory of the Community as a whole, the Member States shall classify as special protected areas those most suited to the conservation of the said species.

2. Keeping in mind the need for protection over the territory of the Community as a whole, the Member States shall take similar steps in

respect of migratory species and their requirements as regards breeding and wintering areas and staging points along their migration routes. In this connection the Member States shall pay particular attention to the protection of wetlands.

3. The Member State shall send the Commission any relevant information to enable it to take appropriate initiatives for the required coordination in order to ensure that the areas referred to respectively in paragraphs 1 and 2 above form a coherent whole which meets the protection requirements of these species over the entire Community territory.
4. In respect of the special protected areas referred to in paragraphs 1 and 2 above, the Member States shall take appropriate steps to prevent pollution or deterioration of the habitat or any disturbances affecting the birds, insofar as these would be significant in relation to the objectives of this Article.

(European Commission, 1977a)

At their Working Group meeting on 28 and 29 July 1977, the Member States gave their views on this revised draft. Ireland maintained its negative approach, entering a reservation regarding the entirety of Article 4(4), which specified the level of protection to be provided to special protected areas, on the basis of "its fear that drainage programmes would be substantially restricted and agricultural development thereby impeded" (CEC, 1977b, p. 5). Several decades later, these same concerns persist in Ireland in respect of raised bog special areas of conservation under the Habitats Directive (Quirke, 2012), highlighting the fact that positions expressed by Member States during negotiations are not necessarily dropped on a directive's being adopted, and may re-emerge during implementation at the national level.

France and Italy offer support for hunting (via habitats)

More generally, at a meeting of the Working Party on the Environment on 8 and 9 September 1977, the French government, with its strong pro-hunting sensitivities, opened up a novel flank in the negotiations regarding habitat protection, arguing that the French government was not in favour of defining a prohibited list of methods of large-scale capture and killing of birds, such as the one then set out in Annex IV (CEC, 1977c).[20] Rather, France argued, it would be better simply to rely on the strict protection of areas of international importance under Article 4 (CEC, 1977c). In other words: don't ban large-scale methods of capture and killing generally, just have a few strictly-protected areas within which such hunting methods could not be used. Such an approach would of course have greatly limited the conservation impact of the directive, since the French approach would have entailed large-scale methods of capture and killing continuing outside protected areas.

Italy offered its support to this argument at a meeting of the Working Party on the Environment on 13 and 14 October 1977, both countries stating that they attached greater importance to the Directive's provisions regarding habitat protection than to the provisions concerning rules for hunting; the other Member States and the Commission stated that they attached equal weight to the habitat protection and hunting rules (CEC, 1977d). In this regard the French and Italians echoed the sentiments – or vice versa – of the then recently-established EU hunting coalition FACE, which stressed in its lobbying efforts the importance of habitat protection: "the destruction of habitats, they argued, accounted for the decline in bird populations" (Meyer, 2010, p. 190). As Stuffmann (1999, p. 17) explains,

> France and Italy [came] under heavy fire from their hunting lobbies which were notably afraid of the general protection status for all bird species with relatively limited exceptions as well as of the explicit prohibition of a number of 'traditional' hunting methods. Thus they insisted very strongly on a reinforcement of the habitat part which of course was completely out of [the] question for the other Member States and would have postponed the adoption of the directive for years. The outcome of this battle was rather positive insofar as the provisions relating to habitat protection were reinforced by the obligation to inform the Commission of the classification of special protected areas in order to enable it to undertake co-ordinative efforts with the perspective to create a coherent network.

On 20 November 1977, the Council's Working Party on the Environment provided a summary of outstanding issues regarding the directive to COREPER (CEC, 1977g). The draft directive attached to this summary reveals that the habitat protection provisions had been strengthened in another way by this time, apparently without much discussion, a line having been added to the end of Article 4(4) reading "Outside the special protected areas, Member States shall also make every effort to prevent pollution or deterioration of habitats" (CEC, 1977g). This was watered down in the final version of the directive to read, "Outside these protection areas, Member States shall also strive to avoid pollution or deterioration" (Birds Directive, 1979).[21] Thus, already in the Birds Directive the Member States were thinking beyond protected areas to the wider landscape, an issue that would assume much greater significance in the context of the Habitats Directive, where the protection of corridors and stepping stones would be a subject of much discussion.

By the end of November 1977 the text of Article 4 was more or less agreed, and only minor amendments were made to the provision between then and the adoption of the Directive in April 1979. In a report to the Council dated 4 December 1977, COREPER summarised the main outstanding issues, which included the following. Denmark and France had separately proposed text for a Resolution to be adopted at the same time as the Directive – discussed above – concerning, amongst other things, the coordination of special protection areas, their status and characteristics (CEC, 1977f). The Resolution ultimately adopted reveals a lack of clarity regarding how the system of protected areas would work in practice (emphasis added):

> The Council takes note of the Commission's intention of submitting appropriate proposals regarding the criteria for the determination, selection, organization and methods of administration of the special protection areas, and invites the Commission to take into consideration in particular in those proposals *the parts of the areas selected which are to be given intensive protection*, a minimum threshold for those parts, which will enable the objectives of Article 4 to be attained, and the measures to be taken to prohibit hunting and to control other specific activities likely to disturb the birds.

As it would later turn out, the CJEU interpreted the protective provisions of the Birds Directive uniformly across sites; i.e., the CJEU did not apply different standards of protection to different areas of sites, perhaps confounding certain Member States' expectations. Moreover, as late as 1990 an Advocate General of the CJEU highlighted that while, per the Resolution, the Commission had intended to submit appropriate proposals regarding the criteria for the determination, selection, organisation and methods of administration of the special protection areas, it had not yet done so (CJEU, 1990, paragraphs 6 and 35).

By May 1978 the habitat protection provisions of the Directive had been agreed (CEC, 1978), by December 1978 political agreement had been reached regarding the entire Directive (Meyer, 2011, p. 84; European Commission, 1979a) – during a German Presidency of the Council – and on 2 April 1979 the Directive was formally adopted, accompanied by a separate Council Resolution of the same date, as originally proposed by the Commission as a compromise in response to French and Italian demands (CEC, 1977g, 1977f). John Temple Lang's original drafting of the key site protection provision (European Commission, 1976, Article 4(3)) had barely changed in over two years of negotiations (Birds Directive, 1979, Article 4(4), first sentence).

Site selection and protection: the species to be protected

A first formal discussion regarding the Annexes to the proposed directive was held at a meeting of the Working Party of the Environment, composed of ornithologists, on 13 and 14 October 1977 (CEC, 1977e). At that time, Annex I to the Commission's original proposal, which listed species requiring the designation of special protection areas pursuant to Article 4, comprised 62 species in total (CEC, 1977e). Per the European Commission's (1976, Article 4(2)) proposal, migratory species not listed on Annex I would, of course, also require the designation of special protection areas. The Member States appear to have had remarkably few negative comments regarding the Commission's proposed species list for Annex I (perhaps owing to advance consultations with the Commission), with only Germany, Italy and France raising objections.[22] Interestingly, when one considers the history of the later Habitats Directive in respect of which the Member States sought to greatly reduce the Commission's proposed lists of species requiring Natura 2000 sites, here the Member States proposed *supplementing* the Commission's Annex I list by the addition of 16 species.[23] Following negotiations, the end result was that the Commission's original proposed list of 62 Annex I species grew to a final list of 74 species in the version adopted in 1979 (Birds Directive, 1979). The list has since grown enormously, following amendments pursuant to EU enlargement, amongst other things. By 1985, for example, the number of species listed on Annex I had almost doubled to cover 144 species (Wils, 1994; European Commission, 1985), and the version of Annex I for the EU 28 comprises 193 species and sub-species (Codified Birds Directive, 2009),[24] covering, amongst other areas, important countries for birds such as Spain and Poland, as well as Arctic areas, which of course were not within the territories of the nine EU Member States in 1979 (Temple Lang, 2012c).

Interestingly, future changes to the Annex I list of species requiring the classification of SPAs were left to a Committee comprising Member States and chaired by the Commission, which would operate by way of qualified majority voting (Stuffmann, 1999; Birds Directive, 1979, Articles 15 and 17). However, by ensuring that key hunting annexes (e.g., the list of huntable species in Annex II) were not covered by the qualified majority voting procedure before this Committee, France – and the hunters' organisations (Gammell, 1999) – ensured that such hunting annexes would be extremely difficult to amend, requiring a unanimous decision in the Council: a "rather strange logic", according to Stuffmann (1999, p. 17), and "something I reflect they may have come to regret!" according to Gammell (1999, p. 8).[25]

The question of the species permissible to hunt had proven particularly controversial during the negotiations, and was the subject of the "last battle" between Member States (Stuffmann, 1999, p. 17). As Meyer (2010) explains, "What caused a major delay was the conflict about the trade in birds between the French government, on the one hand, and the Danish and British governments, on the other. In an attempt to build up a bargaining chip, as contemporary observers suspected, the French government sought to impose stronger restrictions on the trade in birds", on the purported basis that the ability to trade encouraged excessive hunting. However, the real reason appears to have been that trade in hunted waterfowl was popular in northern countries, which "were of the opinion that there was no justification at all to prohibit the marketing of legally taken birds" (Stuffmann, 1999, p. 17). But by seeking to restrict trade in certain hunted species, the French of course gained valuable leverage to secure their own aims in terms of the species that could legally be hunted. In December 1978[26] a compromise was finally reached, under which it would remain legal to hunt the skylark in France and Italy (Meyer, 2010). As Gammell (1999, p. 8) puts it, "to have the Annexes finally adopted, the skylark was sacrificed to become a species which might be hunted, where the Ortolan Bunting was protected", France having sought the hunting of both (CEC, 1978). Proving that old habits die hard, however, and that protection on paper is not the same as protection in practice, it emerged years later that former French President François Mitterrand had dined illegally on ortolan as part of a grand final meal in 1995 (Paterniti, 1998).

Nevertheless, as Stuffmann (1999, p. 17) reflects, "after only two years the Commission's draft of 1976 was adopted as a Council directive with very minor modifications rather reinforcing the important conservation issues than weakening them".

Early implementation

Having drafted the key habitat protection parts of the Birds Directive, John Temple Lang of the Commission's Legal Service played an important role in the directive's early implementation. While most Member States proved slow to react to their new obligations, Temple Lang (1983, p. 176) had a clear vision of what had been achieved: "A wholly new situation, the importance of which has not yet been realised, has come about as a result of [the Birds Directive]", he wrote, continuing, "For the first time, the directive provides a legal basis for a single, common, scientific Community-wide policy on bird conservation, ignoring national frontiers which have, or should have, no biological significance. Also for the first time, the basis for a European policy on nature reserves for migratory birds, forming a network throughout much of Europe, has been created" (Temple Lang, 1982, pp. 23–24), "on the basis of Important Areas using the same criteria everywhere" (Temple Lang, 2012b).

Alistair Gammell, then with the RSPB, recalls:

> I can remember [John Temple Lang] having this conversation with me, when the Birds Directive was . . . finally passed, it had the provision [Article 4] which . . . basically says that Member States shall give protection to special areas for Annex I and migratory species, effectively words like that. And I remember John saying, 'OK, where are these special areas?' He said 'This is an opportunity for the non-government organisations to actually tell the Commission and tell the governments what are these special areas. So it was he that spotted – I guess he'd been working on this all the time – but it was he that spotted that niche, and as a result of that, we set out – 'we' being myself at RSPB and colleagues in BirdLife to produce books which we called Important Bird Areas.
>
> [He] pointed out that the Birds Directive required the Member States to protect special areas, and then the question he asked us was, well, where are these special areas, nobody knows where these special areas are, and there is a fantastic opportunity here for us . . . he was talking with his bird watcher hat on, so to speak, so he was saying it was a fantastic opportunity for us to actually define the nature of the Directive. If we tell people where the areas are, then that will become, effectively, the law. The Commission need help, the Commission don't know where these areas are, so he even suggested that we ask for Commission funding to do these initial studies, which we did, and we got Commission funding to do the first Important Bird Areas studies. So that was a very, very influential discussion, and after that John also was the one to point out, and he certainly led me through this procedure,

> about how you could make an official complaint to the European Union about Member States not having designated areas or not protecting areas that should be designated but . . . had some development on, so he actually advised us about how you send a formal letter of complaint to the European Commission, etc. So all of that is now, as we sit here in 2012, absolutely routine, but if I think back to the 1970s and the very early 1980s, this was completely revolutionary.
>
> *(Gammell 2012)*

Thus, the Commission asked the ICBP and the International Waterfowl Research Bureau (IWRB) to draw up lists of all the areas – which became known as Important Bird Areas (IBA) – in the Member States which deserved special habitat protection (Temple Lang, 1983, 1982).[27] The criteria for identifying areas of international importance for waterfowl and wetlands were well established by the time the Birds Directive was adopted in 1979 (Smart, 1976; Temple Lang, 1982, 1983), and updated criteria were agreed at the first meeting of the Conference of the Contracting Parties to the Ramsar Convention in Cagliari in 1980 (De Klemm and Créteaux, 1995). However, in the case of the Birds Directive, these criteria (slightly adapted) were applied – importantly – not only to waterfowl, which are the principal focus of the Ramsar Convention, but to all Annex I Birds Directive species as well as to regularly occurring migratory species not listed in Annex I (Temple Lang, 1982). Further, these criteria were also applied, significantly, to areas not contemplated by the Cagliari agreement, such as bottlenecks[28] and roosts, in addition to areas traditionally regarded as important such as breeding, wintering and staging areas (Temple Lang, 2012c).

The ICBP and IWRB submitted their first list of IBAs to the Commission and the Member States in 1981 (Osieck and Mörzer-Bruyns, 1981; Temple Lang, 1983). Additional and updated lists of IBAs for Europe followed, and the principle of IBAs has since spread globally, with many regional and national lists now established (BirdLife International, undated). The first European list was quite short, but the most recent list runs to two volumes, over 1,600 pages in total (Heath et al., 2000). In 1998, the CJEU cemented the legal significance of the IBA lists when it held that the lists, although not legally binding on the Member States concerned, can, by reason of their acknowledged scientific value, be used by the Court as a basis of reference for assessing the extent to which a Member State has complied with its obligation to designate special protection areas (CJEU, 1998).[29] And while it seems clear from the arguments in that case that the Commission had up to that point been applying a criterion whereby Member States must designate as SPAs at least half in number and area of the territories listed by IBA 89, the Court went further, holding that "Since it thus appears that the Netherlands has classified as SPAs territories whose number and total area are clearly smaller than the number and total area of the territories suitable, according to IBA 89, for classification as SPAs, the requirements of Article 4(1) of the Directive cannot be regarded as satisfied" (CJEU, 1998, paragraph 72).

The Commission's experience with the Birds Directive – a popular success on its adoption and reportedly Michel Carpentier's favourite directive at the end of his eight years in charge of the ECPS (Caufield, 1981) – evidently gave the Commission a taste for expanding its competences in the direction of additional nature conservation policy-making, with Carpentier using the Directive's example "to bolster his argument that the theoretical trade basis of the environmental action programme need not prevent new conservation measures" (Caufield, 1981, p. 417). As Carpentier himself commented in 1981, "Though we have to interpret the Treaty cleverly we can use it to justify conservation measures. A re-writing of the Treaty might burden us with more restrictions than we have now with our pragmatic approach. The Birds Directive, for example, was accused of being *ultra vires*, but it has been very popular . . . We're trying to extend our programme, not just to deal with pollution" (Caufield, 1981, p. 417). He continued: "And of course the EIA [environmental impact assessment] directive will have an effect on the landscape. I'm surprised at the resistance in Britain to this. It really is a preventive measure, to ensure that the effects of major development projects are known before they get the go-ahead" (Caufield, 1981, p. 417).

The Environmental Impact Assessment (EIA) Directive (1985)

This mention of the EIA Directive brings us to a final piece of the jigsaw that requires discussion before turning to the Habitats Directive and Natura 2000. Having moved from Commissioner Spinelli's *cabinet* to head the 'General Planning and Environmental Improvement' division at the ECPS in 1976, Claus Stuffmann had been thrown immediately into proposing and negotiating the Birds Directive, which proved fraught with difficulties and political sensitivities. Having secured that Directive's adoption "against all odds" (Meyer, 2010, p. 179), Stuffmann went immediately into the process of proposing and negotiating the EIA Directive (Stuffmann, 1981), which would prove at least as difficult and controversial, and whose genesis overlapped with negotiations regarding the Birds Directive (Roberts and Shaw, 1982; Wood, 1982).

The US having introduced environmental impact assessments via its National Environmental Policy Act 1969, the Commission began to consider the issue by funding several studies of EIA (Wood, 1982), before stating, as early as 1975, in a scoping document for the Second Environmental Action Programme: "The Commission believes that [EIA] procedures should be introduced throughout the Community on the basis of common evaluation criteria defined and adopted under the second programme" (European Commission, 1975, p. 5). By the time the Second Environmental Action Programme (1977, p. 36) was adopted two years later there had been a degree of backpedalling, the programme stating: "In order to determine whether and to what extent the regulations of the Member States need to be harmonized, and whether this is necessary at Community level, the Commission will undertake the necessary studies [of EIA procedures]" and then "draw up appropriate proposals" for the Council.

The Commission issued its first preliminary draft directive in 1978 (Wood, 1982; Roberts and Shaw, 1982) and, following twenty such drafts and detailed consultation, the Commission tabled a formal proposal to the Council in June 1980 (Wood, 1982; European Commission, 1980b). The draft Directive was, comments Wood (1982, p. 92), "reported to have been the most discussed draft directive ever issued by the Commission". Thus, while the EIA directive was expected by some to be adopted in 1982/3 (Wood, 1982; Roberts and Shaw, 1982), it in fact took until June 1985 (Environmental Impact Assessment (EIA) Directive, 1985). The Commission had high hopes for the instrument; as Environment Commissioner Stanley Clinton Davis commented in July 1987, "perhaps the area in which Community policy is going to make the greatest difference in the development world over the next few years will be through implementation of the environment impact assessment directive which comes into force in July 1988" (Clinton Davis, 1987).

Thus, between 1976 and 1985, Claus Stuffmann and his division were involved, amongst other things, in proposing and negotiating two of the most important early EU environmental measures – the Birds Directive and the EIA Directive – both of which remain in force today.[30] As we have seen, the Birds Directive put in place, *inter alia*, a strict protection regime for habitats with which, as Temple Lang (1983) commented, it would be difficult to comply without introducing environmental impact assessments. However, the Birds Directive did not specifically require such assessments, and hence did not specify the mechanics for such assessments. In contrast, the EIA Directive was very much a process directive, setting out procedures to be followed before projects likely to have significant effects on the environment could be approved at the national level. With the Commission having taken this step-wise approach from the outcome-oriented Birds Directive to the process-driven EIA Directive, the stage was set for the Habitats Directive. However, the adoption of the Birds Directive in 1979 coincided with the EU's entering a period of "Europessimism" and "Eurosclerosis": as Moravcsik (1991, p. 33) argues, "the early 1980s marked the apogee of Europessimism. Disputes over agricultural policy, the accession of Spain and Portugal, and budgetary reform festered. Ideas of splitting the EC into Subgroups proliferated: Europe *à deux vitesses*, a two-tier Community". It would take almost a decade before the Commission would be in a position to propose the EU's next major leap into nature conservation policy-making by way of the Habitats Directive.

Summary

With the Birds Directive, a process that began in the late 1960s and early 1970s with public protests and questions in the European Parliament regarding the hunting of migratory birds resulted, in the end, in protective legislation that went much further. As Meyer (2011, p. 84) comments, "[t]he Birds Directive was originally motivated by Northern European resentment against the hunting of songbirds in Southern Europe", but ended up introducing, in addition to its hunting control measures, "remarkably strict rules on habitats protection valid in all member states. It provided an important precedent for EC action in the area of nature protection and paved the way for the Habitats directive of 1992 and the Natura 2000 Programme". The EU's second *State of the Environment* report makes clear that the Commission was well aware of this fact before the Birds Directive had even been adopted:

> The importance of this directive lies not so much in its detailed lists of menaced species or of species which can be hunted, much though these will improve the situation in practice. It lies in the establishment of the general principle that birds are to be protected by conserving their habitats in all our Member States and that this policy should be based on sound scientific evidence. This is a firm basis on which to build for the future.
>
> *(European Commission, 1979b, p. 48)*

Notes

1 Thus excluding introduced species and rare visitors, according to the Explanatory Memorandum.

2 The Council's working group spent nearly twenty days discussing the subject, and the Environment Council itself devoted four sessions to the directive (Stuffmann, 1999, p. 16).

3 Mireille Jardin, one of France's negotiators, comments, "it was actually, if I remember well, more than reservations as far as France was concerned, but a real opposition to all which dealt with hunting methods" (Jardin, 2013).

4 Pagh (1999) goes on to criticise this attitude as ill-founded.

5 While he does not name names, Stuffmann (1999, p. 16) was perhaps referring to Denmark (amongst others) when he noted the delaying tactic adopted during negotiations regarding the Birds Directive of "emphasising certain 'green' positions as conditions *sine qua non* knowing very well that they were totally unacceptable to other parties".

6 He continues: "Also because at that time, or some years before, we had listed 26 Ramsar Sites, and they were protected, not directly but indirectly, by the planning legislation, nature conservation legislation, and we felt that we could implement the directive, the habitat protection provisions of the directive to a large extent by the Ramsar Sites, and in fact there were at that time, we had 26 Ramsar Sites with an area of almost 6,000 km^2, so considerable areas compared to the size of Denmark, which is only around 45,000 km^2, although many of the areas were in the sea, outside the land territory, but we didn't feel that the restrictions were so important, and we could gain something in connection with the species protection. I think that those were the kind of deliberations we had, and that led us to be rather positive during the negotiations, although we had our constant fights with the Ministry of Agriculture, which was very sceptical to the species protection provisions, especially as water birds were concerned because hunting of water birds was a traditional activity in Denmark, also in that respect we had most of our problems" (Koester, 2012).

7 As he wrote to the EEB in May 1978, there is the "difficulty of translating essentially British ways of doing things into Community wide measures without extending the jurisdiction of the Community into areas which are of purely domestic concern and are thus essentially inappropriate for Community action. Frankly I do not want the Community moving too far into an area like this one which I regard as being very much one for Member States" (cited in Fairbrass, 2000, p. 22).

8 This comes as no surprise, given that John Temple Lang drew up recommendations which formed the basis for Ireland's Wildlife Act 1976 (Temple Lang, 1983, p. 170; IELA, undated), before drafting key parts of the Birds Directive proposal for the Commission (Meyer, 2010; Temple Lang, 2012a).

9 How this expressed position could be reconciled with Secretary of State Shore's negative views is unclear.

10 As the UK government noted in evidence before the House of Lords in 1989, "the EC Directive on the protection of wild birds has not proved a happy precedent. A rigid interpretation of legal issues in Brussels has made many countries hesitant about how to implement the Directive. The UK has a good record in this field but we have had to expend disproportionate effort and time in explaining ourselves to Brussels" (Department of the Environment, 1989).

11 Given subsequent discussions, Denmark was certainly one (e.g., see CEC, 1977c).

12 The Commission evidently disagreed with this final point, because in its modified Directive proposal of 3 August 1977, further to the European Parliament's opinion of 14 June 1977 (European Parliament, 1977), the Commission had dropped Article 100 as a base, but was now proposing Articles 235 and Article 43 (European Commission, 1977b). To underline its belief in the common agricultural angle to the proposal, the Commission proposed inserting a new recital in the preamble, to read: "Whereas the disturbance of biological balance could have negative consequences for the common market, in particular for agricultural production" (European Commission, 1977b, p. 3). The Working Party on the Environment met again on 8 and 9 September 1977 to discuss, amongst other things, the appropriate legal base for the proposal. The Commission and the Council agreed on Article 235 as a legal base, and agreed to meet to discuss whether the Treaty contained other articles that could be used as bases (CEC, 1977c). Evidently they could not agree on any additional bases, since the directive was adopted on the basis of Article 235 alone (Birds Directive, 1979).

13 Hubert David, first Secretary General of the EEB, believes that Germany may have had constitutional issues regarding the Birds Directive; that is, the appropriate division of competences between the federal and regional levels of government in Germany (David, 2012), and this is confirmed by Stuffmann (1999).

14 The Paris Summit (1972, paragraph 14) declaration stated that the Heads of Government "agreed that in order to accomplish the tasks laid out in the different action programmes, it was advisable to use as widely as possible all the provisions of the Treaties including Article 235 of the EEC Treaty".

15 The question of the validity of Article 235 as a legal base for the Birds Directive was tested before the French Conseil d'Etat in 1990, a hunting association having raised the issue – the Conseil d'Etat held that the Directive had indeed been competently adopted by the Council (Wils, 1994).

16 Germany and Denmark most likely had in mind: the IUCN and its Ramsar Convention (1971) on wetlands of international importance; UNEP and its proposed convention on migratory species of wild animals (for which Germany was the principal driver; this instrument became the Bonn Convention on its adoption in 1979 (Caddell, 2005)); and the Council of Europe and its proposed convention on the conservation of European wildlife and natural habitats (which ultimately became the Bern Convention on its adoption in 1979).

17 Stuffmann (1999) characterises this end point as a positive outcome in respect of habitat conservation.

18 On the early history of the Bern Convention, see Ribaut (2004).

19 It is clear that certain Member States did not think that the new habitat protection provisions were going to make much difference. For example, giving evidence before a House of Commons Select Committee in 1977, a UK official stated that he did not "see the Directive making a great deal of difference . . . [M]ember States [have] to take requisite measures to maintain or restore a sufficient diversity of habitats. We are already doing

this in Britain . . .We would not see this Directive as imposing any greater obligation in Britain than we are already doing" (cited in Fairbrass and Jordan, 2001, p. 10).

20 Which then included, for example, snares, limes, hooks, nets, semi-automatic or automatic weapons, live birds used as decoys (such as blinded or mutilated birds), etc.

21 Note that this obligation has, however, proven enforceable before the CJEU: see the judgment in Case C-418/04 in which Ireland was held to have failed to transpose and apply this obligation in practice (CJEU, 2007).

22 Germany entered a reservation regarding the Brent goose (*Branta bernicla*), which was subsequently excluded; France entered a reservation regarding the Red-crested pochard (*Netta rufina*) and the Ruff (*Philomachus pugnax*), which were subsequently excluded from the version adopted in 1979 (the Ruff was reinserted during a later revision of the Annex), and the Great snipe (*Gallinago media*), which was nevertheless retained; and Italy entered a reservation regarding the Ruff (*Philomachus pugnax*) (see above in this note), and the Great northern diver (*Gavia immer*), which was nonetheless retained.

23 France entered a reservation in respect of the proposed addition of the continental population of the Golden plover (*Pluvialis apricaria apricaria*), which was nonetheless included on the final version of Annex I, and Italy proposed adding the Shag (*Phalacrocorax aristotelis*), but this was resisted by other delegations for the purpose of the version of the Annex adopted in 1979 (CEC, 1977e). The Mediterranean Shag (*Phalacrocorax aristotelis desmarestii*) was later included in Annex I as part of a revision process.

24 Another source gives this as 195 species for the EU-27 (EEA, 2007).

25 Gammell (1999, pp. 10–11) comments, "in the 20 years of the Birds Directive, BirdLife has never attacked legal shooting and has never tried to reopen the question of what should or should not be on Annex II. An historic compromise was reached . . . in 1978 and whilst the hunters did not get to shoot everything they desired, nor did bird conservationists get certain species protected that we believe should have been. Compromises are like that. It is a matter of sadness to me that others have continually sought to reopen this issue rather than deal with the real issue that diminishes the number of birds for both hunter and conservationist alike, how to stop the destruction of Europe's habitats".

26 Following what Stuffmann (1999, p. 17) called "an extremely tense discussion, in particular at the last decisive Council session in December 1978, [which] led to a surprisingly complicated compromise stipulating three groups of species: those free for marketing throughout the Community, those which Member States could admit to trade under certain conditions and those for which a final decision was postponed for two years".

27 For a detailed discussion of the early years of this process, see Temple Lang (1982, pp. 14–18).

28 "[P]laces where migrating birds congregate, even if they do not land ([e.g.] Straits of Messina, Bosphorus, Kalo in S. Sweden, Orgambideska in the Pyrenees, Gibraltar, hirundine roosts)" per Temple Lang (2012b).

29 Interestingly, in 1993, of the then twelve EU Member States, ten Member States did not accept the IBA list as an official candidate list; Italy's position was unclear; Belgium did not object to the list; and Denmark's position is worthy of note: it did not accept the list as an official candidate list of sites, but had nonetheless designated 102 of the 118 sites listed, and had undertaken to review the other 16 sites with a view to their possible designation (European Commission, 1993).

30 Albeit since codified.

References

Archer, C. 2000. Euroscepticism in the Nordic region. *Journal of European Integration*, 22, 87–114.

Bern Convention 1979. Convention on the conservation of European wildlife and natural habitats, concluded in Bern on 19 September 1979 (UN Treaty Series I-21159).

Birdlife International undated. *Important Bird and Biodiversity Areas (IBAs)* [Online]. Available: www.birdlife.org/worldwide/programme-additional-info/important-bird-and-biodiversity-areas-ibas [accessed 30 June 2017].

Birds Directive 1979. Council Directive 79/409/EEC of 2 April 1979 on the conservation of wild birds, *OJ L 103*, 25 April 1979, p. 1, since codified as Directive 2009/147/EC of the European Parliament and of the Council of 30 November 2009 on the conservation of wild birds. *OJ L 20*, 26 January 2010, p. 7.

Caddell, R. 2005. International law and the protection of migratory wildlife: an appraisal of twenty-five years of the Bonn Convention. *Colorado Journal of International Environmental Law and Policy*, 16, 113–156.

Caufield, C. 1981. Environment: Britain lags, Europe leads. *New Scientist* (13 August), 416–417.

CEC 1973. Minutes of the 251st meeting of the Council of the European Communities held in Brussels on Thursday 19 and Friday 20 July 1973. 1586/73 (PV/CONS 23).

CEC 1977a. Note: Proposal for a Council Directive on bird conservation [R/3265/76 (ENV 145)] – Outcome of the meeting of the Working Party on the Environment on Friday 11 March 1977. Council of the European Communities. Brussels, 22 March 1977 (29.03). R/633/77 (ENV 46) (AGRI 174).

CEC 1977b. Note: Proposal for a Council Directive on bird conservation [R/3265/76 (ENV 145)] – Outcome of the meeting of the Working Party on the Environment on Thursday 23 and Friday 24 June 1977. Council of the European Communities. Brussels, 7 July 1977 (13.07). R/1601/77 (ENV 95) (AGRI 435).

CEC 1977c. Note: Proposal for a Council Directive on bird conservation [R/3265/76 (ENV 145)] – Results of the meeting of the Working Party on the Environment on Thursday 8 and Friday 9 September 1977. Council of the European Communities. Brussels, 15 September 1977 (20.09). R/2048/77 (ENV 119) (AGRI 568).

CEC 1977d. Note: Proposal for a Council Directive on bird conservation [R/3265/76 (ENV 145)] – Results of the meeting of the Working Party on the Environment on Thursday 13 and Friday 14 October 1977. Council of the European Communities. Brussels, 19 October 1977 (26.10). R/2441/77 (ENV 138) (AGRI 663).

CEC 1977e. Note: Subject: Proposal for a Council Directive on bird conservation [R/3265/76 (ENV 145)] – Outcome of the meeting of the Working Party on the Environment, composed of ornithologists, on 13 and 14 October 1977. Council of the European Communities. Brussels, 20 October 1977 (25.10). R/2440/77 (ENV 137) (AGRI 662).

CEC 1977f. Note: Subject: Proposal for a Council Directive on bird conservation [R/3265/76 (ENV 145)] – Report from the Permanent Representatives Committee to the Council (document concerning the Articles). Council of the European Communities. Brussels, 4 December 1977 (08.12). R/3077/77 (ENV 170) (AGRI 823).

CEC 1977g. Note: Subject: Proposal for a Council Directive on bird conservation [R/3265/76 (ENV 145)] – Results of the meeting of the Working Party on the Environment on Thursday 22 and Friday 23 September 1977. Council of the European Communities. Brussels, 29 September 1977 (11.10). R/2158/77 (ENV 126) (AGRI 594).

CEC 1978. Note: Subject: Proposal for a Council Directive on bird conservation [R/3265/76 (ENV 145)] – Report by the Permanent Representatives Committee to the Council. Council of the European Communities. Brussels, 22 May 1978 (26.05). R/1214/78 (ENV 74) (AGRI 345).

CEC 1979. Council Resolution of 2 April 1979 concerning Directive 79/409/EEC on the conservation of wild birds. Council of the European Communities. *OJ C 103*, 25 April 1979, p. 6.

Cichowski, R. A. 2007. *The European Court and civil society: litigation, mobilization, and governance*. Cambridge: Cambridge University Press.

CJEU 1990. Opinion of Advocate General Van Gerven of 5 December 1990, *Commission* v. *Germany*, C-57/89, EU:C:1990:432.

CJEU 1998. Judgment of the Court of Justice of the EU of 19 May 1998, *Commission* v. *Netherlands*, C-3/96, EU:C:1998:238.

CJEU 2007. Judgment of the Court of Justice of the EU of 13 December 2007, *Commission* v. *Ireland*, C-418/04, EU:C:2007:780.

Clinton Davis, S. 1987. Environment and development: extracts from a speech by Commissioner Stanley Clinton Davis at the Placemakers' Luncheon Club, London, 23 July 1987, IP/87/318.

Codified Birds Directive 2009. Directive 2009/147/EC of the European Parliament and of the Council of 30 November 2009 on the conservation of wild birds (codified version). *OJ L 20*, 26 January 2010, p. 7.

Council Legal Service 1977. Note: Subject: Proposal for a Council Directive on bird conservation (R/3265/76 (ENV. 145)) (Brussels, 25 July 1977, R/1867/77 (JUR 95) (ENV 106)).

Cramp, S. 1977. *Bird conservation in Europe: a report*. Prepared for the Environment and Consumer Protection Service, Commission of the European Communities. London: HMSO.

David, H. 2012. Interview with Hubert David, founder and first Secretary General of the European Environmental Bureau, conducted by Andrew L. R. Jackson on 3 August 2012.

De Klemm, C. and Créteaux, I. 1995. *The legal development of the Ramsar Convention*. Gland, Switzerland: Ramsar Convention Bureau.

Department of the Environment 1989. Memorandum by the Department of the Environment dated 23 January 1989, headed "House of Lords European Communities Committee: Sub Committee F (Environment). Enquiry into habitat and species protection".

EEA 2007. *Number of habitats types and species in the annexes of the Birds and Habitats Directives. EU 27 – February 2007* [Online]. Available: https://bd.eionet.europa.eu/activities/Natura_2000/pdfs/Nb_annexes_Directives-2007.pdf [accessed 30 June 2017].

Environmental Impact Assessment (EIA) Directive 1985. Council Directive 85/337/EEC of 27 June 1985 on the assessment of the effects of certain public and private projects on the environment, *OJ L 175*, 5 July 1985, p. 40, since codified as Directive 2011/92/EU of the European Parliament and of the Council of 13 December 2011 on the assessment of the effects of certain public and private projects on the environment (codification), *OJ L 26*, 28 January 2012, p. 1. Codified version since amended by Directive 2014/52/EU, *OJ L 124*, 25 April 2014, p. 1.

European Commission 1975. Environment: new guidelines for Community action. Industry and Society, Brussels, 1 July 1975, No. 26/75.

European Commission 1976. Proposal for a Council directive on bird conservation (COM(76) 676 final). Brussels, 23 December 1976, R/3265/76 (ENV. 145), attaching a letter from the Commission to the Council dated 20 December 1976 with COM(76) 676 final as its enclosure.

European Commission 1977a. Departmental Working Paper for the Group Environment of the Council, 28–29 July 1977; Ref: Proposal for a Council Directive on Bird Conservation, Brussels, 26 July 1977, SEC(77)2855.

European Commission 1977b. Modifications of a proposal for a Council Directive on bird conservation. Presented by the Commission to the Council on 3 August 1977 pursuant to the second paragraph of Article 149 of the EEC Treaty. *OJ C 201*, 23 August 1977, p. 2.

European Commission 1979a. The European Community and the environment. European File (4/79, February 1979).

European Commission 1979b. *State of the environment: second report*, Brussels – Luxembourg: ECSC – EEC – EAEC.

European Commission 1980a. Progress made in connection with the environment action programme and assessment of the work done to implement it. Communication from the Commission to the Council. COM(80) 222 final, Brussels, 7 May 1980.

European Commission 1980b. Proposal for a Council directive concerning the assessment of the environmental effects of certain public and private projects. COM(80) 313 final, Brussels, 11 June 1980.

European Commission 1985. Commission directive 85/411/EEC of 25 July 1985 amending Council Directive 79/409/EEC on the conservation of wild birds. *OJ L 233*, 30 August 1985, p. 33.

European Commission 1993. State and perspectives of the classification of Special Protection Areas by Member States. ORNIS 93/7.

European Parliament 1977. Resolution embodying the opinion of the European Parliament on the proposal from the Commission of the European Communities to the Council for a Directive on bird conservation (14 June 1977). *OJ C 168*, 11 July 1977, p. 28.

Fairbrass, J. 2000. EU and British biodiversity policy: ambiguity and errors of judgement. CSERGE Working Paper GEC 2000–4: Centre for Social and Economic Research on the Global Environment, University of East Anglia and University College London.

Fairbrass, J. and Jordan, A. 2001. European Union environmental policy and the UK government: a passive observer or a strategic manager? *Environmental Politics*, 10, 1–21.

First Environmental Action Programme 1973. Declaration of the Council and of the European Communities and of the representatives of the Governments of the Member States meeting in the Council of 22 November 1973 on the programme of action of the European Communities on the environment. *OJ C 112*, 20 December 1973, p. 1.

Gammell, A. 1999. The contribution of BirdLife International to the Birds Directive and the management of birds and their habitats in Europe. 20 years with the EC Birds Directive. Proceedings from a conference on the Council's Directive on the Conservation of Wild Birds, 18–19 November 1999, Elsinore, Denmark. The Danish National Forest and Nature Agency.

Gammell, A. 2012. Interview with Alistair Gammell, formerly of the RSPB, conducted by Andrew L. R. Jackson on 31 July 2012.

Habitats Directive 1992. Consolidated version of Council Directive 92/43/EEC of 21 May 1992 on the conservation of natural habitats and of wild fauna and flora. *OJ L 206*, 22 July 1992, p. 7. Consolidation date 1 July 2013.

Heath, M. F., Evans, M. I., Hoccom, D. G., Payne, A. J. and Peet, N. B. (eds) 2000. *Important Bird Areas in Europe: priority sites for conservation, Volume 1 Northern Europe; Volume 2 Southern Europe*. Cambridge: BirdLife International.

IELA undated. *Dr. John Temple Lang, Patron of the Irish Environmental Law Association* [Online]. Available: www.iela.ie/council-members/dr-john-temple-lang/ [accessed 30 June 2017].

Jardin, M. 2013. Personal communication: e-mail dated 12 April 2013 from Mireille Jardin, member of French government delegation which negotiated the Birds Directive, to Andrew L. R. Jackson.

Jordan, A. 1998. EU environmental policy at 25: the politics of multinational governance. *Environment*, 40, 14–20.

Koester, V. 2012. Interview with Veit Koester, senior negotiator for the Danish government in respect of the Habitats Directive, conducted by Andrew L. R. Jackson on 22 August 2012.

Meyer, J. H. 2010. Saving migrants: a transnational network supporting supranational bird protection policy in the 1970s. *In:* Kaiser, W., Gehler, M. and Leucht, B. (eds) *Transnational networks in regional integration: informal governance in Europe 1945–83.* Basingstoke: Palgrave Macmillan.

Meyer, J. H. 2011. Green activism: the European Parliament's Environmental Committee promoting a European environmental policy in the 1970s. *Journal of European Integration History*, 17, 73–85.

Moravcsik, A. 1991. Negotiating the Single European Act: national interests and conventional statecraft in the European Community. *International Organization*, 45, 19–56.

Osieck, E. R. and Mörzer-Bruyns, M. F. (eds) 1981. Important Bird Areas in the European Community, unpublished report by International Council for Bird Preservation European Community Working Group for the EEC Environment and Consumer Protection Service. London: International Council for Bird Preservation.

Pagh, P. 1999. Denmark's compliance with European Community environmental law. *Journal of Environmental Law*, 11, 301–319.

Paris Summit 1972. Declaration of the first summit conference of the enlarged Community: Meetings of the Heads of State or Government, Paris, 19–21 October 1972. French (VO) and English versions. *Bulletin of the European Communities*, No. 10.

Paterniti, M. 1998. The last meal. *Esquire* (1 May 1998).

Quirke, J. 2012. Quirke Report, arising from 'Peatlands forum 2012', held from 28th February–2nd March 2012, at the Hodson Bay Hotel, Athlone, Ireland.

Ramsar Convention 1971. Convention on wetlands of international importance especially as waterfowl habitat (UN Treaty Series I-14583), amended in 1982 and 1987.

Ribaut, J.-P. 2004. How the Bern Convention came into being. *Naturopa*, 101, 4.

Roberts, P. W. and Shaw, T. 1982. Environmental impact assessment: retrospect and prospect. *Planning Outlook*, 24, 85–92.

Second Environmental Action Programme 1977. Resolution of the Council of the European Communities and of the representatives of the governments of the Member States meeting within the Council of 17 May 1977 on the continuation and implementation of a European Community policy and action programme on the environment. *OJ C 139*, 13 June 1977, p. 1.

Smart, M. 1976. Proceedings of the international conference on conservation of wetlands and waterfowl, Heiligenhafen, December 1974. Annex II, Recommendations for criteria to be used in identifying wetlands of international importance. Slimbridge: International Waterfowl Research Bureau.

Stuffmann, C. 1981. Europe and the environmental assessment. *In:* Cairns, W. J. and Rogers, P. M. (eds) *Onshore impacts of offshore oil: proceedings of the International Conference on Oil and the Environment, Scotland 1980, held in the Pollock Halls, University of Edinburgh, 28 September–1 October.* Essex: Applied Science Publishers.

Stuffmann, C. 1999. The upcoming of the Birds Directive: 20 years with the EC Birds Directive. Proceedings from a conference on the Council's Directive on the Conservation of Wild Birds, 18–19 November 1999. Elsinore, Denmark: The Danish National Forest and Nature Agency.

Temple Lang, J. 1982. The European Community Directive on bird conservation. *Biological Conservation*, 22, 11–25.

Temple Lang, J. 1983. The Wildlife Act and EEC bird conservation measures. *In:* Blackwell, J. and Convery, F. J. (eds) *Promise and performance: Irish environmental policies analysed.* Dublin: The Resource and Environmental Policy Centre.

Temple Lang, J. 2012a. Interview with John Temple Lang, ex-official with the European Commission's Legal Service, conducted by Andrew L. R. Jackson on 17 August 2012.

Temple Lang, J. 2012b. Personal communication: e-mail dated 1 August 2012 from John Temple Lang, ex-official with the European Commission's Legal Service, to Andrew L. R. Jackson.

Temple Lang, J. 2012c. Personal communication: e-mail dated 16 August 2012 from John Temple Lang, ex-official with the European Commission's Legal Service, to Andrew L. R. Jackson.

Treaty of Rome 1957. Treaty establishing the European Economic Community. United Nations Treaty Series.

Wils, W. P. J. 1994. The Birds Directive 15 years later: a survey of the case law and a comparison with the Habitats Directive. *Journal of Environmental Law*, 6, 219–242.

Wood, C. 1982. The impact of the European commission's directive on environmental planning in the United Kingdom. *Planning Outlook*, 24, 92–98.

Wurzel, R. 2013. Member States and the Council. *In:* Jordan, A. and Adelle, C. (eds) *Environmental policy in the EU: actors, institutions and processes*, 3rd edn. Abingdon: Routledge.

5

TOWARDS AN EU HABITATS DIRECTIVE

Early EU plans to extend EU nature conservation law and policy

By May 1980 – just one year after the Birds Directive was adopted – the Commission was already communicating to the Council its view that new EU nature conservation legislation was needed to preserve habitats "in order to guarantee the protection of biotopes of Community importance on a permanent basis" (European Commission, 1980c, p. 8). Significantly, the Commission had a clear, detailed plan by this time – a full eight years before a formal proposal for the Habitats Directive would be forthcoming – covering the establishment of a network of protected areas to supplement the network under the Birds Directive, the need for adequate EU financing, the need for impact assessment rules, as well as a recognition that EU action in this area was likely to encounter "certain difficulties":

> The Commission hopes that certain hesitations which have become apparent as to the scope of the Community's powers [in the field of nature conservation] which have hindered the adoption of important and urgent measures will be overcome in the future.
>
> . . . [I]t should be emphasized that it is becoming increasingly clear that it will be possible to avoid the extinction of several species only if greater efforts are made to preserve the habitats of these species; indeed, the scale of the measures to be taken, especially by the Commission, is greater than was originally proposed.
>
> It is therefore on this aspect that future action must be primarily concentrated; it is necessary to identify the requirements both in terms of the total area needing protection and of the administration of the different zones, and to establish a framework which will guarantee the permanent protection of those biotopes which are considered to be of Community importance.

> To achieve this end, the enactment of general rules and the evaluation of the impact of certain developments on the habitat of the endangered species will not suffice. It will also be necessary to have available all the means which can make it possible to cope with the pressures exerted by economic development, including those which arise from Community policies. Such means would be incomplete without the requisite funds.
>
> [. . .]
>
> As far as the protection and maintenance of wild species is concerned, the [Birds] Directive of April 1979 will be supplemented by the creation and maintenance of a network of sufficiently large protected zones which will be protected and administered under common rules and principles. If this aim is to be achieved by informing and interesting the public, the Community will doubtless have to introduce legislation and provide funds.
>
> The procedures for assessing the environmental impact of public and private developments must expressly require the examination of their effects on the fauna and flora.
>
> At international level it will be necessary for the Community and the Member States to ratify as soon as possible the [Bern] Convention of the Council of Europe and the International Convention on the Conservation of Migratory Species.
>
> *(European Commission, 1980b, pp. 36–37)*

Thus, already by 1980 the Commission had a well-developed picture and plan of how a future Habitats Directive might look. However, these ambitions were evidently constrained in practice by some Member States' continuing reluctance to countenance additional EU nature conservation measures in the immediate wake of the Birds Directive. The Third Environmental Action Programme (1983, p. 13) ultimately contained the following aspirational wording, which gave deference to such views:

> The main problem [in respect of fauna and flora] is the conservation of habitats where their gradual, irreversible disappearance in many cases constitutes the chief threat to the survival of species. While it is recognized that local, regional and national responsibilities are decisive in this case, a Community framework is becoming essential if greater cohesion is to be given to such efforts. Such a framework would ensure that a network of properly protected biotopes, sufficient in both extent and number, and interlinked in a rational fashion, was set up and maintained. The network should be designed in such a way as to guarantee — as far as the habitat is concerned — the survival of all species native to the Community. This would be made much easier if it were possible to use Community financial resources, and in particular those destined for the protection of the environment. One cost-effective use of these resources may be to grant support to voluntary organizations, within a framework of appropriate rules, to manage nature reserves.

Notwithstanding these ambitions, the creation of such a network was not secured during the life of the Third Environmental Action Programme (1983).

Relationship with the Bern Convention[1]

In 1975, the European Commission (1975, p. 14) commented, with a hint of inter-institutional rivalry, that the intergovernmental Council of Europe – which had twice as many member countries as the EEC at that time – is "less active in the environmental field than the Communities".[2] Yet just as the EEC was celebrating the formal adoption of the Birds Directive in April 1979, the Council of Europe appeared to have gone one better, securing agreement in respect of the Bern Convention on the conservation of European wildlife and natural habitats, which opened for signature in September 1979 and entered into force in June 1982 (Lasén-Díaz, 2010; European Commission, 1979).

The Bern Convention – which was being negotiated at the same time as the Birds Directive, complicating matters somewhat given the overlap in subjects (Ribaut, 2004)[3] – was ambitious in aims and scope, and "visionary" in approach, according to Stanley Johnson (Johnson, 2012). The Convention applied to listed plants, birds, and other animals; to the habitats of those listed species; and to unlisted "endangered natural habitats", and aimed "to conserve wild flora and fauna and their natural habitats, especially those species and habitats whose conservation requires the co-operation of several States, and to promote such co-operation", with particular emphasis given to "endangered and vulnerable species, including endangered and vulnerable migratory species" (European Commission, 1979). However, at the time of its coming into force the Convention did not specifically require protected areas to be designated. As Ribaut (2004, p. 4), who was involved in the negotiations for the Convention, comments, "Among the substantive discussions, those concerning the protection of habitats (Article 4) were undeniably the most important and the most difficult . . . Once again, common sense and a flexible approach prevailed, resulting in an Article 4 which may appear very general (excessively so perhaps), but which has proved to be effective, thanks to the case-law which has gradually been built up in this area".[4]

Lasén-Díaz (2010, p. 186) notes that, "The relationship with the EU, a contracting party to the Bern Convention since 1982, has been a dynamic one, having evolved and changed over the years". Indeed, as highlighted in Chapter 4, the EU's Second Environmental Action Programme (1977) envisaged the prospect of EU legislation to implement the Bern Convention even before the Convention's signature, noting that "in so far as action is necessary at Community level in order to ensure that the instrument is satisfactorily applied, the Commission will submit appropriate proposals to the Council". Following the entry into force of the Bern Convention, the EU's regulation of nature conservation issues was certainly narrower in scope than the Council of Europe's, albeit that the EU's more limited regime was backed up by much more sophisticated legal architecture to ensure compliance, and would soon begin slowly expanding with the adoption in 1981 of a Regulation relating to imports of products from whales and other cetaceans (Whales Regulation, 1981), closely followed by the first CITES Regulation (1982) on trade in endangered species, and by the Seal Pups Directive (1983), discussed below.

Stanley Johnson and the European Parliament: 1979–1984

As well as marking the signature of the Birds Directive and the Bern Convention, 1979 would mark another important milestone in the history of the Habitats Directive. That year, Stanley Johnson, who had been working with the ECPS from 1973, left the European Commission having been elected as an MEP in the first direct European parliamentary elections (Johnson, 2009). Crucially, he took up his position as an MEP on the basis of a *congé de convenance personnelle* (leave on personal grounds) from the Commission (Johnson, 2011). This later gave him the option, on leaving the Parliament in 1984, of reinstatement to a position within the Commission (Staff Regulations, 1968, Article 15).

Once at the Parliament, Johnson obtained the post of vice-chair of the Parliament's Environment Committee, a role that allowed him to pursue an active environmental agenda, albeit that the Parliament's powers were relatively limited in those days (Johnson, 2009, 2011). One of Johnson's major achievements during these years as an MEP, with the help of the Dutch MEP Hemmo Muntingh, who was similarly passionate about nature conservation issues (Muntingh, 2013), related to the adoption of EU legislation banning the import of skins of baby seals. As Johnson (2011) explains, "In order to get that through, we had to build first a huge degree of pressure on the Commission, who didn't want to propose it at all". The main Commission official involved was Johnson's then ex-colleague Claus Stuffmann, negotiator of the Birds Directive, who "was not anxious to do this", according to Johnson (2011). There was countervailing pressure from the Canadian government, there were people within the Commission who did not think the EU should be involved in regulating the trade in seal products at all (Johnson, 2011), and Stuffmann would of course at the time have been heavily involved in negotiating the controversial Environmental Impact Assessment (EIA) Directive (1985). Ultimately, however, the Commission relented, and EU legislation was proposed and adopted (Seal Pups Directive, 1983).[5]

As Johnson (2011) reflects,

> one of my main activities when I was in the Parliament, as it related to the Commission, was actually trying to force the Commission to do things the Commission didn't want to do . . . [T]his question of the EU seal ban . . . was one of the first examples where the European Parliament really forced the Commission to do something it didn't want to do. Now, that's not at all the same thing as saying that the Commission lost the power of initiative. The Commission had then and still has the power of initiative. Of course if the Commission hadn't made a proposal, there wouldn't have been a seal ban. But we did manage to get the Commission to a point where it was under such pressure that it felt it had to make a proposal.

This episode sheds light on Johnson's unconventional but effective campaigning style, and reveals some of the tactical lessons he picked up regarding how to secure a desired legislative outcome. In respect of the former, the seals campaign involved rallies in Brussels and Luxembourg, at one of which helium balloons in the shape of baby seals were floated up level with the floor of a building in Brussels in which the Member States' Ministers were meeting, such that they were surrounded (Johnson, 2011).[6] And in respect of the latter, as Johnson (2011) comments, "if you are an effective parliamentarian, you have to get the Commission on your side, that's obvious; but you also have to get the Member States of the Council on your side. The way sometimes you do that is by working with the *grand public* in the Member States". Thus, having been an active policy entrepreneur with the Commission from 1973–1979, Johnson, plus other activist MEPs such as Hemmo Muntingh, began a phase of similar activism within the European Parliament, which slowly began building momentum towards a broad-based EU instrument on nature conservation.

For example, in May 1980 the Parliament adopted a Resolution on the World Conservation Strategy,[7] calling on the Commission to "submit proposals concerning possible ways of improving the internal structure of the Community with regard to the latter's policy on nature conservation" (European Parliament, 1980b). The following month the Parliament adopted a Resolution regarding the Bern Convention, in which the Parliament urged the Council and the Commission to give a high priority to a Community policy on nature and the environment, requested the Commission to accelerate the implementation of projects in the Second Environmental Action Programme relating to natural habitats, and to submit proposals for Directives on the protection of plants and vertebrate animals in the Community (European Parliament, 1980a). By July 1987 – by which time Stanley Johnson was back working for the Commission – the European Parliament (1987, p. 122) was calling for "land and marine parks and nature reserves to form a Community network of protected areas which is representative of the natural resources of the various geographical regions of the Community". Welcoming this Resolution, Environment Commissioner Stanley Clinton Davis assured the Parliament that "the contents of the report have been very fully noted, both by myself and by my staff, and they will serve as an important guide for the Commission's work in this important field of nature conservation" (Clinton Davis, 1987a, p. 288).

Treaty revision: the Single European Act

In addition to this direct activism in respect of nature conservation, Stanley Johnson's years at the European Parliament coincided with moves towards the first major EU Treaty revision since the Treaty of Rome was signed in 1957. If the late 1970s and early 1980s could be described as periods of "Europessimism" and "Eurosclerosis", by the early 1990s the EU had entered a period of "optimism and institutional momentum" (Moravcsik, 1991, p. 19). The source of this change in atmosphere, argues Moravcsik (1991), was the Single European Act, one of the grand bargains that punctuate the EU's history, signed by political leaders in February 1986, but owing its origins to initiatives in the early 1980s.

Altiero Spinelli – former EU Commissioner and a founding father of the EU's environmental policy, as described in Chapter 3 – played an important role. Having been elected to the European Parliament along with Stanley Johnson in the first direct elections in 1979, Spinelli – ever the committed federalist – wrote to his fellow MEPs in June 1980 to seek support for an initiative to further democratise the EU, and to deepen integration by overcoming institutional barriers (Ward, 2003; Lodge, 1984; Cardozo and Corbett, 1986). Eight MEPs responded to this initial call (Gamble, 1998), and the nine met for the first time on 9 July 1980 in *Au Crocodile*, a restaurant in Strasbourg (Cardozo and Corbett, 1986). One of these nine founding members of the so-called Crocodile Club was Stanley Johnson (Booker and North, 2003). The Crocodile Club grew rapidly in size, and would eventually comprise around 180 members (D'Ancona, 1990), almost half of the MEPs of the then European Parliament. The Club's efforts culminated in a Parliament resolution of September 1983 on the substance of a preliminary draft treaty on European Union (European Parliament, 1983), as well as a draft Treaty produced by the European Parliament (1984) itself, which issued in February 1984 and contained a short provision granting the EU explicit competence in environmental matters.

While the Parliament's draft would not ultimately be adopted, it led to the creation, during a dynamic French Presidency of the EU under François Mitterrand, of a committee of personal representatives of the Member States (known as the Dooge Committee, after its Irish Chair), tasked with investigating possible Treaty changes (Jacobs, 2000).[8] The work of the Dooge Committee coincided with the appointment in January 1985 of a new Commission President, Jacques Delors, who had been freed from his job as France's Minister for Finance by a reshuffle (Moravcsik, 1991). Delors immediately set to work, announcing in his maiden speech plans to complete the common market by 1992 (Moravcsik, 1991). During the course of 1985 the Council endorsed this goal and tasked the Commission with drawing up detailed proposals; that job fell to a Briton, Lord Cockfield, then Internal Market Commissioner, who produced a "celebrated White Paper", published in June 1985, containing almost three hundred specific proposals aimed at liberalising the common market (Moravcsik, 1991, p. 40).

Delors then began linking the idea of market liberalisation with the need for qualified majority voting (rather than unanimity) in EU decision-making, arguing that the common market could not be achieved without such voting reform; significantly, he also began proposing extensions in the EU's substantive competences (Moravcsik, 1991). An Intergovernmental Conference was convened, beginning in September 1985, to consider amendments to the Treaty of Rome. A draft of what became the new Treaty – the Single European Act (SEA) – was written during the first month of the conference, the subsequent details were worked out at five meetings between October and December 1985, and the SEA was signed in February 1986 (Moravcsik, 1991).

The SEA sought to relaunch the EU project by setting a deadline of 1992 for the completion of the common market, linking this market liberalisation with procedural reforms designed to ease decision-making within the EU, just as Delors had sought (Moravcsik, 1991). Since 1966, qualified majority voting had in practice been limited within the EU by the so-called Luxembourg compromise, under which France had asserted the right to veto a proposal by declaring that a vital or very important issue was at stake (Moravcsik, 1991). The SEA overcame the Luxembourg compromise by greatly expanding the use of qualified majority voting, though principally on matters relating to the common market. A price paid for securing these reforms was the expansion of Structural Funds aimed at less developed parts of the EU (more below); as Moravcsik (1991, p. 43) comments, this was "a side payment to Ireland [and] the Southern nations in exchange for their political support".

While the content of the SEA did not meet the European Parliament's aspirations,[9] the new Treaty did grant the EU explicit competence in environmental matters, as the Parliament's draft had sought.[10] In this regard, Moravcsik (1991, p. 46) argues that "the Commission might be credited with having quietly slipped some new EC functions, such as environmental and research and development programs, into the revised treaty. But these were functions that the EC had been handling under indirect authorization for a number of years, and there was little opposition from member states to extending a concrete mandate to cover them". This relative lack of opposition might at first glance seem surprising: although the EU had indeed been exercising certain environmental policy-making functions since 1973, the precise ambit of the EU's environmental competences remained a bone of contention in advance of the SEA negotiations in 1985. For example, as Leann Brown (1995, p. 14) notes, "although the Danes had reluctantly agreed to the directive protecting wildbirds, they argued that the EC lacked legal competence to legislate in areas such as protecting wildlife habitats and, as a result, it was understood that no further wildlife legislation should be adopted unless it was directly related to trade".[11]

The question therefore arises: how were disputes over the EU's environmental competences settled via the SEA, and how were Member States' earlier objections overcome? Lodge (1994) and Leann Brown (1995) provide convincing accounts. By seeking to expand the ambit of the EU's competences via the SEA, Leann Brown (1995, p. 14) argues, Commission President Delors created tactical opportunities for "coalitions, bargains and trade-offs to be struck across the issues of market reform and institutional reform. Combining market enhancement, institutional reform and environmental provisions would make the single market more attractive to those concerned about the democratic deficit and environmental protection". This did not mean that extending EU competences in environmental matters went without discussion – these functions were not "quietly slipped" in, as Moravcsik (1991, p. 46) claims. Indeed, as Lodge (1994, p. 35) highlights, "The potential for dispute was particularly large on the environmental front. This potential owed as much to the usual disagreement over whether the Council should go over to majority voting in such matters as to the fact that EC intervention in this field remained contested. Once again, however, the overarching demands of the Single Market's realization impelled action".

The Commission proposed four new Treaty articles regarding EU competence in environmental policy matters, containing general aims, principles and specific measures, with unanimity preserved for defining aims and principles, and qualified majority voting to apply for implementing measures (Lodge, 1994). True to form, and fearing a lowest common denominator approach, Denmark was keen to ensure that it would not be forced to lower its own domestic environmental standards by EU environmental law (Lodge, 1994). On the other side of the coin were certain laggard states: Greece, Ireland and the UK, which "considered stringent legislation costly" (Lodge, 1994, p. 35). Supported by Greece, Ireland proposed an amendment regarding the need to balance environmental and economic considerations (Lodge, 1994), and both Denmark and Greece objected to majority voting, albeit for different reasons (Lodge, 1994).

The provisions finally adopted represented a compromise between competing interests: the EU was given a broad environmental policy mandate, and the general principles of prevention, rectification at source, polluter pays, integration of environmental protection requirements into other EU policies, and subsidiarity were included (Single European Act, 1986, Article 25, inserting Article 130r(1), (2) and (4)). With a nod to the concerns of Ireland and Greece, it was agreed that in preparing action relating to the environment the EU was to take account of, amongst other things, costs and benefits, and the economic and social development of the EU as a whole and the balanced development of its regions (Single European Act, 1986, Article 25, inserting Article 130r(3)). New Article 130s provided that, as a general rule, EU environmental measures were to be adopted by unanimity, following consultation with the European Parliament and the Economic and Social Committee, but that the Council could define matters on which decisions were to be taken by a qualified majority (Single European Act, 1986, Article 25). And to reflect Denmark's concerns, new Article 130t provided that EU environmental measures would not prevent any Member State from maintaining or introducing more stringent protective measures compatible with the Treaty (Single European Act, 1986, Article 25). Importantly, the SEA also introduced an additional legal base for *inter alia* environmental measures – new Article 100a, under conditions of qualified majority voting – in respect of the establishing and functioning of the common market (Single European Act, 1986, Article 18). Unsurprisingly, this later led to the appropriate legal base for EU environmental proposals being a common subject of discussion/dispute, since qualified majority voting applied in respect of Article 100a, whereas unanimity generally applied in adopting other EU environmental measures.

Thus, rather than the SEA merely representing a constitutionalisation of existing environmental functions, the SEA went further and gave the EU broad competences in previously contested environmental policy matters, such as nature conservation. As Leann Brown (1995, p. 24) argues, "In the end, the substantive content of the

[SEA] legislation was the product of Commission-brokered compromise. The most contentious issues were 'side-stepped' so that the overall initiative might move forward. All parties to the final decision making were eager that this 'fringe' area bring positive incentives rather than confrontation to the SEA negotiations". In other words, in the high politics arena of Treaty negotiations, turf disputes over the ambit of the EU's environmental remit were subsumed by the political imperative to move forward with EU integration, complete the common market, and reform the EU's decision-making procedures.

Legal actions under the Birds Directive

As the Crocodile Club's initiatives in the European Parliament were beginning to bear fruit in the direction of Treaty revisions, the German Karl-Heinz Narjes, then Commissioner for the Environment, noted in an April 1983 speech reviewing ten years of EU environmental policy that enforcement was proving a particular problem, and that he had therefore given instructions that "use should be made of all the legal remedies available in the event of infringements of Community law" (European Commission, 1983b, p. 2). By September 1984[12] this task was principally in the hands of a German lawyer, Ludwig Krämer, who headed the 'Application of Community Law' unit (hereafter the Legal Unit) within DG XI (Environment, Consumer Protection and Nuclear Safety), which unit was attached directly to the then Director General Athanase Andreopoulos (European Commission, 1984b), who had taken over from Michel Carpentier in 1981 on the ECPS's being 'upgraded' to a Directorate General (Caufield, 1981; European Commission, 1981).[13] Such direct attachment gave the Legal Unit considerable autonomy and power (Williams, 1994).

To explain: by 1984 DG XI had been divided into two Directorates: Directorate A was in charge of *Protection and Improvement of the Environment*; and Directorate B was charged with *Protection and Promotion of Consumer Interests* (European Commission, 1984b). Each Directorate was headed by a Director, and comprised several units; heads of unit would report to the Director of Directorate A or B, as appropriate, and the Directors of the two Directorates would in turn report to the Director General. However, the Legal Unit fell outside this Directorate structure and instead reported directly – and hence had direct access – to the Director General. In contrast, the heads of units within the Directorates would gain access to the Director General only with the approval of the Director of the relevant Directorate (Williams, 1994). In deciding whether to grant such access, such Directors would "be weighing up [the proposed action] in the light of political requirements and the various policy and technical issues" he or she happened to be responsible for at the relevant time (Williams, 1994, p. 355). As head of the directly-attached Legal Unit, Ludwig Krämer had no such barrier to overcome.

Soon after taking office as head of the Legal Unit, Krämer gave an early demonstration of the power of the Birds Directive. In July 1984 the UK government granted outline planning permission to Scottish Malt Distillers to drain a peatland named Duich Moss on the island of Islay in Scotland's Inner Hebrides, famous for its malt whisky (Batten, 1987). The distillers intended to drain the peatland, build an access road, and begin peat extraction for use in whisky production (Haigh, 1991; Batten, 1987). Permission was granted in spite of objections from the Nature Conservancy Council (the NCC, the government's nature conservation body), the RSPB and others, which objections were based on the site's importance for the Greenland white-fronted goose (*Anser albifrons flavirostris*) (Parry, 1985; Stead, 1985b; JNCC, 2001), a then declining Annex I species. In winter, 4–5% of the world population of this endangered species was found at Duich Moss (Haigh, 1991), yet the UK had not designated the site as a special protection area (SPA) under the Birds Directive.

The RSPB lodged a complaint with the Commission, and in September 1984 the Commission wrote to the UK government requesting that Duich Moss be designated as an SPA and given appropriate protection (Batten, 1987). However, the Commission's concerns were reportedly not passed on to the Scottish Secretary owing to an administrative error, and in December 1984 the government confirmed detailed planning consent (Parry, 1985; Batten, 1987). Between March and May 1985 the Commission wrote to the UK government a further three times, and on 1 July 1985 the UK responded, arguing that peat extraction was compatible with the maintenance of the site's conservation importance (Batten, 1987). On 31 July the Commission replied, asking for a postponement of works for two months to allow the UK's claims to be investigated; however, the UK rejected this request (Batten, 1987). Construction work began in August 1985, but was halted almost immediately when demonstrators from Friends of the Earth staged a sit down protest in front of the bulldozers (*The Guardian*, 1985). At a subsequent public meeting on the island, the naturalist David Bellamy and then director of Friends of the Earth, Jonathan Porritt, were amongst the speakers shouted down by islanders, who drew parallels with the clearances of the 19th century and urged the environmentalists to "go back to the backyards of England where you belong" (Stead, 1985c).[14]

In October 1985, in an unprecedented move, Ludwig Krämer paid a personal visit to Duich Moss (Batten, 1987). As Haigh (1991) records,

> As a matter of courtesy, the [UK's] Foreign Office was informed of [Krämer's] intended visit and was thereby thrown into a state of mild panic. Nothing like this had ever happened before in the EC. Did the official have the right to come? Was his visit desirable? Could he be stopped if he insisted on coming? One can imagine the questions being asked. There is no doubt that the official could visit Islay as a holiday maker to watch birds. But he was coming as a representative of the EC institution that is the guardian of the treaty, to investigate a complaint by a private body against a member state for failure to fulfil an obligation under the treaty. The treaty gives no guidance whatever on the right to visit. In this case, the British government put a motor car at the disposal of the official and had him accompanied to Islay.[15]

Krämer produced a report recommending that infringement proceedings be commenced against the UK (Haigh, 1991), and the Commission took the first steps in that regard (Ezard, 1986). The UK ultimately backed down, following "discreet talks" between Commissioner Stanley Clinton Davis and Scottish Secretary Malcolm Rifkind (Norton-Taylor, 1986), and the end result was that the distillery agreed to use an alternative source of peat on the island (Stead, 1987),[16] such that Duich Moss was duly protected by the Birds Directive (Haigh, 1991).

The realisation that the Birds Directive had teeth was a wake-up call for the EU's Member States. The experience certainly was not a welcome one from the UK government's perspective, as Alistair Gammell explains: "Duich Moss was an awful experience for the British government. They didn't enjoy it very much. They are not used to being questioned and found wanting" (Brooke, 1989, p. 55). And despite the positive conservation outcome in the case of Duich Moss, "the embarrassment caused to ministers by the demonstrated teeth of the [Birds Directive]" arguably had an "unwelcome spin-off" in the form of potential negative impacts on negotiations regarding the Habitats Directive (Brooke, 1989, p. 55), as explained in more detail in Chapter 7.

It was not merely the UK that had begun to realise the full impacts of the Birds Directive. In December 1986 an Advocate General of the CJEU handed down two Opinions in respect of alleged breaches of the Birds Directive by Belgium and Italy (CJEU, 1986a, 1986b), ultimately leading in July 1987 to the first judgments under the Birds Directive, which went against the countries concerned and were "very firm" and gave "a real boost to our efforts to ensure that member countries protect Europe's natural heritage", in the words of Commissioner Stanley Clinton Davis (CJEU, 1987a, 1987b; Clinton Davis, 1987a, p. 287; Clinton Davis, 1987b). Judgments against Germany, the Netherlands and France in respect of the Birds Directive were not far behind, handed down in September 1987, October 1987, and April 1988, respectively (CJEU, 1987c, 1987d, 1988). The Commissioner threw his weight behind these enforcement efforts, citing Sophocles favourably during a speech in London in April 1987: "'nobody [has] a more sacred obligation to obey the law than those who made the law' . . . The governments of our Community countries would do well to remember that. Certainly, the Commission does not intend to forget it" (Clinton Davis, 1987c). With a supportive Commissioner behind them, Krämer's Legal Unit had evidently been busy: the same speech by Clinton Davis refers to 100 pending environmental infringement cases in respect of only twelve EU Member States (Clinton Davis, 1987c), and by February 1988 the Unit had launched 29 infringement proceedings against Member States seeking to enforce compliance with their obligations to designate SPAs under the Birds Directive (DG XI, 1988).

Financing

Convinced of the need to support EU environmental policy with financial backing by way of a "European Fund for the Environment", the European Parliament took the initiative of entering new budget headings in the EU's 1982 budget, with a view to enabling intervention in support of environmental policy (European Commission, 1983a, p. 3). The Commission used the opportunity presented by these new budget headings to propose a Regulation on action by the EU relating to the environment. As the Commission wrote in its proposal,

> it is becoming increasingly clear that Community support for the conservation of certain sensitive areas of Community importance is needed . . . It is above all a question of encouraging the competent authorities or non-governmental organizations to take appropriate steps to ensure the protection of such areas concerning which the Community has special responsibilities arising from the community Directive on the conservation of wild birds, and the Berne and Bonn Conventions and the Protocol to the Barcelona convention [for the Protection Of The Mediterranean Sea Against Pollution].
>
> *(European Commission, 1983a, p. 4)*

The Commission's draft instrument therefore proposed that EU co-funding could be provided for, amongst other things, "measures to protect the natural environment in sensitive areas of Community interest" (European Commission, 1983a, Article 2(1)(b)). However, this was amended and restricted during the course of negotiations relating to the Regulation, such that the wording in the version finally adopted in 1984 applied only to co-funding of measures under the Birds Directive: "projects providing an incentive and aimed at contributing towards the maintenance or re-establishment of seriously threatened biotopes which are the habitat of endangered species and are of particular importance to the Community, under [the Birds Directive]" (CEC, 1984, Article 1(1)(c)). Certain Member States were likely motivated to agree this amendment due to their unease at the idea of extending the EU's competences into additional areas of nature conservation policy; and perhaps also by the Commission's statement in its proposal to the effect that "Generally, such damage [to sensitive areas] can be prevented only by imposing restrictions on local economic activities", such that EU financial intervention would therefore be necessary (European Commission, 1983a, p. 4). One would imagine that some Member States (at least) would not have been keen to extend any such restrictions beyond what was strictly required as a matter of EU law.

The Commission's draft Fourth Environmental Action Programme of October 1986 effectively reopened this debate by stating that "the 'nature conservation' section of the Community's ACE Regulation (EEC) No 1872/84 should be expanded and extended to help the achievement of the above objectives [the protection of flora and fauna]; it would be neither logical nor desirable to restrict its scope to species covered by the 'Birds Directive'; the Commission will make appropriate proposals" (European Commission, 1986a, p. 31). The Commission acted quickly, issuing its draft Regulation to the Council only two months later, in December 1986. In respect of nature conservation, the draft provided that EU co-funding could be made available for:

> (e) projects providing an incentive and aimed at contributing towards the protection, maintenance or re-establishment of areas of particular Community-wide importance for the conservation of nature and especially of seriously threatened biotopes which are the habitat of endangered species and are of importance to the Community;
>
> (f) projects providing an incentive and aimed at implementing programmes on the conservation or restoration of populations of species in danger of extinction in the Community.
>
> *(European Commission, 1986b, Article 1(1)(e) and (f))*

In the draft instrument's Explanatory Memorandum, the Commission justified this proposed change on the basis that "In a number of urgent cases, this restriction [allowing funding only in respect of the Birds Directive, insofar as nature conservation was concerned] has made it impossible for the Community to grant support to protect the natural habitat of other severely endangered species in the Community, such as the monk seal, the brown bear, the sea turtle or certain bats, even though no-one denies that an effective, rapid Community campaign is essential to the survival of these species in the Community" (European Commission, 1986b, p. 5). However, the Member States again rejected the Commission's proposal, with the instrument finally adopted in July 1987 again restricting EU co-funding in the nature conservation field principally to measures in respect of the Birds Directive,[17] and this in spite of the new EU environmental competences provided by the SEA, which had come into force on 1 July 1987 (Sands, 1990). Nonetheless, the final version of the Fourth Environmental Action Programme (1987) retained the above-quoted aspirational wording from the 1986 draft,[18] and so the issue remained very much on the Commission's agenda.

Indeed, DG XI's Nature Unit was creative in seeking to take action in respect of organisms other than birds, and did not lose sight of the species mentioned above, as Pierre Devillers of Belgium's IRSNB explains:

> one thing that is important in the genesis of the Habitats Directive is that Richard [Geiser of the Nature Unit] was very conscious of the need to do something for other organisms, especially for mammals, and he launched a series of actions using as best he could some of the dispositions of the Treaty that permitted the Commission to look at more than the birds, and so he initiated some action plans for certain mammals and that he did mostly, if not entirely, with our institute. The first he did was the monk seal [(Anselin et al., 1988)], and also the sea turtles [(Kunkel et al., 1989)] . . . because he was very aware of the plight of the monk seal particularly in Greece, and so we were very, very involved with that, and of course that also meant contact with the Bern Convention. And we tried to go as far as we could . . . I wouldn't say it was stretching the Treaty but it was certainly using it to its fullest possibilities.
>
> *(Devillers, 2013)*

Stanley Johnson returns to the Commission in 1984

Shortly before the European parliamentary elections of June 1984, Stanley Johnson moved back to the European Commission, returning from his five-year *congé de convenance personnelle* as an MEP (Johnson, 2011). He rejoined the Commission at the same senior grade he had held on his departure for the Parliament, and assumed a dual role very soon after his return: head of the 'Waste Management and Prevention of Noise Pollution' unit, and adviser on intersectoral environmental issues to Anthony John Fairclough, then Director of Directorate A (Protection and Improvement of the Environment) but soon to be promoted to acting Director General of DG XI on the early retirement of Athanase Andreopoulos, who had not been well-suited to the job (Johnson, 2012; Krämer, 2011).[19] "It was quite a strange couple of years [from May 1984 to January 1987] in a way", reports Johnson (2011), "because I came back being vice-chairman of the Parliament's environment committee. I didn't have the kind of overarching role which I'd had before".[20] Strange for Johnson those two plus years may have been – that is, from his return to the Commission in June/July 1984 to the appointment of a new Director General to DG XI, Laurens Jan Brinkhorst, in January 1987 (*Japan Economic Newswire*, 1987), at which point Johnson would drop his role as head of the Waste unit and act solely as main adviser to Brinkhorst (Johnson, 2011) – but they turned out to be important years for EU nature conservation policy.

A British Commissioner for the Environment, Stanley Clinton Davis, took up office in January 1985 (Brown, 1984), and Stanley Johnson was given responsibility for drafting the Fourth Environmental Action Programme, to cover the period from 1987–1992; that is, the "golden era" of EU environmental policy described in Chapter 6. As Johnson (2011) explains, "It was my job to write the whole thing, and indeed I did. Of course I got input from everybody else in the DG". Not only did Johnson and Clinton Davis share a common country of origin, they also shared interests; as Johnson (2011) comments, "Clinton Davis I regarded as wonderful because he was particularly keen on the things I was interested in", which of course included nature conservation.[21]

Others in DG XI were not the only ones seeking to provide input to the draft Fourth Environmental Action Programme. For example, of the many conservation conferences taking place across Europe in this period, at a British Association of Nature Conservationists' (BANC) conference in Cambridge in 1985 the need for an EU level instrument on species and habitat conservation was widely discussed (Everett, 1987; Hepburn, 1994). The issues were recorded in detail in a 1987 journal article, in which Sue Everett, a former chair of BANC, called specifically for a Habitats Directive (Everett, 1987). Hepburn (1994, p. 2) recalls that "This was based upon discussions held at and subsequent to the BANC conference on international wildlife treaties at Cambridge in 1985 and, in particular, encouragement from the RSPB and others with experience of the Birds Directive. This fortuitous event, combined with judicious lobbying by BANC, ensured that these views were not lost on the European Commission". As Alistair Gammell explains,

"Myself and one or two others . . . lobbied to put [a proposal for a directive on habitats and species protection] in [the Fourth Environmental Action Programme], although there were Commission officials in DG XI . . . who were . . . very easy to lobby . . . They wanted to do it and we wanted it in" (Gammell, 2012).

This aspect of DG XI's work was given powerful political impetus around this time by the agreement of the SEA Treaty changes in February 1986. Knowing that the Commission could rely on the new environmental competence provisions agreed in the SEA, Stanley Johnson worked on his draft of the Fourth Environmental Action Programme throughout 1986 (Johnson, 2011), and the Commission submitted its final draft to the Council on 15 October 1986, during a British Presidency of the Council of the EU, containing the following wording drafted by Johnson (European Commission, 1986a, p. 31; Johnson, 2012):

> 5.1.2 More than six years have now elapsed since the Council adopted its Directive and resolution on the conservation of wild birds (79/409/EEC). It is of course essential to ensure that this Directive and resolution are fully implemented in the Member States . . . But this alone is not enough; the time is now ripe for the Community and Member States to make a major new thrust in the field of nature conservation.
>
> 5.1.4 What essentially is needed is a Community instrument aimed at protecting not just birds but all species of fauna and flora; and not just the habitats of birds, but the habitat of wildlife — animals and plants — more generally. Such a comprehensive framework should ensure that, throughout the Community, positive measures are taken to protect all forms of wildlife and their habitat; such measures should be aimed at the three main objectives of the World Conservation Strategy:
>
> — the maintenance of essential ecological processes and life support systems,
> — the preservation of genetic diversity,
> — the sustainable utilization of species and ecosystems.
>
> 5.1.5 The Commission will make appropriate proposals on these lines. The Commission is also working on the preparation of a comprehensive list of sites throughout the Community that are protected under the various categories of protected areas. Such a list is a necessary basis for the consistent implementation of the kind of framework instrument discussed above.
>
> 5.1.6 It seems clear from recent reports that the implementation of the Berne Convention in the Member States of the Community, as elsewhere, is poor. A comprehensive framework of nature protection measures, at Community level, would undoubtedly help to improve the situation of endangered species of plants and animals within the Community, whilst at the same time helping to achieve the three-fold objective of the World Conservation Strategy.[22]

An EU instrument with these aims would naturally be a significantly more complex undertaking than the Birds Directive, given the former's much broader ambit. However, by 1987 the Commission already had some data to work with.

The CORINE biotopes project[23]

In addition to the lists of species annexed to the Bern Convention, the Commission had launched, in 1985, the CORINE (Coordination of Information on the Environment) programme (Devillers, 2013). One project within CORINE was CORINE biotopes, which was "designed to assemble reliable and consistent information about the location and status of ecosystems, habitats and species in need of protection and to make this information accessible to policy-makers" (Moss and Wyatt, 1994, p. 329); or, in other words, per Pierre Devillers – a scientist from the Institut Royal des Sciences Naturelles de Belgique (IRSNB) – "it was essentially doing, if you like, Natura 2000, it was doing exactly that" (Devillers, 2013).

The CORINE biotopes project – which was "essential in the early process" of the Habitats Directive, according to Devillers (2013) – was under the direction of Gunter Schneider of DG XI, with Michel Cornaert the main official responsible under Schneider (Devillers, 2013). Schneider had worked as Claus Stuffmann's deputy head of division – with special responsibility for "economic aspects and land use" – from 1980 to 1988, at which time Schneider moved to head up his own division, 'Economic aspects, employment and statistics' (European Commission, 1980a, 1988). The relations between Schneider's CORINE team and Stuffmann's Nature Unit were reportedly not particularly good, but there were good contacts between the CORINE team and the Council of Europe (Devillers, 2013). At this time – that is, from the launch of CORINE in 1985 – there were three separate habitat classification projects running in parallel in Europe, as well as a fourth project that had already resulted in a report. The first, already published, was the Council of Europe's *Classification of European Ecosystems* designed by Géhu (1984);[24] second, as part of the implementation of the Birds Directive, Stuffmann's team had developed a "very short typology of habitats . . . the kind of typology that bird watchers do" (Devillers, 2013); third, there was the CORINE biotopes project, which involved identifying important nature conservation sites and describing them; and fourth, there was ongoing work between the Council of Europe and the Commission – overseen by Professor Noirfalise of the Agricultural Science Faculty, Gembloux, Belgium – to produce a map of the natural vegetation of the member countries of the European Community and the Council of Europe, work that would ultimately be published in 1987 (Council of Europe and European Commission, 1987).[25]

As Devillers and Devillers-Terschuren (1996) record,

> Throughout the development of the [CORINE] classification great efforts were made to establish or retain compatibility, in the sense of possibilities of one to one conversion, between the CORINE classification and two

other Europe-wide projects. These were, on the one hand, the Council of Europe *Classification of European Ecosystems* designed by J. M. Géhu (1984), and, on the other hand, the *Carte de la végétation naturelle des Etats membres des Communautés européennes et du Conseil de l'Europe* that was being prepared by Professor Noirfalise simultaneously with the CORINE project.

However, it was the task of seeking to ensure compatibility between the Birds Directive's simple typology and the developing CORINE initiative that would see Devillers become involved in CORINE biotopes in the first place, as he explains:

> Richard [Geiser] asked me to try to bridge the two, to try to convince Cornaert to take onboard what was already done for the Birds Directive, because he [Geiser] was very worried about the incoherence that it would be for the Commission to have two systems of describing, and he was very aware of that, and so was Michel Cornaert, after all CORINE was Coordination, Information, so the idea was to do the two together.
>
> *(Devillers, 2013)*[26]

Having started out in this way, Devillers soon became involved with broader discussions regarding coherence with the Council of Europe's two ongoing projects, leading to some "difficult discussions" in Strasbourg, finally resulting in a compromise, allowing Devillers to begin work on a typology of EU habitats as part of CORINE biotopes, work that would continue throughout much of the Habitats Directive negotiations, before being published in 1991 (Devillers, 2013).[27] This work, explains Devillers (2013), "differed only from something like Natura 2000 in the sense that it was not legally binding, it was just to be a scientific inventory". The CORINE team established lists of animals and plants to be taken into consideration by policy-makers, as well as a typology of habitats. This typology was, explains Devillers (2013), originally strictly

> a tool to describe the habitats identified, but very . . . early in the process, I proposed – and it was quickly accepted . . . that it also be used as a selection criterion. The idea was that we could not indefinitely make longer and longer lists of species, that it was not that all species had to be protected and all populations; all representative populations, we had to have a geographical basis, so all representative populations should be protected, nothing should fall through the net, but we could not do that by listing all the species of Europe, including all the unglamorous creepy-crawlers and little plants. And so I proposed to use habitat types as surrogates, and therefore to make sure that the list of habitats was detailed enough, went into fine enough details to make sure that every distinctive community was identified, and that therefore every distinctive community would have to be represented in the network in sufficient quantity in every geographical area, so that we would be sure that all the species would be covered, or as sure as possible

> that all the species would be covered. So that was the philosophy, and that's why the typology of habitats became extremely detailed, exactly to ensure that [element of surrogacy].

At this time, the work was not being done with a Habitats Directive in mind; rather, the CORINE team simply saw CORINE biotopes as

> the best thing we could do. We felt that it was a way to get it through the back door, if you like. We did not think that it was possible to get a binding instrument, but we felt that notoriety is essential, and that if we made a scientific inventory of what needs to be conserved, it would bring a strong leverage on what the Member States would have to do. So essentially we were doing something like the Habitats Directive but without thinking of an instrument. CORINE ran into trouble later on, there was very strong pressure from the Member States to find other things to reduce its importance and its significance . . . but it's clear that it came from the fact that many Member States were afraid of CORINE. And we were told, in fact, by representatives of Member States, that we were going much too far because, OK, we were admitting that it was not a binding instrument but in reality we were making it such that it would be difficult not to regard it as that. So it was our wish, and it's fundamental to what happened with the Habitat Directive.
>
> *(Devillers, 2013)*

Notes

1 For additional accounts of the Habitats Directive's relationship with the Bern Convention, see Jones (2012) and Coffey and Richartz (2003).

2 This claim might be up for some debate, given the Council of Europe's extensive own-initiative activities in the environmental field in the 1960s and 1970s.

3 Such complexities would have included, for example, the species to be listed and the prohibited methods of capture and killing.

4 This reference to 'case-law' relates to the Convention's system of complaints and case-files, which can result in Recommendations to Parties from the Standing Committee of the Convention: see Lasén-Díaz (2010).

5 As Jans and Vedder (2012, p. 520) explain, the "directive was originally to apply from 1 October 1983 to 1 October 1985, but its duration was extended to 1 October 1989 and in 1989 the Council agreed to extend its application indefinitely".

6 Years later, by this time having left behind his career with the EU institutions, Johnson joined a WTO protest in Seattle dressed as a turtle (*The Evening Standard*, 1999). As his son Boris commented at the time, "It is magnificent . . . It is typically brave of him to take on the Robocop police. He is tremendously keen on the environment, which doesn't make him a lefty" (*The Evening Standard*, 1999).

7 The World Conservation Strategy was drawn up in 1980 by the IUCN in collaboration with UNEP and the WWF. For additional background, see McCormick (1986).

8 Lodge (1994) argues that the SEA was predominantly an initiative of the European Parliament, a view that can be contrasted with that of liberal intergovernmentalist Moravcsik (1991, p. 21), who argues that the SEA reforms primarily "rested on interstate bargains

between Britain, France and Germany". As Lodge (1994, p. 20) counters, "Without the [Parliament's] initiative for the EUT [draft Treaty], the SEA process would not have been launched or been run in the way that it was".

9 MEPs passed a resolution complaining that the SEA "in no way represent[s] the real reform of the Community that our peoples need" (Moravcsik, 1991, p. 45).

10 Lodge (1994, p. 14) argues that the Parliament's draft had "a decisive influence on the SEA's content".

11 As such, it is incorrect to argue that the SEA was merely the constitutionalisation of *de facto* practice: Leann Brown (1995) cites Koppen (1988) as having made this argument.

12 Krämer took on the role of head of the Legal Unit sometime between May and September 1984 (European Commission, 1984a, 1984b).

13 Carpentier was not made head of the newly-created DG when the ECPS was upgraded because the Commission wanted to find a Director General position for a Greek individual, Greece having joined the EU in 1981 (Pleinevaux, 2010; Krämer, 2011). This caused resentment, with the staff of the new DG feeling that Carpentier had been the victim of a serious injustice (Krämer, 2011).

14 The environmentalists were later informed they would be safe if they returned to the island, which gives an indication of the atmosphere at the time (Stead, 1985a).

15 A report by the House of Lords would later note, "The experience of Duich Moss (Islay) in Scotland in 1985, when a Commission official visited the site to investigate a possible case of non-compliance, was regarded by the Scottish Office as having created unwelcome and unnecessary political controversy" (House of Lords, 1989b, paragraph 28). And yet, as Viscount Blakenham reflected, "If that unwelcome visit had not been made the wintering ground of the white-fronted goose is unlikely to have remained protected" (House of Lords, 1989a, p. 1266).

16 The NCC had identified an alternative site nearby, already used by the distillers, which the distillers could use to source their peat, but the distillers had rejected this site on the basis that the supplier at the site could not meet the distillers' requirements, though this was contested by the NCC (Ardill, 1985).

17 "(e) projects providing an incentive and aimed at contributing towards the maintenance or re-establishment of seriously threatened biotopes which are the habitat of endangered species and are of particular importance to the Community under Directive 79/409/EEC; (f) projects providing an incentive and aimed at contributing towards the protection or re-establishment of land threatened or damaged by fire, erosion and desertification" (CEC, 1987, Article 1(1)(e) and (f)).

18 That is, "the 'nature conservation' section of the Community's ACE Regulation (EEC) No 1872/84 should be expanded and extended to help the achievement of the above objectives; it would be neither logical nor desirable to restrict its scope to species covered by the 'Birds Directive'; the Commission will make appropriate proposals" (Fourth Environmental Action Programme, 1987).

19 Fairclough, per Krämer (2011), was brilliant, hard-working and well-liked.

20 Johnson worked as head of the 'Prevention of Pollution and Nuisances' division of the ECPS from 1973–76, but from 1977–79 he worked as adviser to Michel Carpentier, head of the ECPS (Johnson, 2011; European Commission, 1977).

21 Ian Hepburn, the RSPB/WWF coordinator of the EU-wide Habitats Directive campaign, explains that having a British Commissioner with a British adviser, in the form of Stanley Johnson, gave British NGOs an opportunity to go over and "start talking the same language, literally and metaphorically . . . and understand each other rather well" (Hepburn, 2012).

22 See Note 7.

23 For further background, see Moss and Wyatt (1994).

24 For more on Géhu's contribution to European nature conservation, see Evans (2016).

25 This 1987 map was effectively the second edition, the first – covering the Council of Europe Member States – having been published in 1979, with Noirfalise amongst the

authors (Ozenda et al., 1979). The 1987 map formed the basis for the creation of the first indicative map of biogeographical regions for the purposes of the Habitats Directive, which was later extended eastwards: for more detail, see ETC-BD (2006).

26 Devillers (2013) adds, "when I joined the [CORINE] biotope team, at that point, primarily to see to it that it would not conflict with the description of the special protection areas of the Birds Directive, that what they did would work, and that for birds, for instance, their list of important birds would be those in Appendix I and that the descriptions of habitats would be the same as those of the Birds Directive ... it was just to make sure that it did not undermine the Birds Directive, originally" (Devillers, 2013).

27 As Devillers (2013) explains, the typology he developed "incorporated as two digit units the units from the Birds Directive as upper limit, and then accommodated all of Géhu's proposals under those as the hierarchy, and also further developments, etc. And I worked very closely with Professor Noirfalise, who was a friend, and so he was developing at the same time his cartography and we kept in close contact, so that the typology we proposed within the [CORINE] biotope team was a synthesis of that of Géhu, of what Noirfalise was preparing, and of the very little that had to be respected that came from the Birds Directive. That typology took time to finish because it was complicated by accessions – the accession of Spain and Portugal certainly broadened the diversity of habitat types considerably, and so it was finally published in 1991 [(Devillers et al., 1991)] ... [I]t was being finalised all through the preparation of the Habitat Directive".

References

Anselin, A., Van der Elst, M. N., Beudels, R. C. and Devillers, P. 1988. Analyse descriptive et projet préparatoire à une stratégie pour la conservation du Phoque moine de Méditerrannée (Monachus monachus). Rapport à la Commission des Communautés Européennes, Bruxelles.

Ardill, J. 1985. Whisky firm accused over geese haven / Scottish malt distiller rejects alternative site for extracting peat. *The Guardian*, 3 August.

Batten, L. A. 1987. The effectiveness of European agreements for wader conservation. *Wader Study Group Bulletin 49, Supplement/IWRB Special Supplement* 7, 118–121.

Booker, C. and North, R. 2003. *The great deception: the secret history of the European Union*. New York: Continuum.

Brooke, M. 1989. A birthday for the birds. *New Scientist* (4 February), 54–57.

Brown, D. 1984. Delors beats the system with new Commission / EEC Commission president-elect supervises portfolio share-out among his team. *The Guardian*, 10 December.

Cardozo, R. and Corbett, R. 1986. The Crocodile initiative. *In:* Lodge, J. (ed.) *European union: the European Community in search of a future*. London: Macmillan.

Caufield, C. 1981. Environment: Britain lags, Europe leads. *New Scientist* (13 August), 416–417.

CEC 1984. Council Regulation (EEC) No. 1872/84 of 28 June 1984 on action by the Community relating to the environment. Council of the European Communities. *OJ L 176*, 3 July 1984, p. 1.

CEC 1987. Council Regulation (EEC) No 2242/87 of 23 July 1987 on action by the Community relating to the environment. Council of the European Communities. *OJ L 207*, 29 July 1987, p. 8.

CITES Regulation 1982. Council Regulation (EEC) No 3626/82 of 3 December 1982 on the implementation in the Community of the Convention on international trade in endangered species of wild fauna and flora. *OJ L 384*, 31 December 1982, p. 1.

CJEU 1986a. Opinion of Advocate General Da Cruz Vilaça of 2 December 1986, *Commission* v. *Belgium*, Case 247/85, EU:C:1986:458.

CJEU 1986b. Opinion of Advocate General Da Cruz Vilaça of 2 December 1986, *Commission* v. *Italy*, Case 262/85, EU:C:1987:340.

CJEU 1987a. Judgment of the Court of Justice of the EU of 8 July 1987, *Commission* v. *Belgium*, Case 247/85, EU:C:1987:339.

CJEU 1987b. Judgment of the Court of Justice of the EU of 8 July 1987, *Commission* v. *Italy*, Case 262/85, EU:C:1987:340.

CJEU 1987c. Judgment of the Court of Justice of the EU of 17 September 1987, *Commission* v. *Germany*, Case 412/85, EU:C:1987:370.

CJEU 1987d. Judgment of the Court of Justice of the European Union of 13 October 1987, *Commission* v. *Netherlands*, Case 236/85, EU:C:1987:436.

CJEU 1988. Judgment of the Court of Justice of the EU of 27 April 1988, *Commission* v. *France*, Case 252/85, EU:C:1988:202.

Clinton Davis, S. 1987a. European Parliament, Verbatim report of proceedings, 9–10 July 1987, Strasbourg.

Clinton Davis, S. 1987b. Mr Clinton Davis welcomes Court judgements on Bird Directives, 10 July 1987, IP/87/288.

Clinton Davis, S. 1987c. Vital need to apply Community environment law – extracts from a speech given by Commissioner Stanley Clinton Davis at the conference of the Institution of Environmental Health Offices, London, 28 April 1987, IP/87/161.

Coffey, C. and Richartz, S. 2003. *The EU Habitats Directive: generating strong responses*. Berlin: Institute for European Environmental Policy.

Council of Europe and European Commission 1987. *Map of the natural vegetation of the member countries of the European Community and the Council of Europe*. 2nd edn. Explanatory noted by A. Noirfalise. Luxembourg, Office for Official Publications of the European Communities.

D'Ancona, H. 1990. Italy and its contribution to the making of Europe. *In:* Bruyning, L. F. and Leerssen, J. T. (eds) *Yearbook of European Studies. 3 Italy – Europe*. Amsterdam: Editions Rodopi B.V.

Devillers, P. 2013. Interview with Pierre Devillers of Belgium's IRSNB conducted by Andrew L. R. Jackson on 25 April 2013.

Devillers, P. and Devillers-Terschuren, J. 1996. A classification of Palaearctic habitats. Council of Europe, Strasbourg: Nature and environment, No 78.

Devillers, P., Devillers-Terschuren, J. and Ledant, J.-P. 1991. *CORINE biotopes manual. Vol 2. Habitats of the European Community*. Luxembourg: Office for Official Publications of the European Communities.

DG XI 1988. Internal DG XI note dated 18 February 1988 from the Legal Unit to the Director of Directorate B, the adviser to the Director General, the Nature Unit, and the Economics Unit. Subject: Draft of a directive on habitats protection. Ref: Version of FFH directive of 15.2.1988.

Environmental Impact Assessment (EIA) Directive 1985. Council Directive 85/337/EEC of 27 June 1985 on the assessment of the effects of certain public and private projects on the environment, *OJ L 175*, 5 July 1985, p. 40, since codified as Directive 2011/92/EU of the European Parliament and of the Council of 13 December 2011 on the assessment of the effects of certain public and private projects on the environment (codification), *OJ L 26*, 28 January 2012, p. 1. Codified version since amended by Directive 2014/52/EU, *OJ L 124*, 25 April 2014, p. 1.

ETC-BD 2006. *The indicative Map of European Biogeographical Regions: methodology and development*. Paris: ETC-BD.

European Commission 1975. Non-governmental organizations for the protection of the environment in the Community (X/280/75-E). *Industry and Society*, Brussels, 27 May 1975, No. 21/75.

European Commission 1977. Directory of the Commission of the European Communities (September 1977).

European Commission 1979. Commission Communication to the Council on the results of the negotiations for the conclusion of the Convention on the Conservation of European Wildlife and Natural Habitats. Proposal for a Council Decision concerning the conclusion of the Convention on the conservation of European wildlife and natural habitats. COM(79) 414 final, Brussels, 17 July 1979.

European Commission 1980a. Directory of the European Communities (October 1980).

European Commission 1980b. Progress made in connection with the environment action programme and assessment of the work done to implement it. Communication from the Commission to the Council. COM(80) 222 final, Brussels, 7 May 1980.

European Commission 1980c. Rational land use: Its role in Community environment policy. Communication from the Commission to the Council. COM(80) 253 final, Brussels, 7 May 1980.

European Commission 1981. Organigramme de la Commission des Communautés Européennes (December 1981). Luxembourg: Office des publications officielles des Communautés Européennes.

European Commission 1983a. Proposal for a Council Regulation on action by the Community relating to the environment (ACE). COM(82) 849 final, Brussels, 11 January 1983. Submitted by the Commission to the Council on 13 January 1983.

European Commission 1983b. Ten years of European Community environment policy. Address by Mr Narjes to the DIHT, Bonn, 21 April 1983. IP(83) 139, Brussels, 21 April 1983.

European Commission 1984a. Directory of the Commission of the European Communities (May 1984).

European Commission 1984b. Directory of the Commission of the European Communities (September 1984).

European Commission 1986a. Draft for a resolution of the Council of the European Communities on the continuation and implementation of a European Community policy and action programme on the environment (1987–1992). COM(86) 48 5 final. Submitted by the Commission to the Council on 15 October 1986. *OJ C 70*, 18 March 1987, p. 3.

European Commission 1986b. Proposal for a Council Regulation on action by the Community relating to the environment. COM(86) 729 final, Brussels, 19 December 1986.

European Commission 1988. Directory of the Commission of the European Communities (January 1988).

European Parliament 1980a. Resolution embodying the opinion of the European Parliament on the proposal from the Commission of the European Communities to the Council for a Decision concerning the conclusion of the convention on the conservation of European wildlife and natural habitats (17 June 1980). *OJ C 175*, 14 July 1980, p. 18.

European Parliament 1980b. Resolution on the World Conservation Strategy (WCS), 20 May 1980. *OJ C 147*, 16 June 1980, p. 27.

European Parliament 1983. European Parliament Resolution concerning the substance of the preliminary draft Treaty establishing the European Union (14 September 1983). *OJ C 277*, 17 October 1983, p. 95.

European Parliament 1984. *Draft Treaty establishing the European Union (14 February 1984)*. Luxembourg: Office for Official Publications of the European Communities.

European Parliament 1987. Resolution on the establishment and conservation of Community nature reserves (10 July 1987). *OJ C 246*, 14 September 1987, p. 121.

Evans, D. 2016. Jean-Marie Géhu and nature conservation in Europe. *Documents Phytotosociologiques*, 3e série Vol. 8, 227–232.

Everett, S. 1987. Strengthening protection for Europe's wild flora and wildlife habitats. *ECOS: A Review of Conservation*, 8.

Ezard, J. 1986. Legal move on Islay / European Commission to demand protection for geese on Inner Hebrides island. *The Guardian*, 8 March.

Fourth Environmental Action Programme 1987. Resolution of the Council of the European Communities and of the Representatives of the Governments of the Member States meeting within the Council of 19 October 1987 on the continuation and implementation of a European Community policy and action programme on the environment (1987–1992). *OJ C 328*, 7 December 1987, p. 1.

Gamble, C. J. 1998. Godfather of the European Union: Altiero Spinelli. *European Business Review*, 98, 135.

Gammell, A. 2012. Interview with Alistair Gammell, formerly of the RSPB, conducted by Andrew L. R. Jackson on 31 July 2012.

Géhu, J. M. 1984. Classification of European ecosystems. Report SN-VS(S4)3 to the European Committee for the Conservation of Nature and Natural Resources. Council of Europe, Strasbourg.

Haigh, N. 1991. The European Community and international environmental policy. *International Environmental Affairs*, 3, 163–180.

Hepburn, I. 1994. European wildlife policy – primed for action? *ECOS: A Review of Conservation*, 15, 1–2.

Hepburn, I. 2012. Interview with Ian Hepburn, English RSPB/WWF campaigner who headed an EU-wide NGO campaign in support of the Habitats Directive during the instrument's negotiation and was seconded to DG XI's Nature Unit for a period during the negotiations, conducted by Andrew L. R. Jackson on 16 August 2012.

House of Lords 1989a. House of Lords' debate on 14 November 1989 regarding "Habitats and Species Protection: ECC Report", columns 1248–1279.

House of Lords 1989b. House of Lords' Select Committee on the European Communities. 15th report: Habitat and Species Protection. Ordered to be printed 18 July 1989.

Jacobs, F. 2000. The European Parliament. *In:* Lynch, P., Neuwahl, N. and Rees, W. (eds) *Reforming the European Union: from Maastricht to Amsterdam*. Harlow: Pearson Education Limited.

Jans, J. H. and Vedder, H. H. B. 2012. *European environmental law: after Lisbon*. Groningen: Europa Law Publishing.

Japan Economic Newswire 1987. Former Netherlands Premier takes up Tokyo EC post in April. *Japan Economic Newswire*, 26 January.

JNCC 2001. SPA description: Eilean na Muice Duibhe (Duich Moss), Islay. Joint Nature Conservation Committee.

Johnson, S. 2009. *Stanley I presume*. London: Fourth Estate.

Johnson, S. 2011. Interview with Stanley Johnson conducted by Christian van de Velde on 18 October 2011. Ref: INT185. HISTCOM.2: Histoire interne de la Commission Européenne 1973–1986. Project coordinator: Université catholique de Louvain.

Johnson, S. 2012. Interview with Stanley Johnson, former European Commission official and former MEP, conducted by Andrew L. R. Jackson on 16 July 2012.

Jones, G. 2012. The Bern Convention and the origins of the Habitats Directive. *In:* Jones, G. (ed.) *The Habitats Directive: a developer's obstacle course?* Oxford: Hart Publishing.

Koppen, I. J. 1988. The European Community's environment policy. From the summit in Paris, 1972, to the Single European Act, 1987. *Working Paper 88/328*. Florence: European University Institute.

Krämer, L. 2011. Interview (in French) with Ludwig Krämer conducted by Christian van de Velde on 7 October 2011. Ref: INT194. HISTCOM.2: Histoire interne de la Commission Européenne 1973–1986. Project coordinator: Université catholique de Louvain.

Kunkel, P., Van der Elst, M. N., Beudels, R. C. and Devillers, P. 1989. Sea turtles in the European Community. Parts 1 and 2. Rapport à la Commission des Communautés Européennes, Bruxelles.

Lasén-Díaz, C. 2010. The Bern Convention: 30 years of nature conservation in Europe. *Review of European Community & International Environmental Law*, 19, 185–196.

Leann Brown, M. 1995. Advocacy coalitions and the "greening" of the Single European Act. *Fourth Biennial International Conference of the European Community Studies Association, 11–14 May 1995*. Charleston, South Carolina, USA.

Lodge, J. 1984. European Union and the first elected European Parliament: the Spinelli initiative. *Journal of Common Market Studies*, 22, 377–402.

Lodge, J. 1994. Negotiating the Single European Act in the European Community. *In:* Zartman, I. W. (ed.) *International multilateral negotiation: approaches to the management of complexity*. San Francisco: Jossey-Bass Publishers.

McCormick, J. 1986. The origins of the World Conservation Strategy. *Environmental Review*, 10, 177–187.

Moravcsik, A. 1991. Negotiating the Single European Act: national interests and conventional statecraft in the European Community. *International Organization*, 45, 19–56.

Moss, D. and Wyatt, B. K. 1994. The CORINE biotopes project: a database for conservation of nature and wildlife in the European community. *Applied Geography*, 14, 327–349.

Muntingh, H. 2013. Interview with Hemmo Muntingh, former MEP, conducted by Andrew L. R. Jackson on 18 March 2013.

Norton-Taylor, R. 1986. Diplomacy may preserve geese's grazing ground / Inner Hebrides geese versus whisky dispute. *The Guardian*, 26 April.

Ozenda, P., Noirfalise, A., Tomaselli, R. and Trautmann, W. 1979. Vegetation map of the Council of Europe Member States. Strasbourg: Council of Europe.

Parry, G. 1985. Imminent destruction of wildlife site blamed on negligence of Government / Scottish island of Islay. *The Guardian*, 15 June.

Pleinevaux, C. 2010. Interview (in French) with Claude Pleinevaux conducted by Christian van de Velde on 8 October 2010. Ref: INT238. HISTCOM.2: Histoire interne de la Commission Européenne 1973–1986. Project coordinator: Université catholique de Louvain.

Ribaut, J.-P. 2004. How the Bern Convention came into being. *Naturopa*, 101, 4.

Sands, P. 1990. European Community environmental law: legislation, the European Court of Justice and common-interest groups. *Modern Law Review*, 53, 685–698.

Seal Pups Directive 1983. Council Directive 83/129/EEC of 28 March 1983 concerning the importation into Member States of skins of certain seal pups and products derived therefrom. *OJ L 91*, 9 April 1983, p. 30.

Second Environmental Action Programme 1977. Resolution of the Council of the European Communities and of the representatives of the governments of the Member States meeting within the Council of 17 May 1977 on the continuation and implementation of a European Community policy and action programme on the environment. *OJ C 139*, 13 June 1977, p. 1.

Single European Act 1986. Single European Act. *OJ L 169*, 29 June 1987, p. 1.

Staff Regulations 1968. Regulation (EEC, Euratom, ECSC) No 259/68 of the Council of 29 February 1968 laying down the Staff Regulations of Officials and the Conditions of Employment of other Servants of the European Communities and instituting special measures temporarily applicable to officials of the Commission, as amended and unofficially consolidated. *OJ L 56*, 4 March 1968, p. 1.

Stead, J. 1985a. Friends of the geese get safe passage / Friends of the Earth to continue campaign to protect geese on Islay. *The Guardian*, 12 August.

Stead, J. 1985b. Geese feel the nip of whisky / Battle for control of peat bog on the Scottish island of Islay. *The Guardian*, 16 July.

Stead, J. 1985c. Whisky island vents its anger on conservationists / Friends of the Earth director Porritt shouted down at meeting on Island of Islay. *The Guardian*, 6 August.

Stead, J. 1987. Whisky men yield ground to geese. *The Guardian*, 2 July.

The Evening Standard 1999. Johnson Snr's turtle protest in Seattle. *The Evening Standard*, 2 December.

The Guardian 1985. UK news in brief: Whisky moor work halted / Work on a peat moor on the Scottish island of Islay. *The Guardian*, 1 August.

Third Environmental Action Programme 1983. Resolution of the Council of the European Communities and of the representatives of the Governments of the Member States, meeting within the Council, of 7 February 1983 on the continuation and implementation of a European Community policy and action programme on the environment (1982 to 1986). *OJ C 46*, 17 February 1983, p. 1.

Ward, I. 2003. *A critical introduction to European Law*. London: Reed Elsevier.

Whales Regulation 1981. Council Regulation (EEC) No 348/81 of 20 January 1981 on common rules for imports of whales or other cetacean products. *OJ L 39*, 12 February 1981, p. 1.

Williams, R. 1994. The European Commission and the enforcement of environmental law: an invidious position. *Yearbook of European Law*, 14, 351–399.

6

THE HABITATS DIRECTIVE PROPOSAL

Three phases in the history of EU environmental policy

Second phase (1987–1992): the "golden era"

With hindsight, comments Ward (1997, p. 178), "the period between the Single European Act (SEA) of 1987 and the Maastricht Treaty in 1992 could be seen as [a] uniquely favourable era for strengthening Community environmental policy". Various factors combined to contribute what was arguably a "golden era" (Ward, 1997, p. 178): the emergence of, and an increasing awareness of, transnational environmental problems; widespread public support for environmental action; a growing awareness of the potentially damaging impacts of the common market; a growth in support for Green political parties; the entry into force in 1987 of a clear legal base for EU environmental policy (in the Single European Act, 1986); and the momentum provided by numerous influential civil servants (*fonctionnaires*) within DG XI, supported by two progressive EU Environment Commissioners: the Briton Stanley (now Lord) Clinton Davis, who was Commissioner from 1985 until 1988, and the "flamboyant" Italian Carlo Ripa di Meana (Mason, 1992), who was Commissioner from 1989 until 1992 (Ward, 1997; Hey, 2005; Markham, 1989).

Before taking up post as Environment Commissioner, Ripa di Meana had been Commissioner for Culture, and this culture job had suited him well (Henningsen, 2012). As such, he had not wanted the environment portfolio (per Ripa di Meana, cited in Schoon, 1991). Nevertheless, on appointment as Environment Commissioner, Ripa di Meana "at once made it plain that he intended to use the Commission's legal powers and his own considerable political skills to introduce a radical programme of environmental improvement in the Community" (*The Observer*, 1991). The combined result of the above-mentioned factors was a notable expansion in EU environmental policy-making during this period.[1] But there were good reasons for this expansion: at a European summit in February 1988 it was agreed that spending under the Structural Funds would double between 1989 and 1993, most of which would be "spent in predominantly rural regions of the less developed parts of the Community; in Southern Europe, Ireland and parts of the United Kingdom" (House of Lords, 1989, paragraph 40). The result was likely to be heavy infrastructural investment and development that risked damaging the environment unless properly regulated (see House of Lords, 1989, paragraphs 40–41).

In addition to expanding the EU environmental *acquis*, Ripa di Meana – nicknamed variously Carlo the Ripper (White and Wolf, 1991) and the Grim Ripa (Fraser, 1992) by the UK press – was particularly active in his pursuit of the effective implementation and enforcement of EU environmental law, to the annoyance of certain Member States. He is said to have formed a "potent partnership" with Laurens Jan Brinkhorst, the Dutch Director General of DG XI from 1987 to 1994 (*The Observer*, 1991). A third prominent member of this potent team was Ludwig Krämer, introduced in the previous chapter, who headed DG XI's Legal Unit from 1984–1995 (European Commission, 1984, 1994, 1995). On taking over as head, Krämer (2011) explains that he stated publicly that the Commission would examine each and every complaint submitted to it, and the number of complaints soon went up from a tiny number to several hundred. Krämer reportedly gained the nickname 'the Sheriff' because of the uncompromising stance he took towards Member States in breach of EU law (Wurzel, 2002, p. 66).[2] As *Environment Watch* commented, "Krämer's unit had its heyday between 1989 and 1992 as the environment commissioner at the time, Carlo Ripa di Meana, aggressively and publicly pursued member states over alleged infringements of environmental legislation" (cited in Wurzel, 2002, p. 67). Other influential members of DG XI during these years, of particular importance in the context of this book, are Stanley Johnson, who (to recap) joined DG XI from the European Parliament in 1984, having left the Environment and Consumer Protection Service (ECPS) to serve as an MEP from 1979, and took up where he left off as an active policy entrepreneur (Johnson, 2012, 2011); and Claus Stuffmann, former *cabinet* member of Altiero Spinelli and head of the unit responsible for nature conservation at the ECPS and later DG XI from 1976 to 1994 (European Commission, 1995, 1976).

Introduction to the Habitats Directive proposal

In January 1987 Laurens Jan Brinkhorst was appointed as Director General of DG XI (*Japan Economic Newswire*, 1987). Brinkhorst had history regarding EU environmental policy: e.g., having served as State Secretary for Foreign Affairs in the government of the Netherlands from 1973–1977, in this capacity Brinkhorst spoke strongly in favour of developing a Community environment policy, directly challenging France's recalcitrance at the difficult first meeting of the Environment Council in July 1973 at which political agreement was reached regarding the First Environmental Action Programme (1973) (CEC, 1973, pp. 23–24; and see Chapter 3).

As Brinkhorst's main adviser at DG XI, "I was really able to push things through", comments Stanley Johnson (Johnson, 2011, 2012). Yet, even after the new environmental powers in the SEA came into force in July 1987, DG XI was aware that nature conservation negotiations would remain a challenge – and not only with the twelve Member States,[3] but also internally within the Commission. As DG XI had commented in February 1987,

> like other EEC policies, environmental policy has to work with the need to find compromise between 12 Member States, whose priorities and depth of commitment often vary widely. This tends to lead to a lowest common denominator approach, with legislation moving at the pace of the 'slowest' member government; a similar compromise has to be found within the Commission itself, where different departments also have different priorities. Officials responsible for Community agriculture or industry policies may see efforts to protect the countryside and wildlife habitats or to impose stricter pollution requirements as a threat to the success of their own policies.
>
> *(DG XI, 1987, p. 63)*

Nonetheless, DG XI and others were convinced of the need for an EU nature conservation instrument going beyond birds, and by August 1987 the policy was beginning to take shape, with Commissioner Clinton Davis commenting, "We would like . . . to use a Community instrument, probably a directive, to provide for the EEC-wide notification and protection of sites of special interest, whether in terms of species, habitat or landscape" (Clinton Davis, 1987). Separately, also in August 1987, DG XI's Legal Unit launched the first stage in legal proceedings against Germany in respect of the Birds Directive for allegedly failing to protect bird habitats from coastal protection works in Leybucht bay, a case that would play an important recurring role during the negotiations regarding the Habitats Directive (Freestone, 1996). Two months later, in October 1987, the Council adopted the Fourth Environmental Action Programme (1987), retaining Stanley Johnson's language regarding the need for a new EU nature conservation instrument,[4] thus clearing the path for the Commission to bring forward proposals for a Habitats Directive.

As Alistair Gammell comments, "I think governments, my explanation is that governments put . . . in [the wording about a new EU instrument on nature conservation] with the full expectation that they wouldn't do anything about it. So they did, governments allowed it to remain in, but I'm pretty certain that they never intended to do anything about it" (Gammell, 2012). This is not impossible; after all, in the Resolution accompanying the Fourth Environmental Action Programme, the Council and Member States merely "[noted] the action programme set out in the Annex to this resolution and [approved] in general the guidelines" (Fourth Environmental Action Programme, 1987, p. 4). Moreover, in terms of the priority areas on which the EU should concentrate, the Resolution accompanying the Action Programme referred to: "(g) measures to protect and enhance Europe's natural heritage, in particular . . . the protection, as laid down in the Treaty, of areas of importance in the Community, *inter alia*, in the framework of [the Birds Directive], or of areas which are particularly sensitive environmentally, and encouragement of the revival of areas which are environmentally impaired" (Fourth Environmental Action Programme, 1987, p. 3). This was hardly a direct endorsement of the Commission's proposal, setting to one side the ambiguity of the paragraph.[5] Indeed, as the UK government would later argue, "[a possible Habitats Directive] had not been one of the 22 priorities identified explicitly in the Resolution of 19 October 1987" (House of Lords, 1989, paragraph 38).

Nevertheless, having successfully negotiated the Fourth Environmental Action Programme (1987) through the Council on behalf of the Commission, Stanley Johnson – as main adviser to DG XI's Director General – found himself in a position to progress the idea of a Habitats Directive, albeit that this caused certain internal difficulties within DG XI. While Johnson held an overarching senior role as adviser, he no longer had responsibility for any policy sector (Johnson, 2012). The head of the unit that would naturally be responsible for the Habitats Directive – 'Agriculture, nature conservation and relations with other policies' (European Commission, 1988a) (the Nature Unit) – was Claus Stuffmann, who (as detailed in Chapters 3–4) had earlier negotiated the Birds Directive and the EIA Directive on behalf of the Commission. Both Directives had proven controversial, and Stuffmann would of course also have been involved at this time with their implementation, including multiple legal proceedings and the dispute with the UK government over Duich Moss detailed in Chapter 5. For some or all of these reasons, Stuffmann was reportedly reluctant to proceed with the idea of a new nature conservation directive, despite the commitment in the Fourth Environmental Action Programme (Johnson, 2012; Sundseth, 2012). This was not, of course, because Stuffmann was not keen on nature conservation; on the contrary, in fact, as Pierre Devillers of the IRSNB highlights: "both Claus Stuffmann and Richard Geiser were extremely dedicated conservationists and extremely proactive". The Commission was "a force for conservation in Europe" at that time, adds Devillers (2013), "and it's very largely those two men that did it. I think that will never be too much recognised". Notwithstanding these convictions, it seems that Stuffmann did not think he would be able to get a Habitats Directive proposal through the Commission, not to mention the Council (Johnson, 2012). As Kerstin Sundseth, a *stagiaire* in DG XI at the time, recalls, "There was a little bit of humming and hawing [regarding] whether the Commission dared to launch something as major as this when the Birds Directive was . . . not under attack, but was in the news, but the argument was, look, many Member States have already signed up to the Bern Convention, we should be coherent and continue" (Sundseth, 2012).[6]

In the circumstances, Stanley Johnson decided to take matters into his own hands, and began the process of producing a draft Habitats Directive, causing a "good deal of friction" within DG XI, given that Johnson did not hold the policy brief for nature conservation matters (Johnson, 2012). This dynamic between Stanley Johnson and Claus Stuffmann was an important feature of the early history of the Habitats Directive (Henningsen, 2012), and it would take all the skills and experience of both to see the Directive successfully adopted. As Sundseth reflects, "I really do think that if Stanley hadn't been there we would never have seen the Habitats Directive. He just went ahead, he was determined that . . . they had committed to it [in the Fourth Environmental Action Programme] and that they should follow it through" (Sundseth, 2012). Johnson having set the ball in motion in this way, Claus Stuffmann would then play a pivotal role in the successful negotiation of the Habitats Directive, just as he had with the Birds Directive: as John Temple Lang comments, "Without [Claus Stuffmann] I doubt if either [the Birds Directive or the Habitats Directive] would have been adopted. He patiently and professionally steered the ships through many shoals . . . It was an achievement of the best kind that Commission officials can do" (Temple Lang, 2013).

First draft of the Habitats Directive and Annexes: January 1988 to formal proposal in August 1988

The first draft: Devillers and Temple Lang

The official starting signal for producing a draft Habitats Directive was given in mid-January 1988 (DG XI, 1988o), and the first draft was produced for Johnson very rapidly by Pierre Devillers of the IRSNB[7] and John Temple Lang, who at the time was still with the Commission's Legal Service, having about a decade before played a key role in drafting the Birds Directive (Temple Lang, 2012), as detailed in Chapters 3–4. As Devillers (2013) explains, Stanley Johnson "met John Temple Lang for lunch somewhere and [Stanley] asked [John] if he thought it could be done, and John called me,[8] and because of the existence of the CORINE biotope habitat list, I told him indeed yes I was sure it could be done. And so [John] told Stanley and I remember very well that on the Friday [15 January 1988][9] Stanley said he wanted the draft by the Monday and John spent the weekend with me, here in my home, and we drafted it during that weekend".

The first draft was modelled on the Birds Directive, with Temple Lang doing the drafting given his legal skills, though it was obvious to Devillers and Temple Lang that the Habitats Directive was going to be a much more complicated job than the Birds Directive had been, because of the wide variety of habitats and species concerned and the multifarious problems those species and habitats faced (Temple Lang, 2012). A major difference between the Birds Directive and this new draft instrument was that, whilst the Birds Directive covered, *inter alia*, habitat conservation (via protected areas) and, in addition, species protection (inside and outside protected areas), the first draft of the Habitats Directive related only to habitat protection and did not contain a parallel species protection regime (IRSNB, undated). As Devillers (2013) explains, he and Temple Lang did "not put into [this first draft] anything about taking [i.e., species protection], and that came from the allergy I had developed to the taking problems under the Bird Directive, and I didn't want to restart with snails or with frog legs what we had been doing with thrushes and pâté and the rest".[10]

A few additional points are worthy of note (all from IRSNB (undated)): first, in terms of site designations, the draft envisaged the designation of (at least) the one hundred most important areas in the EU and the five most important areas in each region[11] for each of the species and habitats to be listed in Annexes (Article 4(2)).[12] Second, the draft gave the Commission no greater role than it had under the Birds Directive in respect of site designations; that is, the Commission was merely to be provided with information by the Member States to allow the Commission to perform a coordinating function (Article 4(3)). Third, in terms of site protection, the draft contained a provision based on Article 4(4) of the Birds Directive, but going a little bit further:

> In respect of the protection areas referred to in paragraphs 1 and 2 above, Member States shall take appropriate steps to avoid pollution or deterioration of habitats or any disturbances affecting the flora and fauna, in so far as these would have significant undesirable effects on the species or types of habitats for which the area has been classified. They will consider in particular the creation of nature reserves, the upkeep and management of other areas in accordance with the ecological needs of the species and types of habitat concerned, the re-establishment of destroyed biotopes, the creation of biotopes. Outside the special protection areas, Member States shall also strive to avoid pollution or deterioration of habitats.
>
> *(Article 4(4))*

Further, the draft envisaged the Annexes covering bird species in addition to other species of fauna and flora, with a cross-reference to the Birds Directive providing that obligations satisfied under Article 4 of the Birds Directive would fulfil site protection obligations under the Habitats Directive (Article 4(5)). Fourth, the draft proposed settling the long-running dispute over nature conservation financing[13] by providing that species recovery plans for Annex I species and projects concerning the conservation, management, maintenance, or reestablishment of habitats or species of flora or fauna in protected areas designated pursuant to Article 4 would be eligible for EU co-funding of up to 75% as a general rule, and up to 100% in the case of species or habitats threatened with extinction in the EU – higher levels of co-funding than was the general rule under the then applicable Regulation.[14]

Revising the first draft with Stanley Johnson

Devillers and John Temple Lang presented this draft to Stanley Johnson in the latter's office on Monday 18 January 1988, as Devillers (2013) explains: "on the Monday John and I went to Stanley's office and . . . we worked on the floor with various pieces of paper and scissors and started to put things together and to jot the missing pieces and we probably [finished] it that evening, it's certainly not much longer than that. I think by the next day anyway it was in Stuffmann's division". Indeed, the revised draft that was passed to Stuffmann – dated Monday 18 January 1988 – drew heavily on material from the Bern Convention to supplement the draft produced over the weekend (DG XI, 1988s). Johnson then immediately sought comments from Stuffmann's Nature Unit (DG XI, 1988q). The draft as revised on the Monday had developed quite a bit from the original version of the weekend. One major change concerned the addition of species protection measures: the revised draft contained a substantial section on this subject, modelled largely on the Bern Convention, containing, for example, rules about deliberate capturing and killing of certain species, etc. (DG XI, 1988q, Articles 5–9). Making changes on the Monday, Johnson had insisted on inserting these species protection measures, saying: "we've got to, it's a question of coherence with the Council of Europe, we have to cover what Bern covers" (Devillers, 2013). The EU hunting coalition FACE was therefore correct on the substance (but not quite on the timing) in stating that "our information of about a year ago – end '87, early '88 – [was] that the Commission was preparing a Directive on habitat protection, and that, to the surprise of many observers – including bodies such as the E.E.B. [European Environmental Bureau] – the extension to the conservation/management of flora and fauna species, probably took place only in February or March '88" (FACE, 1989, p. 1).[15]

In terms of site designations, the revised draft proposed obliging Member States to classify territories sufficient in number and size to ensure the survival of species and habitats to be listed in Annexes; this was backed up by site selection criteria in an Annex, including a '100 most important areas in the EU' formula as one criterion (DG XI, 1988q, Annex V). The revised draft no longer envisaged listing birds in an Annex, such that the directive would in general terms cover only other fauna and flora and their habitats. In terms of site protection, Article 4(4) remained in the revised draft, but the draft now included an interesting new idea: "Article 16: In carrying out their obligations under this directive, Member States shall apply the provisions of Articles . . . of [the EIA Directive] in respect of programmes and projects susceptible of having a significant effect on the territories classified as special protection areas under this directive or under [the Birds Directive]". The EIA Directive had yet to come into force at this time – it would do so on 3 July 1988 (Environmental Impact Assessment (EIA) Directive, 1985) – but here the draft proposed requiring projects *and programmes*[16] to undergo impact assessment, in supplement to the bare protective provision set out in Article 4(4). Finally, it is in this 18 January 1988 draft that Annexes of specific species and habitats were mentioned for the first time, and here the influence of the Bern Convention and the CORINE biotopes project is evident. In terms of the species and habitats requiring the designation of protected areas, the draft states: "ANNEX I Plants – identical to latest Berne list . . . ANNEX II Animals – Berne . . . ANNEX IV Types of habitat to be protected – to be completed (see for example Section 4.4 of Technical Handbook Volume I of CORINE Biotopes Project)" (DG XI, 1988q).[17]

This Devillers/Temple Lang/Johnson draft of 18 January 1988 was passed immediately to Stuffmann's Nature Unit, which gave its initial reflections on 19 January, amongst which was a proposal to delete the financing provisions altogether from the draft instrument (DG XI, 1988q, p. 6). As Devillers (2013) reports, the Nature Unit commented (in person) "that [the draft directive] was certainly a very good text, a very good idea, but that it was very visible that it had not been seen by lawyers, there were things that were totally impossible surely for the Legal Service of the Commission, which I thought was very amusing at the time . . . John [Temple Lang] was a very high officer in the Legal Service, but at the time it could not be known that it was him who had drafted it".[18]

The Nature Unit begins work on the draft

This 18 January 1988 draft formed the base material on which Johnson and the Nature Unit began to work, with Johnson assisted in the early stages principally by his *stagiaire* Kerstin Sundseth, and from April 1988 by Henriette Bastrup-Birk, a Danish Commission official who later played a key role in negotiating the instrument as part of Stuffmann's Nature Unit (Johnson, 2012). Johnson (2012) reports that he retained control over the draft instrument by typing it out on his personal Amstrad computer. There was, he explains, a general reluctance on the part of others to get involved in heavy retyping; in other words, there was "what you might call a presumption that you would think twice before changing a text in a typeface you couldn't alter without retyping the whole thing" (Johnson, 2012). Thus, others within the Commission could suggest changes, but Johnson retained control of the working draft at this time (Johnson, 2012).

Following a meeting with the Nature Unit on 21 January 1988, Johnson produced a revised draft of the directive (DG XI, 1988m). Article 1, which recorded the aims of the directive, had been amended to provide "The aim of this directive is to conserve natural habitats and wild flora and fauna (other than birds)", though the question of whether or not birds should be covered by the directive remained open, as evidenced by a handwritten comment on the list of Annexes and by the fact that this draft retained the idea of applying the EIA Directive to protected areas under this new directive *and* under the Birds Directive (DG XI, 1988m). The issue of financing was now proposed to be dealt with by way of a simultaneous amendment to the ACE Regulation (1984) (DG XI, 1988m), and an interesting development had taken place in respect of the directive's site protection obligations. Two new paragraphs were proposed for Article 4, reading:

> 4.9 Outside the special protection areas, Member States shall also strive to avoid pollution or deterioration of natural habitats.
>
> Member States shall take the necessary measures to conserve characteristic or representative habitats in all of the regions where they naturally occur.
>
> Version 1
>
> [They will furthermore take all necessary measures to ensure the protection of landscape features, especially those mentioned in Annex VIII, of major

importance for the conservation of the species mentioned in Annex I whose conservation cannot be assured solely through the designation of special protection areas.]

Version 2

[Furthermore, to ensure the maintenance or re-establishment of a network of areas of biological importance, Member States shall take the necessary measures to protect landscape features, mentioned in Annex VIII, of major importance for the conservation of the species listed in Annex I.]

4.10 [Member States may allow derogations from the provisions of paragraph [4.9], if the impact on landscape features can be counterbalanced or if the measures of derogation are required by overriding public interest, provided, however, that compensatory measures are taken.]

(DG XI, 1988m)

In other words, in this draft, special protection areas themselves were not to be the subject of any form of derogation; however, a derogation could be applied in respect of important areas *outside* special protection areas, which the directive would also require to be protected. It is also interesting to note that, as early as January 1988, the Commission had in mind the concepts of "overriding public interest" and "compensatory measures" in their draft derogation, which concepts would re-emerge later in Article 6(4) of the final version of the directive to be applied in respect of Natura 2000 sites under the Habitats and Birds Directives (Habitats Directive, 1992).

Although the draft directive was already highly detailed by the end of January 1988,[19] several major questions remained outstanding. For example, whether the Directive should remain limited to the European territory of the EU (as was the case with the Birds Directive). After all, as a DG XI official noted, "One should not forget the great number of endangered species in the overseas territories!" (DG XI, 1988g). Also, the possibility of Member States derogating from the provisions of Article 4 (the main site protection provision) remained open (DG XI, 1988g). Several decisions had, however, been taken: the representatives of four federations of NGOs were to be invited to discuss the draft directive: the WWF, IUCN, EEB and FACE (DG XI, 1988t).[20] Moreover, a decision had been taken that the UK government's nature conservation advisory body, the Nature Conservancy Council, was to work on producing Annexes listing species requiring the designation of special protection areas, and Belgium's IRSNB (Devillers) was to work on a list of habitats as well as site selection criteria (DG XI, 1988g).

Rapid progress and the question of consultation

The Fourth Environmental Action Programme (1987) had only been adopted in October 1987, yet by February 1988 Commissioner Clinton Davis was already promising an audience that "I will soon present a first Directive of general scope with the aim of protecting habitats, fauna and flora" (Clinton Davis, 1988). That said, much remained to be done. By this time the Nature Unit led by Claus

Stuffmann was fully engaged in the process, and on 8 February the Unit sent out an ambitious draft work programme for the directive, comprising the following actions for 1988: 15 February: Dispatch of draft directive to other Directorates General (DG) of the Commission for a first consultation and to the Environment Commissioner's *cabinet*; 16 February: Working group meeting on the species Annexes of the draft directive (with representatives of the WWF, IUCN, the UK government's Nature Conservancy Council, the French National Museum of Natural History's SFF,[21] and the IRSNB); 8 March: Consultation meeting with representatives of DGs VI (agriculture), XII (science, research and development), XIV (fisheries) and XVI (regional policy); 15 March: Dispatch of the draft directive to Member States' Permanent Representations with a view to discussing; 18 April–22 April: Consultation meeting with national experts; 25 April–29 April: Consultation with representatives of NGOs: WWF (International and UK), EEB, IUCN, FACE; June: Dispatch of draft directive for formal consultation of other DGs; 30 June: Transmission of proposed directive to Commission (DG XI, 1988n).

Despite the proposal here of consultations with Member State representatives, it seems that no formal[22] consultations were held with governments before the instrument was proposed to the Council in August 1988 (DG XI, 1988c, 1988j; Wildlife Link, 1988b).[23] The reason for this is unclear, but it seems that the proposal to consult Member State representatives may have come from the Nature Unit, while Stanley Johnson was in practice running the day-to-day process. This interpretation is given some support by the following comment from Claus Stuffmann in 1989, who said the draft directive was "pulled out of the Commission's drawers in the space of three months, without consulting national administrations" (cited in Bader and May (1992), in turn cited in Kautz, 2002, p. 54). The UK government would later argue that the proposal for a directive was not one of the priorities identified explicitly in the Resolution of 19 October 1987 that accompanied the Fourth Environmental Action Programme (1987), and that more extensive prior consultation would therefore have been appropriate (House of Lords, 1989). The Commission's response was that "there had been extensive discussion in Council working groups at earlier stages – for example during the preparatory stages of the Fourth Environmental Action programme – so [more extensive discussion with Member States before the launch of the proposal] had not been necessary", and that "during the working group discussion of the Fourth Environmental Action Programme, no Member State had recorded a 'reserve' against this particular item, unlike others" (House of Lords, 1989, paragraphs 38 and 66).

Comments from DG XI's Economics Unit

In letters dated 10 and 11 February 1988, Gunter Schneider's 'Economic aspects, employment and statistics' unit of DG XI (European Commission, 1988a) (the Economics Unit) – which held responsibility for the CORINE biotopes project (Devillers, 2013) – expressed concerns regarding the draft directive. In particular, the Economics Unit did not believe that adequate financial provisions were available to meet the aims of the draft directive, and argued that the draft did not set realistic objectives in certain respects (DG XI, 1988h). The Unit noted that the financial capacity of Member States was inversely proportional to biodiversity richness (i.e., the most biodiverse regions of the EU tended to have least capacity), and that financing of a different order of magnitude to that available in the ACE Regulation's (1984) budget would be required (DG XI, 1988h).[24] Further, the Unit argued that a step-wise approach to deadlines for site designations would be required, since very short timescales could not be met by all countries (the draft directive had proposed 3 years for the designation of sites): some, such as Denmark, could perhaps meet a short timetable,[25] the Economics Unit noted, but for others even 10 years would be insufficient (DG XI, 1988h). Finally, the Economics Unit suggested naming the new network of protected areas "OIKOS",[26] and proposed merging the Birds Directive with the new directive, with the former to be repealed on the entry into force of the new directive (DG XI, 1988i).

At this point, DG XI's work was given added impetus by the European Council's decision, at its meeting of 11–13 February 1988, to double spending under the Structural Funds between 1989 and 1993 (House of Lords, 1989), completing part of the deal done to secure agreement on Treaty changes via the SEA. Most of these Funds would be spent in rural regions of Southern Europe, Ireland and parts of the UK, including "heavy investment . . . on new roads, bridges, sewers and energy supplies, factories, tourist developments, and changes to established patterns of agricultural production" (House of Lords, 1989, paragraph 40). Claus Stuffmann expressed particular concern at the speed of development envisaged, noting that, whatever their benefits, the Structural Funds were creating "a development which is faster perhaps than can readily be controlled in environmental aspects" (House of Lords, 1989, paragraph 41). Giving evidence before the House of Lords in the UK, the Commission argued that "such potential problems with the Structural Funds were themselves an important argument for a new Habitats Directive granting the Community internal competence for protection of all Community flora and fauna, rather than simply for birds, as hitherto" (House of Lords, 1989, paragraph 44).

Returning to the draft directive, the Economics Unit's comments were taken on board in the next version of the directive, dated 15 February 1988, which called for the "basic structure" of the OIKOS network to be established by the designation of sites within two years from the entry into force of the directive, as well as by the immediate designation of remaining undesignated sites under the Birds Directive (DG XI, 1988k, Article 4.5). Within three years of the new directive's entry into force, the Commission, assisted by a Standing Committee on Nature Conservation, would then prepare a Joint Programme for the strengthening and development of the network, and this Programme would be reviewed every three years (DG XI, 1988k, Article 4.7). Amongst other things, the Programme would specify deadlines by which further areas needed to be designated (DG XI, 1988k, Article 4.7). Thus, slowly but certainly, the Commission began to propose greater powers for itself in respect of site designations than had been the case in respect of the Birds Directive. In addition, the revised draft retained the possibility of

Member States derogating from the protection provided to important "distinct natural features" *outside* the special protection areas: "Member States may allow derogations . . . if the impact on distinct natural features can be counterbalanced or if the derogations are required by overriding public interest, provided, however, that they do not have adverse effects on the sustainability of the biological systems connecting the special protection areas" (DG XI, 1988k, Article 4.11).[27]

The Legal Unit provides comments: hands off the Birds Directive

On 18 February 1988 the Legal Unit of DG XI gave its views on the draft directive, expressing particular concern regarding the idea of amalgamating the Birds Directive with the new Habitats Directive (DG XI, 1988l). Many of the Legal Unit's then 29 infringement proceedings in respect of site designations under the Birds Directive "have already had a positive result", the Unit argued, and the introduction of new deadlines and a "dynamic" (stage-by-stage) system of site designations would "obviously call in question the legal basis of the existing 29 infringement proceedings against Member States and the work that has been done till now by Member States and the Commission. Furthermore by introducing new deadlines the pressure that the Commission has been exercising to force Member States to classify and protect habitats, will be radically reduced" (DG XI, 1988l, pp. 1–2). Moreover, the Legal Unit foresaw a "reduction of the level of protection provided for by [the Birds Directive] since the draft provides for derogations and has not such a clear wording as [the Birds Directive]" (DG XI, 1988l, p. 2). In addition, the Legal Unit highlighted the imprecision of the obligation to establish the "basic structure" of the network of protected areas within two years (DG XI, 1988l, p. 3). The Legal Unit's conclusions were clear: bird species should be removed from the draft directive and a new paragraph should be inserted into the habitat protection provision stating that "The obligations of this article are without prejudice to the provisions of [the Birds Directive]" (DG XI, 1988l, p. 2).

Natura 2000 coined; and contact with the Council of Europe

By 26 February 1988, the OIKOS network had been renamed the NATURA network in the draft directive (DG XI, 1988f, Article 4.3), and by May 1988 this had been further amended to NATURA 2000 (DG XI, 1988e), a name that "seems to be invented by a cheap PR-advertising firm", according to the EU hunting coalition FACE (1989, p. 2).[28] At about this time, Stanley Johnson paid a visit to the Council of Europe in Strasbourg, which held responsibility for the Bern Convention, the "mother" of the Habitats Directive (Jaffeux, 2013). As Johnson (2012) explains,

> The Bern Convention was undoubtedly an extremely important part of my thinking at this stage, because it was perfectly clear that the Bern Convention did, at the level of the Council of Europe, address the two things that I was most interested in, which was habitat and species, and it was a fine

> document . . . [Y]ou could say to yourself, well, what is the value-added to having an EU directive? Well the value added from my point of view was that it meant that all the mechanisms for implementation and enforcement which the EU could bring to bear would then be available. So even if we produced an instrument which actually mimicked the Bern Convention, there would still be a value added in terms of its implementation.

However, in Strasbourg, the then Director of Environment and Local Authorities at the Council of Europe "was rather worried about the EU adopting its own instrument" on nature conservation (Johnson, 2012). There had always been a sensitivity in terms of whether the EU would in some way usurp functions of the Council of Europe, particularly as the EU's membership grew in size (Johnson, 2012). This sensitivity was evidently acute in respect of the Bern Convention because, at one point in the new EU directive's passage, the above-mentioned Director at the Council of Europe went so far as to ask Johnson not to proceed with the draft directive (Johnson, 2012). The relationship between the Bern Convention and the new directive would remain complex and difficult for some time, as further detailed in Chapter 7.[29]

The genie leaves the bottle: draft circulated to NGOs and others

In March and April 1988 the Commission began receiving written comments on the draft directive from NGOs (nature conservation and hunting) and others such as the Council of Europe, Stanley Johnson having circulated the draft directive and its annexes for comments. As Alistair Gammell explains:

> when the draft was complete, Stanley, with our [RSPB] support, sent it out to about 200 NGOs around Europe . . . on the grounds that the Commission had been criticised a lot for putting . . . what the governments described as half-baked ideas; ill-researched and half-baked ideas to the Council. And so we said . . . – it can really be described as no more than an excuse – . . . that we didn't want to put a half-baked idea . . . to the Commission and to the European Council, we wanted to make sure it was fully informed, and therefore we sent it round to about 200 NGOs to ask for their opinion as to whether or not this was a good idea and whether it could be improved or not.
>
> *(Gammell, 2012)*

Comments came back from a range of organisations between April and June 1988, including, for example, the hunting coalition FACE (FACE, 1988a, 1988b), the Italian bird protection NGO LIPU (1988), a Dutch bird protection NGO (NVTBVV, 1988), the Max Planck Institute for Behavioural Physiology (MPI, 1988), the European Environmental Bureau (EEB, 1988), and Wildlife Link, an umbrella organisation of UK conservation NGOs of which Stanley Johnson was Vice-Chair (Wildlife Link, 1988a, 1988c). Clearly, the genie of a broad-based EU nature conservation instrument was out of the bottle.

While this type of inclusive consultation might perhaps be regarded as good practice today, at the time it was unconventional to say the least. As Gammell (2012) explains, "I had Nicholas Ridley, who was the British Minister . . . shout at me that I'd broken every rule in the book, and I had to point out that I was actually an employee of an NGO and as far as I was aware I hadn't broken any rules whatsoever . . . [T]he governments and some people in the Commission were absolutely apoplectic. I have no idea what they said to Stanley". Kerstin Sundseth, Johnson's *stagiaire* at the time, does not recall this particular episode but she remembers more generally "a lot of furious phone calls" (Sundseth, 2012). However, as a senior DG XI colleague from the time reflects, "Stanley is and was not a person who finds it terribly important to respect the rules" (Henningsen, 2012).

Jacques Delors' cabinet intervenes

By 27 June 1988, DG XI had a final version of the Habitats Directive and Annexes ready, and the complete draft was circulated that day to the rest of the Commission with a view to its being considered formally on 6 July 1988 (DG XI, 1988d). By this time the Commission's Legal Service had already given a favourable opinion, and the other Directorates General of the Commission had indicated their agreement, with the exception of DG VI (agriculture), which was still being consulted (DG XI, 1988d). The draft evidently remained in a state of flux: several revised versions were circulated to the rest of the Commission in the first half of July, and the date of the Commission's consideration was repeatedly pushed back.

On 13 July 1988 the formidable *cabinet* (Ross, 1994) of Commission President Jacques Delors intervened, and from this point on Stanley Johnson conducted the final internal Commission negotiations alone (DG XI, 1988o). As Johnson (2012) explains, the draft was

> blocked by the gentleman in the [Delors] *cabinet*, who was a Frenchman, who was in charge of looking after the proposals coming up to do with the agricultural sector, and his name was . . . [Michel-Jean] Jacquot,[30] and I remember him telephoning me at one stage, saying that the *cabinet*, the President, was not minded to allow this proposal to be adopted by the Commission, because it was really just much, much too detailed.[31] Now, if you remember, if you go back to the implementation of the Birds Directive, there had been . . . right at the beginning, when the Birds Directive was first adopted, which was 1979 . . . a colossal row with the French over the listing of . . . what the French call the ortolan, which you know French hunters absolutely love to hunt. So it was a very, very sensitive area, and I think Jacquot took one [look] at all these species and all these habitats [in the draft annexes] and thought 'Oh my God, the French farmers are going to rise up in arms'.

The President's *cabinet* highlighted that the Council had not said 'yes' to the Fourth Environmental Action Programme (1987); rather, the Council had noted the proposals (Delors' cabinet, 1988). Further, the President's *cabinet* stressed that under the Single European Act (1986) it was necessary to show that the objectives of an EU environmental proposal could be better attained at EU level than at the level of the Member States; in the *cabinet*'s view, this demonstration of subsidiarity was lacking from the Explanatory Memorandum for the proposal (Delors' cabinet, 1988).[32] Nevertheless, the *cabinet* was clearly not altogether against the idea of a habitats directive, as evidenced by the *cabinet*'s detailed, constructive comments on the Explanatory Memorandum for the instrument, including proposing wording to tackle the subsidiarity point mentioned above (Delors' cabinet, 1988). Delors' *cabinet* was also keen for references to be inserted into the Explanatory Memorandum highlighting "provisions in [the] draft directive permitting derogations etc. to allow for flexibility" (Delors' cabinet, 1988, p. 6).

Annexes dropped from the proposal; and departure of Commissioner Clinton Davis

Stanley Johnson and DG XI took these comments on board, and rapidly produced a revised version of the Explanatory Memorandum that was amended to note, amongst other things, that the draft directive requires "an appropriate balance to be struck between conservation and economic interests and [provides] for the possibility of derogations from the general obligations in certain specific cases" (DG XI, 1988r, p. 4). More significantly, Johnson took a tactical decision to drop the annexes of species and habitats from the proposal altogether, the idea being that they could be re-inserted later in the process. As Johnson (2012) explains,

> So this was a unilateral decision I took, to say, 'OK, at least get this show on the road, let's get it to the European Parliament anyway, even without Annexes, so that at least we've got a start . . .'. I thought, 'OK, better to get the process actually officially launched rather than have it die a death in the Commission, by trying to get permission to agree Annexes' . . . And then I said to myself, 'OK, we will get the Annexes, we'll introduce them or we'll somehow get them introduced at the stage of the Parliamentary discussion'. OK, that was the thought I had, let the thing move.

As it later turned out, the European Parliament would refuse to consider the draft directive without seeing its detailed Annexes, but Johnson's plan was simply to get a proposal onto the Commission's table and thereafter the Council's, and it was a plan that worked.

On 18 July 1988, DG XI commenced legal proceedings against Spain in respect of the Birds Directive, alleging that Spain had failed to protect the Santoña marshes in Cantabria in various ways, including by allowing a new road to be built across the marshland and allowing industrial estates to be established on land that formed part of the marshes (CJEU, 1993). Just two days later, on 20 July 1988, a final draft of the Habitats Directive and annexes (minus the species and habitats annexes) was transmitted by DG XI to the rest of the Commission, and the draft was agreed at a

Commission meeting on 26 July 1988 (European Commission, 1988b), subject to a handful of minor amendments.[33] As planned, most of the annexes were left out of the proposal, the idea being that these would be left to the Committee envisaged by Article 23 of the draft directive to define, by way of a cooperative procedure between the Commission and the Member States (DG XI, 1989b). Just four days before the directive proposal was agreed by the Commission, Commissioner Stanley Clinton Davis – an appointee of the UK's Labour Party – learnt from Conservative Prime Minister Margaret Thatcher that his name was not going to be put forward for the new EU Commission, which would take office at the beginning of 1989 (European Commission, 1988c).[34] Having secured the Commission's agreement to the Habitats Directive proposal, Clinton Davis would thus remain in place until the end of the year, long enough to see negotiations with the Member States begin.

Launching the Directive proposal

At this point, the question of how best, politically, to launch the directive arose. The Director General of DG XI, assisted by the Nature Unit,[35] had a clear plan (emphasis as per the original):

> In view of the fact that the Habitat draft directive constitutes a major proposal in a politically very sensitive area, the timing and the conditions of the launching of the examination process in the Council must be thought through very carefully. It is indeed of crucial importance that the right frame of mind be created before the proper examination of the proposal by the Environmental Group begins. I therefore insist that the proposal be presented by myself or [the Director of Directorate B] at a meeting of high officials (Directors or Director Generals for Nature Conservation) before the detailed discussions take place in the Council. The purpose of that meeting will be to present the objectives and the philosophy of the draft directive in the light of – among other things – experience gained from the application of the Birds Directive.
>
> *(DG XI, 1988o, p. 2)*

This meeting – described by UK government officials as a "réunion de séduction" (DG XI, 1988a) – was arranged for 29 September 1988 (DG XI, 1988o), and is discussed further in the following chapter.

Notes

1 DG XI came to be characterised by some as a "legislative factory" (Ward, 1997, p. 178). Note in this regard the existence of a "heated debate in legal circles about what constitutes the environmental *acquis*" (Jordan et al., 1999, p. 381). As Jordan et al. (1999, pp. 381–382) explain, "Nigel Haigh's definitive manual (1998) and Stanley Johnson's

textbook (Johnson and Corcelle 1995) both arrive at totals of over 500 and 400 measures respectively (see also Table 1). Others, however, such as Krämer (1997) and Macrory and Purdy (1997, p. 37), settle for a much lower figure. This discrepancy is partly a reflection of the unbounded nature of environmental policy, which crosses many other policy areas. It is for this reason that Nigel Haigh includes legislation adopted under Articles 30 and 31 (nuclear radiation) and 43 (agriculture) in his manual. However, the discrepancy also reflects the politics of subsidiarity post-Maastricht. It is revealing in this respect that Ludwig Krämer (1997, p. 9), a DG XI official, contrasts his total with those of Haigh and others 'which are sometimes used to illustrate – and criticise – the amount of legislative activity of the [EU]'. Finally, the discrepancy is also a function of which types of legislation are included in the analysis. The higher figures tend to include amendments and updated items of legislation which now account for an increasing proportion of the environmental *acquis* (Table 1)". Jordan et al.'s (1999, pp. 382–383) definition of the environmental *acquis* steers what seems a reasonable course, and their data support the view that the years 1987 to 1992 were notably productive in terms of EU environmental policy-making compared with the period before 1987, particularly when one focuses on what they term 'major' EU environmental legislation (i.e., excluding "amendments, Decisions which adopt international conventions and other 'non'-environmental statutes" (Jordan et al., 1999, p. 382)). Their data reveal that 74 major pieces of EU environmental legislation were adopted in the 20 years from 1967 to 1986, while 40 major pieces were adopted in the six years from 1987 to 1992 (Jordan et al., 1999, Figure 2 on p. 383). This seems a better measure than Jordan's (1998) claim that between 1989 and 1991 the Environment Council adopted more environmental policies than it had in the previous 20 years.

2 It is important to recognise that it was not only DG XI that sought to enforce EU law systematically around this time: as Várnay (2006, p. 7) explains, "[Commission President Jacques Delors'] Single Market project was transformed into an action programme by [a Commission] White Paper [of June 1985] which envisaged the legislation of almost three hundred directives until 1992. It was realised that if the Member States did not implement these directives in time and in a correct manner, the great project could easily fail, its advantages would not be realised. The Commission ... decided to control the transposition of directives systematically and in case of non-implementation it initiated the [infringement] procedure without any special consideration. It also invited the interested parties to submit complaints which still plays a significant role as a source of information for the Commission about the violation of Community law by the Member States. In order to ensure accessibility for the potential complainants the Commission introduced a complaint form".

3 Greece, Portugal and Spain had joined the EU since the Birds Directive was adopted in 1979.

4 Quoted in Chapter 5.

5 It is not clear if the "inter alia" is attached to "areas of importance in the Community" or instead to "in the framework of [the Birds Directive]".

6 Moreover, as Denmark's Veit Koester, then Chair of the Standing Committee of the Bern Convention, argues, "[T]here is no doubt that one of the reasons why the Commission proposed the [Habitats] Directive was that there were constant problems with the Member States' implementation of the [Bern] Convention, especially as far as the habitat protection provisions of the Convention ... There's no doubt that the Commission was frustrated by the constant complaints [to the Standing Committee] against EU Member States about their poor implementation of the Bern Convention and that was one of the reasons for proposing the Habitat Directive" (Koester, 2012).

7 See Chapter 5 regarding Devillers' role in CORINE. Devillers was also a member of a small informal core group of advisers to Richard Geiser of DG XI's Nature Unit regarding early implementation of the Birds Directive (Devillers, 2013) and was the main architect of an EU classification system for habitats, part of the CORINE

biotopes project (European Commission, 1991; Devillers et al., 1991), and of its "successor" the Palaearctic habitat classification (Devillers and Devillers-Terschuren, 1996), "developed for the Council of Europe". This Palaearctic classification formed the basis for the *Habitats Directive Interpretation Manual* for the EU-15 in 1996, produced to aid the interpretation of Annex I habitat types (Moss, 2008; European Commission, 2007).

8 As Devillers (2013) explains, "I think the reason I was contacted by him [John Temple Lang] was that we knew each other very well, I was his main contact on nature in Brussels, and he was in the habit of getting in touch with me every time he wanted to discuss a point that had to do with nature conservation ... [A]t that time we were birding together at least once a month or twice a month ... we published a paper on raptors together around that time, and ... we were quite often together in the field".

9 Whilst this very first draft is undated in the archive, other documentary evidence suggests that this must have been Friday 15 January 1988. As discussed above, this first draft was amended by Devillers, Johnson and Temple Lang on the Monday (18 January), and the draft was passed to Stuffmann immediately thereafter – the first draft marked up by Stuffmann (removing the funding provision, for example, as discussed above) is dated 19 January 1988 (DG XI, 1988q).

10 When asked whether hunting regulation formed part of the work of the advisory group of Devillers, Gammell, et al., which assisted Richard Geiser with the early implementation of the Birds Directive, Devillers (2013) replied, "unfortunately, very much. Hunting and other forms of taking took a great deal of our time, not to say most of it".

11 The regions being the administrative units designated in the Political Map of the European Community, edited by the Office for Official Publications of the European Communities.

12 Though this first draft of the Directive did not include the Annexes themselves (IRSNB, undated).

13 Under, at that time, ACE Regulation (No 2242/87).

14 See Articles 5 and 6 of ACE Regulation (No 2242/87).

15 That said, the decision to exclude species protection measures from the original draft of the weekend of 16/17 January appears to have been a unilateral decision by Pierre Devillers, based on his Birds Directive experience, and this decision was reversed by Johnson on the Monday (Devillers, 2013), such that the Commission was preparing a directive on habitats (alone) for a matter of only a couple of days.

16 In a speech in August 1987, Commissioner Clinton Davis gave an early indication of the Commission's desire to extend the EIA Directive to cover programmes as well as projects, noting "I do not underestimate the difficulties involved". This would indeed prove difficult – it took a further 14 years before impact assessment of programmes at EU level would be realised, in the form of the Strategic Environmental Assessment Directive (2001).

17 A handwritten comment notes that the IRSNB was updating/reviewing the CORINE list for the purposes of the draft directive, with a deadline of 14 February 1988 (DG XI, 1988m).

18 Indeed, neither Devillers' nor Temple Lang's involvement was revealed at the time (Devillers, 2013).

19 Even going so far as to propose an Annex containing a proforma sign and notice to be used on the ground to indicate a site's status as an EU protected area (DG XI, 1988m). The relevant annex to the directive would later be deleted, on the initiative of the French Presidency of the Council in 1989 (DG XI, 1989a). However, it was reinserted in April 1990 when the Commission was required by the European Parliament to insert annexes formally as part of its proposal (CEC, 1990), only to be deleted again later.

20 The Fédération Europèene des Parcs Naturels was originally to have been the fifth coalition of NGOs invited (DG XI, 1988g), but its name was scored out in a later draft, evidently following an internal discussion (DG XI, 1988t).

21 The Fauna and Flora Secretariat, a unit of France's National Museum of Natural History (Le Muséum National d'Histoire Naturelle) (Jaffeux, 2013).

22 The UK government stated that "EC officials appear to have had contacts with a number of organisations within the UK during the preparation of the Directive (including a short informal session with DOE officials in January 1988). There were however, no formal meetings with the UK government and although the Commission are not obliged to consult we believe that they recognise with hindsight that it would have been more productive to have done so" (Department of the Environment, 1989, paragraph 3). Commenting on what seems to have been this informal session with UK officials, Roy Bunce, then head of the Department's Wildlife and Countryside Division, states: "my first contact was with Stanley Johnson, with this early draft, so called, which I think had been very largely written for him by the RSPB ... We were very concerned about the implications of the first draft as drafted, and Stanley ... turned up with a draft at the end of the day that was virtually the same as what he'd started with. And we felt that our views and concerns had not been taken into account in any way" (Bunce, 2012).

23 Two versions of a Commission file note of a meeting with senior civil servants record subtly different accounts: the first, apparently in draft, records that there was resentment (*ressentiment*) on the part of Member States because NGOs had been consulted and not government representatives (DG XI, 1988c); what seems to be a later draft of the same file note records that there was dissatisfaction (*mécontentement*) because NGOs were consulted *more* than the Member States (DG XI, 1988b).

24 Morillo and Gómez-Campo (2000) state, for example, that "ca. 40% of the EU priority species and >60% of the priority natural habitats [under the Habitats Directive] are found" within Spain's borders.

25 Denmark had proven itself the "honourable exception" in respect of site designations under the Birds Directive, having made a lot of progress with its network of SPAs for birds within four years of the Directive's adoption (Gammell, 1999; Stuffmann, 1999).

26 Oikos is an ancient Greek term meaning 'house, an inhabited place, a habitat' (Hughes, 1975).

27 The paragraph cross-referred to in the 15 February 1988 version of the derogation provision is a typographical error (DG XI, 1988l); subsequent drafts reveal the above interpretation of the intended effect to be correct.

28 It is this 26 February 1988 draft that incorporated for the first time a list of habitats – as Annex IV – to be protected by the designation of special protection areas (DG XI, 1988f).

29 Several interviewees highlighted the fact that, despite the Bern Convention being the "mother" of the Habitats Directive, the Directive's preamble unfortunately omits to mention the Convention. The Convention was, however, mentioned in the Explanatory Memorandum that accompanied the Commission's directive proposal (European Commission, 1988d). The archival history reveals that the Bern Convention was mentioned in the draft preamble at one stage but was removed before the draft directive reached the Council. The reason is unclear.

30 Given as "André Jacquot" during interview (Johnson, 2012) but corrected following further research by the present author. See ELO (2004) for a brief biography of Michel-Jean Jacquot.

31 Briefing the Environment Commissioner in October 1989, the Director General of DG XI wrote: "when DG XI's original proposal – in a complete form – was being deliberated within the Commission ... [y]our predecessor felt constrained to bow to pressure from a number of his colleagues to withdraw the detailed annexes and to make a number of consequential amendments" (DG XI, 1989c).

32 "Cette démonstration de subsidiarité manque fortement" in the original.

33 The amendments agreed at the meeting of 26 July 1988 are recorded in DG XI (1988p). A few additional modifications were made, at the proposal of the Director General of DG XI, assisted by the Nature Unit, before the Commission formally proposed the directive to the Council (see DG XI, 1988o).

34 Potential reasons cited for this decision include: avoiding the embarrassment of sacking only a Conservative commissioner, Lord Cockfield, who had reportedly annoyed

Margaret Thatcher with his commitment to certain EU policies (Palmer and Wintour, 1988); to avoid Stanley Clinton Davis becoming Vice President of the Commission (Palmer and Wintour, 1988); and as an alleged pay-off to Leon Brittan (who would be nominated by Thatcher to the Commission) for his support in the Westland affair, which related to the future of the UK helicopter industry (Webster and Ford, 1988).

35 The letter, signed by the Director General, is marked XI.B.3. (the Nature Unit) and bears the initials of a member of the Nature Unit (DG XI, 1988o).

References

ACE Regulation 1984. Council Regulation (EEC) No 1872/84 of 28 June 1984 on action by the Community relating to the environment. *OJ L 176*, 3 July 1984, p. 1.

Bader, D. and May, H. 1992. *EG und naturschutz, institutionalisierung, regelungslage und verknüpfung zu anderen politikbereichen*. Bonn: Economica Verlag.

Bunce, R. 2012. Interview with Roy Bunce, senior negotiator for the UK government in respect of the Habitats Directive, conducted by Andrew L. R. Jackson on 13 August 2012.

CEC 1973. Minutes of the 251st meeting of the Council of the European Communities held in Brussels on Thursday 19 and Friday 20 July 1973. 1586/73 (PV/CONS 23).

CEC 1990. Outcome of proceedings of Working Party on the Environment on 2 April 1990. Brussels, 24 April 1990, 5885/90, ENV 86. Council of the European Communities.

CJEU 1993. Opinion of Advocate General Van Gerven of 9 June 1993, *Commission* v. *Spain*, C-355/90, EU:C:1993:229.

Clinton Davis, S. 1987. "Environment policy cannot stand alone" – Extracts from a speech by Mr Stanley Clinton Davis at the 30th international seminar, Salzburg, 4 August 1987, IP/87/347.

Clinton Davis, S. 1988. Speech to the European Economic and Social Committee on 2 February 1988.

Delors' Cabinet 1988. Note dated 13 July 1988 from Commission President Delors' cabinet to the cabinet of Environment Commission Stanley Clinton Davis, attaching handwritten comments on draft Explanatory Memorandum for the Habitats Directive proposal. Sent by fax on 14 July 1988.

Department of the Environment 1989. Memorandum by the Department of the Environment dated 23 January 1989, headed "House of Lords European Communities Committee: Sub Committee F (Environment). Enquiry into habitat and species protection".

Devillers, P. 2013. Interview with Pierre Devillers of Belgium's IRSNB conducted by Andrew L. R. Jackson on 25 April 2013.

Devillers, P. and Devillers-Terschuren, J. 1996. A classification of Palaearctic habitats. Council of Europe, Strasbourg: Nature and environment, No 78.

Devillers, P., Devillers-Terschuren, J. and Ledant, J.-P. 1991. *CORINE biotopes manual. Vol 2. Habitats of the European Community*. Luxembourg: Office for Official Publications of the European Communities.

DG XI 1987. *The European Community and the environment*, 3rd edn. Manuscript completed in February 1987. Luxembourg: Office for Official Publications of the European Communities.

DG XI 1988a. DG XI file note dated 9 September 1988. Headed "Présentation au sein du group Environnement du conseil du projet de directive 'Habitats' – 8.9.88".

DG XI 1988b. DG XI file note headed "Note de Dossier. *Reunion des hauts fonctionnaires – 'Protection de la nature'* – 29 September, 1988".

DG XI 1988c. DG XI file note headed "Reunion des hauts fonctionnaires – 'Protection de la nature' – 29 September, 1988" – with handwritten amendments.
DG XI 1988d. Draft directive on the protection of natural and semi-natural habitats and of wild fauna and flora. Communication from Mr Clinton Davis to other members of the Commission. Brussels, 27 June 1988, COM(88) 381. On the agenda of the 926th meeting of the Commission, on Wednesday 6th July 1988.
DG XI 1988e. Draft of habitats directive, headed "FFH-Directive of 17 May 1988".
DG XI 1988f. Draft of habitats directive, headed "FFH-Directive of 26 February 1988", with "FAST – comments" handwritten at the top.
DG XI 1988g. Internal DG XI document headed "Pending issues: (Habitat Directive)", hand dated 26 January 1988.
DG XI 1988h. Internal DG XI letter dated 10 February 1988 from Unit A.4 (Economic aspects, employment and statistics) to the Director of Directorate B and the adviser to the Director General.
DG XI 1988i. Internal DG XI letter dated 11 February 1988 from Unit A.4 (Economic aspects, employment and statistics) to the Director of Directorate B and the adviser to the Director General.
DG XI 1988j. Internal DG XI letter dated 25 October 1988. Headed "Rapport de mission: Object: Mission à la Haye le 24/10/88. Discussion sur la proposition de directive habitats naturels (FFH)".
DG XI 1988k. Internal DG XI note dated 16 February 1988 from the Nature Unit to the Director of Directorate B and the adviser to the Director General, attaching a draft of the Habitats Directive dated 15 February 1988.
DG XI 1988l. Internal DG XI note dated 18 February 1988 from the Legal Unit to the Director of Directorate B, the adviser to the Director General, the Nature Unit, and the Economics Unit. Subject: Draft of a directive on habitats protection. Ref: Version of FFH directive of 15.2.1988.
DG XI 1988m. Internal DG XI note dated 26 January 1988, attaching "Outline of an EEC Directive on the protection of natural habitats and species of wild flora and fauna other than birds" (SPJ version of 26 January 1988). With some handwritten comments.
DG XI 1988n. Internal DGXI note dated 8 February 1988 attaching work programme and draft of Habitats Directive dated 8 February 1988 (XI/B/3).
DG XI 1988o. Note dated 4 August 1988 from the Director General of DG XI (environment) to the head of Commissioner Clinton Davis' cabinet (XI/007867).
DG XI 1988p. Note dated 27 July 1988 from the office of Commissioner Stanley Clinton Davis to the Director General of DG XI. Headed: "Object: Directive 'Habitat'".
DG XI 1988q. Outline of a possible EEC directive on the protection of habitats and species. Version of 18 January 1988, with handwritten amendments by the Nature Unit of DG Environment.
DG XI 1988r. Projet de directive concernant la protection des habitats naturels et semi-naturels ainsi que de la faune et de la flore sauvages. Communication de M. Clinton Davis. Bruxelles, 15 July 1988, COM(88) 381/5. Corrigendum: replacement des pages 1 à 4. On the agenda of the 928th meeting of the Commission, on Tuesday 19 July 1988.
DG XI 1988s. Version of FFH-Directive of 8 February 1988: Outline of an EEC directive on the protection of species of wild flora and fauna and of natural habitats. Marked up with handwritten comments identifying the sources of the draft provisions.
DG XI 1988t. Version of internal DG XI document headed "Pending issues: (Habitat Directive)", hand dated 26 January 1988, marked up with handwritten comments and changes.

DG XI 1989a. DG XI (unit B3) file note dated 19 September 1989, regarding a meeting between DG XI and French representatives to discuss the draft habitats directive, held on 13 September 1989 in Brussels.

DG XI 1989b. DG XI note headed "Conseil Environnement, le 28 novembre 1989, 'Point Habitat', Summary Background Note".

DG XI 1989c. Undated draft briefing note from the Director General of DG XI to the Environment Commissioner, referring to a meeting of the European Parliament on 18 October 1989.

EEB 1988. Letter dated 9 May 1988 from the EEB to DGXI, commenting on the draft directive. Attached to a letter from DGXI to the Nature Conservancy Council dated 27 May 1988.

ELO 2004. Michel-Jean Jacquot. Short biography. *CountrySide*, 67, May 2004. European Landowners' Organization, p. 3.

Environmental Impact Assessment (EIA) Directive 1985. Council Directive 85/337/EEC of 27 June 1985 on the assessment of the effects of certain public and private projects on the environment, *OJ L 175*, 5 July 1985, p. 40, since codified as Directive 2011/92/EU of the European Parliament and of the Council of 13 December 2011 on the assessment of the effects of certain public and private projects on the environment (codification), *OJ L 26*, 28 January 2012, p. 1. Codified version since amended by Directive 2014/52/EU, *OJ L 124*, 25 April 2014, p. 1.

European Commission 1976. Directory of the Commission of the European Communities (September 1976).

European Commission 1984. Directory of the Commission of the European Communities (September 1984).

European Commission 1988a. Directory of the Commission of the European Communities (January 1988).

European Commission 1988b. Protecting the Community's wildlife heritage: Commission proposes creation of special protection areas. Press release P/88/92 of 26 July 1988.

European Commission 1988c. Statement by Mr Stanley Clinton Davis. Press release IP/88/478, 22 July 1988.

European Commission 1988d. Translation of letter dated 16 August 1988 from the Commission to the President of the Council, attaching Habitats Directive proposal COM(88) 381 final. Council document 8149/88, ENV 143, Brussels, 16 September 1988.

European Commission 1991. *CORINE biotopes: The design, compilation and use of an inventory of sites of major importance for nature conservation in the European Community*. Brussels and Luxembourg: ECSC-EEC-EAEC.

European Commission 1994. Directory of the European Commission (16 June 1994).

European Commission 1995. Directory of the Commission of the European Communities (October 1995).

European Commission 2007. *Interpretation Manual of European Union Habitats*. EUR 27. July 2007.

FACE 1988a. Note from the Federation of Hunting Associations of the EEC (FACE) dated 29 April 1988, commenting on the draft directive.

FACE 1988b. Note from the Federation of Hunting Associations of the EEC (FACE) dated 31 May 1988, commenting on the draft directive.

FACE 1989. Memorandum by the Federation of Hunting Associations of the EEC (FACE), dated 26 January 1989, headed "Preliminary comments and observations on the proposal for a Council Directive 'on the protection of natural and semi-natural habitats and of wild fauna and flora' COM(88) 381 final".

First Environmental Action Programme 1973. Declaration of the Council and of the European Communities and of the representatives of the Governments of the Member States meeting in the Council of 22 November 1973 on the programme of action of the European Communities on the environment. *OJ C 112*, 20 December 1973, p. 1.

Fourth Environmental Action Programme 1987. Resolution of the Council of the European Communities and of the Representatives of the Governments of the Member States meeting within the Council of 19 October 1987 on the continuation and implementation of a European Community policy and action programme on the environment (1987–1992). *OJ C 328*, 7 December 1987, p. 1.

Fraser, J. 1992. Joy as 'Grim Ripa' goes back to Rome. *The Glasgow Herald*, 29 June.

Freestone, D. 1996. The enforcement of the Wild Birds Directive: a case study. *In:* Somsen, H. (ed.) *Protecting the European environment: enforcing EC environmental law*. London: Blackstone Press.

Gammell, A. 1999. The contribution of BirdLife International to the Birds Directive and the management of birds and their habitats in Europe. 20 years with the EC Birds Directive. Proceedings from a conference on the Council's Directive on the Conservation of Wild Birds, 18–19 November, Elsinore, Denmark. The Danish National Forest and Nature Agency.

Gammell, A. 2012. Interview with Alistair Gammell, formerly of the RSPB, conducted by Andrew L. R. Jackson on 31 July 2012.

Habitats Directive 1992. Consolidated version of Council Directive 92/43/EEC of 21 May 1992 on the conservation of natural habitats and of wild fauna and flora. *OJ L 206*, 22 July 1992, p. 7. Consolidation date 1 July 2013.

Henningsen, J. 2012. Interview with Jørgen Henningsen, former Director at DG Environment, conducted by Andrew L. R. Jackson on 24 August 2012.

Hey, C. 2005. EU environmental policies: A short history of the policy strategies. *In:* Scheuer, S. (ed.) *EU environmental policy handbook: a critical analysis of EU environmental legislation*. Brussels: European Environmental Bureau.

House of Lords 1989. House of Lords' Select Committee on the European Communities, 15th Report: Habitat and Species Protection. Ordered to be printed 18 July 1989.

Hughes, J. D. 1975. Ecology in ancient Greece. *Inquiry*, 18, 115–125.

IRSNB undated. Draft directive in DG Environment's archive hand-marked "1er Texte directive (IRSNB)". From the archive folder "Directive Habitats – Preparations à 1988".

Jaffeux, H. 2013. Personal communication: e-mail dated 18 April 2013 from Henri Jaffeux, ex French government official who worked on implementing the Habitats Directive, to Andrew L. R. Jackson.

Japan Economic Newswire 1987. Former Netherlands Premier takes up Tokyo EC post in April. *Japan Economic Newswire*, 26 January.

Johnson, S. 2011. Interview with Stanley Johnson conducted by Christian van de Velde on 18 October 2011. Ref: INT185. HISTCOM.2: Histoire interne de la Commission Européenne 1973–1986. Project coordinator: Université catholique de Louvain.

Johnson, S. 2012. Interview with Stanley Johnson, former European Commission official and former MEP, conducted by Andrew L. R. Jackson on 16 July 2012.

Jordan, A. 1998. EU environmental policy at 25: the politics of multinational governance. *Environment*, 40, 14–20.

Jordan, A., Brouwer, R. and Noble, E. 1999. Innovative and responsive? A longitudinal analysis of the speed of EU environmental policy making, 1967–97. *Journal of European Public Policy*, 6, 376–398.

Kautz, R. 2002. Natura 2000 und eigentum: evaluierung eines naturschutzpolitischen instruments vor dem hintergrund des spannungsverhältnisses zwischen naturschutzanliegen und eingriffen in das eigentumsrecht. Vienna: Instituts für Sozioökonomik der Forst- und Holzwirtschaft an der Universität für Bodenkultur, Wien.

Koester, V. 2012. Interview with Veit Koester, senior negotiator for the Danish government in respect of the Habitats Directive, conducted by Andrew L. R. Jackson on 22 August 2012.

Krämer, L. 2011. Interview (in French) with Ludwig Krämer conducted by Christian van de Velde on 7 October 2011. Ref: INT194. HISTCOM.2: Histoire interne de la Commission Européenne 1973–1986. Project coordinator: Université catholique de Louvain.

LIPU 1988. Letter dated 26 April 1988 from the Liga Italiana Protezione Uccelli to DG XI, commenting on the draft Habitats Directive.

Markham, J. M. 1989. Europe's politicians thinking green as concern grows on environment. *The New York Times*, 12 April.

Mason, D. 1992. Minister plays down boycott by 'flamboyant' EC Commissioner (27 May 1992). *Press Association*, 27 May.

Morillo, C. and Gómez-Campo, C. 2000. Conservation in Spain, 1980–2000. *Biological Conservation*, 95, 165–174.

Moss, D. 2008. EUNIS habitat classification – a guide for users. *European Topic Centre on Biological Diversity*, July.

MPI 1988. Letter dated 11 May 1988 from the Max-Planck-Institut für Verhaltensphysiologie to DG XI, commenting on the draft Habitats Directive.

NVTBVV 1988. Letter dated 10 May 1988 from the Nederlandse Vereniging Tot Bescherming Van Vogels to DG XI, commenting on the draft Habitats Directive.

Palmer, J. and Wintour, P. 1988. PM triggers row with EEC sackings: Heath attacks appointment of 'discredited minister'. *The Guardian*, 23 July.

Ross, G. 1994. Inside the Delors cabinet. *Journal of Common Market Studies*, 32, 499–523.

Schoon, N. 1991. Interview / Woodland 'saviour' with a lonely vision; Britain has until next week to answer claims that it has breached European environmental law. As the deadline nears, Nicholas Schoon interviews Carlo Ripa di Meana, the Environment Commissioner. *The Independent*, 9 December.

Single European Act 1986. Single European Act. *OJ L 169*, 29 June 1987, p. 1.

Strategic Environmental Assessment Directive 2001. Directive 2001/42/EC of the European Parliament and of the Council of 27 June 2001 on the assessment of the effects of certain plans and programmes on the environment. *OJ L 197*, 21 July 2001, p. 30.

Stuffmann, C. J. 1999. The upcoming of the Birds Directive. 20 years with the EC Birds Directive. Proceedings from a conference on the Council's Directive on the Conservation of Wild Birds, 18–19 November, Elsinore, Denmark. The Danish National Forest and Nature Agency.

Sundseth, K. 2012. Interview with Kerstin Sundseth, environmental consultant and former DG XI stagiaire, conducted by Andrew L. R. Jackson on 13 September 2012.

Temple Lang, J. 2012. Interview with John Temple Lang, ex-official with the European Commission's Legal Service, conducted by Andrew L. R. Jackson on 17 August 2012.

Temple Lang, J. 2013. Personal communication: e-mail dated 3 April 2013 from John Temple Lang, ex-official with the European Commission's Legal Service, to Andrew L. R. Jackson.

The Observer 1991. Profile: Playboy hero of the Greens – Carlo Ripa di Meana. *The Observer*, 27 October.

Várnay, E. 2006. The institutionalisation of infringement procedures in EC law – the birth of a Community sanction. *European Integration Studies, Miskolc*, 5, 5–19.

Ward, S. 1997. The IGC and the current state of EU environmental policy: consolidation or roll-back? *Environmental Politics*, 6, 178–184.

Webster, P. and Ford, R. 1988. Kinnock request rejected; Thatcher refuses EEC reappointment; Sir Stanley Clinton Davis. *The Times*, 27 July.

White, M. and Wolf, J. 1991. 'Bully' row threatens EC talks: Environment commissioner accuses PM of intimidation. *The Guardian*, 26 October.

Wildlife Link 1988a. Letter dated 6 June 1988 from WWF UK, on behalf of Wildlife Link, commenting on the draft Habitats Directive and annexes.

Wildlife Link 1988b. Minutes of a meeting of the Habitats Directive group held at 2.30pm on Thursday 15 December 1988 at the Vincent Wildlife Trust, 21 Bury Street, London EC3.

Wildlife Link 1988c. Note of the meeting of the EC Habitats Directive Working Group held at 2pm on Wednesday 21 September 1988 at the Institute of Biology, 20 Queensberry Place, London, SW8.

Wurzel, R. 2002. *Environmental policy-making in Britain, Germany and the European Union: the Europeanisation of air and water pollution control.* Manchester: Manchester University Press.

7

NEGOTIATING THE HABITATS DIRECTIVE I

The Greek and Spanish presidencies (July 1988 to June 1989)

Greek Presidency (July–December 1988)

The Directive in principle

Introduction

The text of the draft Habitats Directive was formally proposed to the Council on 16 August 1988,[1] during a Greek Presidency of the Council. Having secured the successful launch of the proposal, Stanley Johnson gradually played a less prominent role from this point onwards, albeit he engaged in a tour of key capitals with Stuffmann in October/November 1988 (more below), remained informed of developments, and intervened in February 1990 when the proposal came under threat once more from Jacques Delors' *cabinet* (see Chapter 8), before leaving DG XI altogether in May 1990 (Johnson, 2012).

On the Commission side, the key figures in the negotiations (some directly and others in an advisory capacity) once the directive had been formally proposed to the Council were: Claus Stuffmann, head of the Nature Unit, who was "very much in the lead on all the official negotiations of the Habitats Directive" (Bunce, 2012);[2] Henriette Bastrup-Birk, Stuffmann's number two in respect of the negotiations (Neal, 2012), variously referred to by interviewees as the "mother" of the Habitats Directive (Troya, 2012) and "the person who really ended up doing the job", dealing with a lot of the detailed issues relating to the directive (Henningsen, 2012); Catherine Yourassowsky, a Belgian scientific expert and member of the Nature Unit who was the "right-hand" woman of Henriette Bastrup-Birk during the negotiations, working extensively on the text of the directive and its annexes with Bastrup-Birk and Stuffmann;[3] Geert Raeymaekers, who worked with DG XI providing scientific input to the annexes (Raeymaekers, 2012); Reinhard Klein, a

German member of the Nature Unit who dealt with nature conservation financing (Klein, 2012); Richard Geiser, another German member of the Nature Unit, who worked on the species protection aspects of the Birds Directive (Klein, 2012); Pierre Devillers of the IRSNB, who worked closely with the CORINE biotopes team in DG XI; Ian Hepburn, the English RSPB/WWF campaigner who coordinated an EU-wide NGO campaign in support of the directive proposal, and who was seconded to the Nature Unit for a period during the negotiations (Hepburn, 2012); and Antonio Troya, a Spanish government expert seconded to DG XI to act as a liaison during negotiations regarding the Directive (Troya, 2012).

An over-ambitious plan to agree the Directive by November 1988

As the UK government would later note, it seems that Greece at one point hoped (very optimistically) to secure agreement regarding the Directive at the Environment Council meeting in November 1988, and to this end called a series of working group meetings, the first two of which were held on 8 September and 23 September (DG XI, 1988b; Department of the Environment, 1989), somewhat unpicking the Commission's strategy of easing the proposal's path via the proposed "réunion de séduction" of *hauts fonctionnaires* (senior civil servants) scheduled for 29 September.

Amongst the issues covered at the meeting on 8 September, certain delegations expressed the view that the proposal was too complex and needed to be simplified; Germany and France foresaw conflict between the species protection provisions and hunting interests; Spain and Germany expressed serious reservations regarding the absence from the proposal of annexes of species and habitats; to this, DG Environment replied that it would be happy to table draft annexes, and this it did, tabling its existing draft annexes as working documents at the meeting on 23 September (CEC, 1988a; DG XI, 1988f), thus completing a step in Stanley Johnson's strategy.[4] The question of the relationship between the proposal and the Bern Convention was raised, and almost all delegations suggested that the proposal would involve long and intensive negotiations (DG XI, 1988b; CEC, 1988e). They were not wrong: the negotiations would ultimately take more than three years, spanning seven Council Presidencies (Greece, Spain, France, Ireland, Italy, Luxembourg, the Netherlands), albeit the lion's share of the work would be done during the French, Luxembourg and Dutch Presidencies.

NGO preparations

If the Birds Directive was a motor for NGO coordination in Europe (Meyer, 2010; European Commission, 1979; David, 2012), by the time the Habitats Directive was proposed, NGOs were relatively well prepared. Wildlife Link – a coalition of UK conservation NGOs – met on 21 September 1988 to discuss the proposed directive, with Stanley Johnson in attendance as an invited expert,[5] together with experts from the NCC and the Conservation Monitoring Centre of the IUCN, amongst others (Wildlife Link, 1988c).

It was noted at the meeting that certain Member States did not regard the proposed directive as an EU competence (e.g., arguing that national measures would be more appropriate for non-migratory species), and that some countries were concerned that the proposed directive might render the Bern Convention "powerless" (Wildlife Link, 1988c, p. 3). Moreover, it was thought that "Denmark, Netherlands, Spain and the UK would be amongst those raising objections to the Directive" (Wildlife Link, 1988c, p. 3). As such, it was agreed that this was a crucial time to coordinate NGOs in all Member States. The RSPB reported that it had appointed Angus Westgarth-Smith to a newly-created campaigning post in support of the Habitats Directive proposal (ultimately a joint RSPB/WWF position (Wildlife Link, 1988a)), though Westgarth-Smith would be replaced in July 1989 by Ian Hepburn, who played an important role in the directive's history being at one point seconded part-time from this NGO position to assist the Nature Unit of DG XI with the preparation of the Habitats Directive, "much to the disappointment of the UK [government] staff", according to Hepburn (2012), "when [they] saw me sitting at the top table during meetings in Brussels".[6]

Other coalitions of NGOs were also beginning to mobilise, and it was noted that Alistair Gammell of the RSPB had organised a meeting of major coalitions of NGOs in Europe for 18 October 1988 to establish a common stance regarding the directive (Wildlife Link, 1988c). An important milestone was to be a meeting of Environment Ministers on 24 November 1988 at which the Ministers would decide whether to support the directive in principle (Wildlife Link, 1988c), and the NGOs were therefore going to need to act fast.

Wildlife Link resolved to seek a meeting as soon as possible with Roy Bunce of the UK's Department for Environment, preferably before a meeting with then Secretary of State for the Environment Nicholas Ridley. But Bunce had been given clear instructions: "basically the brief was to keep this [directive proposal] in the long grass for as long as possible" (Bunce, 2012). Ridley himself was openly hostile to the proposal, telling Gammell that there would be a habitats directive "over my dead body", prompting Gammell (2012) to reply that he would be sorry to see Ridley go! Gammell feared Ridley's attitude was in part attributable to the UK government's experience in the Duich Moss case under the Birds Directive (see Chapter 5). As Gammell stated in early 1989, "The society and conservation are now paying an awful price in the habitats directive which has been launched in Brussels. Nicholas Ridley has said 'over his dead body', so there is the risk that the new directive will have its teeth removed" (Brooke, 1989, p. 55).

Positions begin to take shape

By the 23 September 1988 meeting between the Commission and Member State officials, positions had begun to take shape. Years of implementing the Birds Directive had opened the Member States' eyes to the potential impacts of EU nature conservation legislation, and it quickly became clear that negotiations would be painstaking, if progress could be made at all. Worryingly from the Commission's perspective, many comments were negative, and several struck at the heart of the proposal. For example, Denmark argued that a policy of conservation based on special protection areas was outdated (CEC, 1988a); the UK remained concerned about economic impacts (DG XI, 1988f); three of the EU's powerhouses – France, Germany and the UK – were anxious to minimise the role of the Commission in general (DG XI, 1988f); Denmark, Greece, France (arguing again from a pro-hunting angle (DG XI, 1988f)) and the UK queried the need for provisions aimed at protecting species, in addition to provisions protecting habitats, arguing that this could lead to a duplication of obligations under the Bern Convention. More generally, DG XI's unofficial notes of the meeting record that the UK and Denmark sought to "hide behind" the Bern Convention, arguing that this instrument would suffice in the nature conservation field (DG XI, 1988f). The Commission responded that species and habitats conservation were inseparable elements of a nature conservation strategy, and that an EU instrument would provide a better guarantee of protection than a less binding international agreement (CEC, 1988a).

Regarding the proposal to amend the EIA Directive to apply impact assessment obligations to special protection areas under the new directive (on which, more below), certain Member States argued that it would not be appropriate to amend the EIA Directive in this way, since that Directive had only just entered into force (CEC, 1988a). The Commission responded that "experience gained in implementing the wild birds Directive had demonstrated the need for prior environmental impact assessment since ex-post corrective measure[s] were often unsuccessful" (CEC, 1988a, p. 8). Spain was concerned that the Canary Islands were to be covered by the Directive, although France – interestingly – argued for the inclusion of its overseas departments and territories (DG XI, 1988f), a view rejected unconvincingly by the Commission, "in view of the very different characteristics of the fauna and flora in those regions compared with those to be found in the European territories of the Member States" (CEC, 1988a, p. 9). Few positive comments emerged from this meeting, although the Greek delegation made the interesting suggestion of distinguishing priority species and habitats from species/habitats requiring less urgent measures, an idea that would later re-emerge and be successfully adopted (DG XI, 1988f). The Commission evidently had a lot of ground to cover if it was to persuade Member States of the merits of the instrument, yet the political atmosphere remained difficult, albeit there were positive developments on the horizon in that regard.

Political developments in the European Parliament, the UK and France

At the end of September 1988 the European Parliament published three reports penned by Hemmo Muntingh – the MEP then most active in the nature conservation field – on the Birds Directive and the Bern, Bonn and Washington Conventions, which together threw significant weight behind the Habitats Directive proposal (European Parliament, 1988a, 1988b, 1988c). As Muntingh explains,

> In the resolutions set out in those reports, Parliament made it clear that it was dissatisfied with the current state of implementation of the conventions [and the Birds Directive], and it made a large number of recommendations. For example, it called on the Commission and the Member States . . . to draw up as quickly as [possible] a Community directive implementing the Bern and Bonn Conventions covering all species of marine and terrestrial flora and fauna present in the wild and their habitats.
>
> *(European Parliament, 1989, p. 25)*

It was not merely in the European Parliament that political developments were beginning to favour the Habitats Directive proposal. Whilst Nicholas Ridley remained a negative influence within the UK government in respect of the proposal throughout 1988 and much of 1989, times in the UK were slowly beginning to change, kickstarted by a landmark speech by Margaret Thatcher to the Royal Society in London on 27 September 1988, in which the Prime Minister reflected that "the health of the economy and the health of our environment are totally dependent upon each other . . . Stable prosperity can be achieved throughout the world provided the environment is nurtured and safeguarded" (Thatcher, 1988). The Prime Minister's speech received a cautious welcome at the time from environmentalists (Ardill and Linton, 1988), with the Royal Society for Nature Conservation maintaining NGO pressure on the government by setting out a five-point plan to achieve the government's vision, including "A lead from Mrs Thatcher on the proposed European habitat directive" (*PR Newswire Europe*, 1988). Significantly, times were also changing in France: in May 1988, Brice Lalonde – ex-head of Friends of the Earth France – was appointed Minister for the Environment (Phillips, 1988; Forsyth, 1988; Lalonde, 2012).

La réunion de séduction

Nevertheless, in late 1988 the Commission was facing into a difficult political wind as regards its proposal. The Director General of DG XI's proposed launch of the proposal at a meeting of senior civil servants took place as scheduled on 29 September 1988, and his speaking notes reveal both the Commission's determination that its proposal should receive a fair hearing notwithstanding early feedback, as well as the Commission's awareness of key points that may arise, reflected in an "arguments to keep in reserve" section of the Director General's notes. Amongst these hypothetical questions that might be posed by Member States were: "What guarantee do we have that the Commission will not continue with the pernickety attitude it has had the tendency to adopt in handling complaints under the Birds Directive?"; and "Why maintain two distinct instruments [i.e., the Birds Directive and the Habitats Directive]?" (DG XI, 1988e).

The Member States restated many of their previously-raised points, and threw a few new ones into the mix (DG XI, 1988d): e.g., the risk of delaying or rendering more difficult the application of the Bern Convention, which was then, countries claimed, entering an important phase (Belgium, Denmark, Luxembourg, UK); the financial, legislative and administrative implications (Belgium, UK, Germany, Luxembourg, the Netherlands), with certain delegations expressing fears of creating a "mega-bureaucracy"; fears of the Commission being granted too prominent a role (Belgium, Germany); reticence to extend the application of the EIA Directive to cover special protection areas under the new directive, so soon after the entry into force of the EIA Directive (Belgium, Spain); dissatisfaction that NGOs had been consulted about the proposal in advance more than the Member States (Belgium, Germany); the possibility of merging the proposed directive with the Birds Directive (Belgium, Luxembourg); the necessity to mobilise EU financial resources to support national nature conservation bodies (Denmark, Ireland, Italy); the necessity to balance nature conservation aims and other policy goals (UK, Ireland), etc. In his response, DG XI's Director General noted that in establishing the new alliance that he wished to see between the Commission and Member States, it would be important to focus on common interests and to leave behind "sterile debates" regarding respective competences and legal points of detail (DG XI, 1988d). In conclusion, the Director General noted that the Commission was ready to take a flexible approach and to listen carefully to the Member States' concerns (DG XI, 1988d).

In this spirit, the Commission continued its discussions with the Member States, and, pending the Environment Council meeting on 24 November 1988 (i.e., at ministerial level), which could make or break the proposal, several tentative decisions were taken by officials at a meeting on 6 October 1988, including (DG XI, 1988g): that the missing annexes, once examined and amended, would be reintroduced to the proposal for adoption by the Council; that it would be desirable to introduce a hierarchy in terms of the species and habitats listed in annexes (cf. Greece's earlier proposal of priority species and habitats); and that, after the meeting of the Environment Council in November, technical groups would be established to consider the criteria for classifying special protection areas and the lists of species and habitats to be annexed to the directive. An appeal was made by the Member States for greater flexibility – presumably compared with their experience with the Birds Directive – in terms of the application of site selection criteria and the envisaged mechanisms for protection. Further, France (joined by Greece) continued to stress the view that protecting habitats would automatically benefit species, the unstated corollary being that it would therefore be unnecessary to introduce rules regulating species protection outside protected areas, which rules would of course impact on hunting.

Johnson and Stuffmann visit the Netherlands, Spain and the UK

The Netherlands

At this point in time, Stanley Johnson and Claus Stuffmann embarked jointly on bilateral negotiations with three Member States – the Netherlands, Spain and the UK – in advance of the crucial Environment Council meeting on 24 November 1988. They visited the Netherlands on 24 October (DG XI, 1988h), Spain on 7/8 November (DG XI, 1988a), and London on 17 November (DG XI, 1988c). The Commission's note of discussions with the Netherlands complicates any characterisation of the Habitats Directive as a "Dutch directive", a perception that appears to have developed in certain quarters over time (Wurzel, 2008, p. 267).[7] Indeed, the Netherlands harboured numerous fundamental concerns regarding the Commission's proposal, and the Dutch authorities expressed their regret that the Commission had not consulted national authorities in advance of proposing the Directive to the Council (DG XI, 1988h). In other words, whilst the Netherlands – like other Member States – certainly played an important role in negotiating the text of the Directive, the central elements of the Directive (Natura 2000 network, parallel species protection regime, etc.) were already in place in the Commission's original proposal, whose history is described in Chapter 6. Moreover, an important 'network' element of the proposal – that is, the protection afforded to features such as corridors outside protected areas – was watered down during negotiations, such that the version finally adopted imposes weaker obligations in this respect than were contained in the Commission's original proposal.

Thus, Liefferink and Zouwen's (2004) argument that the Netherlands "uploaded" the concept of the Dutch National Ecological Network for the purposes of the Habitats Directive, and Leistra et al.'s (2008, p. 27) argument that this Network "proved to be the exemplar for Natura 2000", are not borne out by the history, given that: the Dutch government complained that it had not been consulted regarding the Commission's draft directive before the directive was proposed to the Council; by 1980 the Commission had a clear vision for a network of EU protected areas going beyond birds (European Commission, 1980), while the Dutch Ecological Network formed part of a plan designed in the 1980s and adopted by the Dutch Parliament in 1990, two years after the Habitats Directive proposal (Keulartz, 2009; Verschuuren, 2015); and, most importantly, the protection for important features outside Natura 2000 was weakened during negotiations regarding the Habitats Directive, albeit that the Netherlands repeatedly sought to strengthen this aspect of the Directive. A better characterisation of what happened is that the Netherlands tried, but failed, to upload its concept of a European Ecological Network to the EU level. As Henri Jaffeux comments, "If there is a country that [was] not satisfied out of the [Habitats Directive] negotiation, it is indeed the Netherlands" (Jaffeux, 2013), because the Directive as finally adopted was less ambitious than the Netherlands had sought, in particular in respect of defining and protecting corridors between sites.

Spain

The bilateral meeting with Spain raised a number of interesting issues (DG XI, 1988a), including: the issue of financing, which was central to Spanish concerns, since Spain needed, it said, to develop economically, yet the designation of "vast untouched zones" could create a barrier to such development (DG XI, 1988a, p. 2). As Jørgen Henningsen, a Director in DG XI, explains, Spain was not against the idea of nature protection, "but Spain felt . . . that very large areas that would be covered by this legislation were in Spain, [and] that Spain had actually proportionally and in absolute terms much bigger areas than any other country" (Henningsen, 2012). Further, Spain expressed itself to be categorically opposed to amending the EIA Directive to apply that Directive to special protection areas under the Habitats Directive (DG XI, 1988a). Finally, Spain noted that the list of habitats to be the subject of special protection areas – a short list at this time – was on the one hand too detailed, and on the other contained too many gaps (DG XI, 1988a), an unenviable problem for the Commission to address.

The UK

In preparation for the meeting of the Environment Council on 24 November 1988, and in light of discussions up to that point, following a Working Group meeting on 4 November (Department of the Environment, 1989) the General Secretariat of the Council agreed a list of fundamental issues to be considered by Environment Ministers, which essentially boiled down to whether or not it was a good idea to proceed with the proposed directive at all (CEC, 1988c). Early indications from the UK were not good. On 18 November 1988, Lord Caithness, the UK's junior Environment Minister, gave a speech containing the following remarks, no doubt timed to coincide with the impending meeting of Ministers: "We must keep control of the future of our own countryside and of our wildlife in our own hands", he said, and suggested "the addition of the Brussels bureaucrat to the official pest list" (Caithness, 1988).[8] The day before Lord Caithness made these remarks, Stanley Johnson and Claus Stuffmann had met with UK officials in London to discuss the Habitats Directive proposal (DG XI, 1988c). Noting the UK's "extremely negative attitude", evidenced by several press articles, Johnson and Stuffmann aimed to set in motion a "more constructive dialogue" (DG XI, 1988c).

The UK officials reported that the government remained to be convinced of the need for any EU instrument in the field and, on being reminded of the clear commitment to such an instrument in the Fourth Environmental Action Programme (1987), a senior UK official stated that "the UK had had a change of heart since then" (DG XI, 1988c). Like other Member States, the UK expressed reservations about amending the EIA Directive so soon after its adoption (DG XI, 1988c). Moreover, the UK rejected the notion that the Commission might be granted the power, as the Commission's draft proposed, to identify sites for designation which fell within the EU's "top ten" for a particular species or habitat in circumstances where the country in question had not itself designated the site within two years of the directive's entry into force (DG XI, 1988c). Recognising the strength of resistance to this proposal, the Commission's file note records "we should think how to reformulate this – because it surely won't fly in its present version" (DG XI, 1988c, p. 2). Finally, the UK seemed to be particularly concerned that "sites once designated under the directive would be 'set in concrete' and that it would be impossible to touch them in any way subsequently" (DG XI, 1988c, p. 2). Reacting to this point, Johnson and Stuffmann explained that "It was true that, once designated, the text did not envisage further derogations but we were ready to consider a formula which permitted a certain flexibility as long as it was clear that the protected status of a site was to be maintained" (DG XI, 1988c, p. 2), an early presaging of the Article 6(4) derogation that would ultimately be introduced to the Habitats Directive. As for the UK's fear that things shouldn't be "set in concrete", a Commission aside in its note of the meeting records, "Without action of the kind proposed, sensitive sites are only too likely to be 'set in concrete' – several feet thick!" (DG XI, 1988c, p. 2).

The UK was similarly robust in a meeting with the Wildlife Link coalition of NGOs around the same time (Wildlife Link, 1988a). The government reported that while some Member States would favour a directive in principle, all were opposed to the present draft, "on the ground of its intrusiveness, bureaucracy and inflexibility" (Wildlife Link, 1988a). "This is code for enforceable!" added the RSPB note-taker, adding that "The UK Government's probable unwillingness to negotiate on this draft comes about in part in pique, since they had little to do with the drafting process and the text is not to their liking (because it means something)" (Wildlife Link, 1988a). The government may also have been piqued to be faced with the RSPB, which, the government felt, had had a very strong hand in drafting the directive: the draft seemed to have "been very largely written for [the Commission] by the RSPB", comments Roy Bunce, then head of the UK government's Wildlife and Countryside Division (Bunce, 2012).[9] The UK government was also reportedly "fed up with being hounded by [the Commission's Legal Unit] on matters of opinion" (Wildlife Link, 1988a). According to the RSPB note-taker,

> The UK's position has been to say very little so far at meetings on this Directive in Europe. We knew this from our contacts, but [the senior UK official present] said it as well. I judge this is for two reasons. First they dread their fingerprints being found on the murder weapon, whereas if someone else will do the dirty act that would be preferred by them. Secondly, I sense they genuinely may finally not have resolved their final position. We told [the government] clearly that silence when the draft was under attack, would be construed by us as agreement with the attack and draw the appropriate response from us . . . we do now begin to see the size of the problem . . . We are warning . . . that if we are rebuffed, [the government] can expect quite a rough ride and an issue that will not go away.
>
> *(Wildlife Link, 1988a)*[10]

Make or break Environment Council meeting of 24 November 1988

Ignoring the NGOs' demands, at the ministerial-level meeting on 24 November 1988 the UK – represented by Lord Caithness, since Secretary of State Nicholas Ridley refused to attend EU Council meetings (Travis, 1989; Department of the Environment, 1989) – did its best to derail the directive. Spain and the UK reportedly "felt that a number of preliminary issues needed to be settled before launching a Community scheme of this scale; in particular, there should be appropriate implementation nationally of the Bern Convention and the Directive on wild birds . . . [The] UK thought that Community action might rather be focused on assisting Member States to improve their implementation of the various instruments" (CEC, 1988b, p. 2). An internal Commission file note would later record that only Spain and the UK had expressed themselves in negative terms at this meeting, both "contesting the necessity for EU action in the area" (DG XI, 1989b).[11] The Commission responded that the provisions of the Bern Convention were "too vague and lacked binding force, hence the need for more thorough harmonization at Community level" (CEC, 1988b, p. 3), not to mention the fact that two EU Member States – France and Belgium – were not yet parties to that Convention (Nature Conservancy Council, 1989).

On the issue of impact assessment, some delegations were in favour of compulsory assessments, but Ireland reiterated the view that amending the EIA Directive, as proposed by the Commission, would not be desirable (CEC, 1988b). Many Member States were in favour of reinforcing EU financial support for nature conservation, though some felt this should come from existing funds, such as the ACE Regulation (1984) or the Structural Funds (CEC, 1988b). Ireland stressed the importance of EU financial support, "particularly in view of the number of habitats for protection that were situated in Member States which lacked administrative and financial resources" (CEC, 1988b, p. 6).

In respect of the establishment of annexes of species and habitats, the Council was presented with three options: the first was the establishment of comprehensive annexes to be drawn up by the Council; the second was the progressive establishment of a comprehensive list according to the procedures set down in the draft directive; the third was the establishment in a first stage of a short list of habitats and species agreed by the Council, based on the needs of species or ecosystems that were especially endangered or that were a key factor in terms of general conservation potential; this list would be reviewed regularly and extended by procedures to be agreed (CEC, 1988d). Unsurprisingly, the majority of delegations were in favour of the third option (CEC, 1988b). The Council concluded its discussions by asking COREPER to continue examining the proposal in light of the statements made by Member States (CEC, 1988b). At this point in time the directive's future remained very much in the balance. As a contemporaneous article in *Natural World* magazine argued, "The main danger to this Directive is that it will be ambushed by an alliance of self-interested Member States, who are keener on saving money and pleasing the powerful industrial and agricultural lobbies than in promoting conservation of Europe's natural heritage" (Rose, 1988, p. 8).

Strengthening habitat protection in the Bern Convention

Another risk to the Directive's progress came from a different – perhaps less expected – angle. Just a few days after the Environment Council meeting of 24 November 1988, the Standing Committee of the Bern Convention met in Strasbourg. The Chair of the Committee – Veit Koester of Denmark – had a longstanding relationship with the Bern Convention, having headed the Danish delegation during the instrument's negotiation from 1976–1979 (Koester, 2004). Koester was also prominently involved in early discussions regarding the Habitats Directive, in respect of which Denmark had repeatedly stressed its support for existing instruments such as the Bern Convention, and the need to avoid duplication and overlap. As Koester (2012) explains,

> when Denmark joined the EU [in 1973] . . . it was the official Danish policy, and that was also something that was referred to very openly: 'Do not worry about the EU, for instance nature conservation, land use planning and what belongs to that will remain as national concerns only and there will be no interference from the side of the EU' . . . When the Habitat Directive negotiations started, the Danish position was still the same, namely that nature conservation and land use planning and so forth, they belonged to the national domain and the EU should not interfere. And we could see that the Habitat Directive would have rather strong implications for freedom in respect of nature conservation and land use planning, so the Danish position was at the beginning very negative, and also we have the position that our nature protection and planning legislation was so good that it should rather be copied by all other EU countries than be influenced by the Habitat Directive, so we were very negative at the beginning, because we felt that our legislation was good and we didn't need the EU to interfere.

One of the issues on the agenda of the Bern Convention's Standing Committee in December 1988 was the implementation of Article 4 of the Convention, which concerned habitat protection; the UK delegate reported to the Committee that his country attached considerable importance to the Bern Convention, particularly Article 4 (Council of Europe, 1988). What followed was a concerted, albeit ultimately failed, attempt by certain countries to undermine the need for an EU Habitats Directive by seeking to refine the habitat protection provisions of the Bern Convention (Fernández-Galiano, 2012; Coffey and Richartz, 2003).[12]

Acting on a proposal from the Chair, the Standing Committee instructed the Secretariat to prepare a revised draft of a Resolution that had been presented to the meeting,[13] which was intended to provide a basis for the common interpretation of several expressions provided in the Convention, such as natural habitats, necessary measures, etc. (Council of Europe, 1988). The hope was that Parties to the Convention would, at an extraordinary meeting of the Standing Committee arranged for June 1989, "agree a Resolution on the legal interpretation of the Convention's habitat provisions together with a series of Recommendations on their implementation" (Nature Conservancy Council, 1989, paragraph 19). The UK government was very much in the thick of this approach:

> there are already a number of existing conventions which could in our view secure the same nature conservation objectives [as the proposed Habitats Directive] if properly implemented by all member states. As Lord Caithness made clear at the Council [on 24 November 1988], we believe that the first step should be to invest more effort in making these conventions function more effectively rather than seeking to establish a further overlay. The Standing Committee of the Bern Convention, for example, is currently working up proposals to clarify the interpretation of some of the key articles in the Convention, and to suggest ways of securing their effective implementation. A special session of the Committee is to take place in early June to resolve these matters. The UK government is closely involved in this process.
>
> *(Department of the Environment, 1989, paragraph 18)*

This process would ultimately lead to the adoption of a Resolution and three Recommendations by the Standing Committee of the Bern Convention at its 8th meeting, in June 1989 (Standing Committee, 1989b, 1989c, 1989d, 1989a).[14] Whatever effect these agreements may have had in terms of strengthening the habitat protection provisions of the Bern Convention,[15] they did not have the desired effect of halting the march towards a Habitats Directive. Nevertheless, amongst the Recommendations agreed in June 1989 was one that advised countries to take steps to designate areas of special conservation interest, which ultimately led to the Bern Convention's Emerald Network (Lasén-Díaz, 2010; Standing Committee, 1989c), a site designation and protection regime that operates in parallel to Natura 2000, covering (in principle) all contracting parties and observer states to the Bern Convention, including the EU's Member States.[16] Ironically, therefore, given that the two systems are now complementary, what became the Emerald Network was initially born of an attempt to arrest the development of the Habitats Directive.[17]

It did not take long for the outcome of the Environment Council's meeting of 24 November 1988 to be portrayed in the UK as a rejection of the Habitats Directive. By 14 December, the House of Lords' European Communities' Committee was busily convening an enquiry into the future of habitat protection in the EU, "Following the rejection by the Environment Council, on 24 November 1988" of the Commission's proposal (House of Lords, 1988). The UK government maintained that reports of its having blocked or opposed the directive were misinformed: "The UK has at no stage formally opposed the principle of a Directive, and has at no stage been isolated in Community discussions.

We have in fact been right in the mainstream of Community thinking" (Department of the Environment, 1989, paragraph 14). The Wildlife Link coalition of NGOs was similarly quick to organise, holding a meeting with a senior UK government official as well as an internal meeting on 15 December 1988 (Wildlife Link, 1988b). The message the NGOs subsequently put out to the media was not one of defeat: as Huw Davies (1988) reported before the year was out, "Next year the European environment ministers in Brussels will start debating the draft [Habitats Directive]", which would be a "major test of the government's commitment to nature conservation" according to Alistair Gammell of the RSPB. "Ironically", continued the article, "the proposed directive is a very British initiative. It was largely drafted by a senior EEC official and former British MEP, Stanley Johnson, and endorsed by the outgoing Environment Commissioner and former MP, Stanley Clinton Davis" (Huw Davies, 1988). The outgoing Commissioner, in the final weeks of his term, expressed his pride at having secured the Commission's agreement to propose the directive, but also his fears for its future: "I am very proud of the fact that my colleagues agreed to [the] habitats directive but I cannot hide from you my apprehension that there is a good deal of opposition to it in the Council. It is not very worthy opposition, it is opposition which is designed to erode the purpose of the habitats directive, based upon anxieties alleged in relation to the birds directive. I think it is a reflection of the difficulties which emerged from that directive" (Clinton Davis, 1988).

Spanish Presidency: January–June 1989

Green shoots?

The new year brought a new Commissioner for the Environment, the Italian Carlo Ripa di Meana (European Commission, 1988b), as well as a new Presidency of the Council, held by Spain for the first half of 1989. Ripa di Meana would prove a flamboyant and dynamic Environment Commissioner but nature conservation was not his environmental priority, which was very much the urban environment, an area that he felt "should be a much more important part of the Commission's environmental agenda" (Henningsen, 2012).[18] However, support for the Habitats Directive proposal emerged from an unlikely corner: despite having reportedly expressed itself in negative terms at the meeting of Environment Ministers on 24 November 1988, Spain performed a *volte-face* on taking over the Presidency, revealing that the Habitats Directive would be one of its environmental priorities during its six-month Presidency (DG XI, 1989b). In briefing the new Environment Commissioner for a meeting with Spain, DG XI's Nature Unit noted that the Spanish Presidency could usefully focus on advancing the technical work needed to establish the annexes to the Habitats Directive; as the Nature Unit explained, at its meeting of 24 November 1988 the Council had favoured producing short annexes that would be capable of being operational in a short space of time. The establishment of such annexes could be worked upon, the

Nature Unit suggested, by an *ad hoc* group of experts who could work in parallel to the Council's Working Group on the Environment, which would focus on the directive's text (DG XI, 1989b).

From SPAs to SACs

The EU hunting coalition FACE provided its preliminary comments on the directive proposal on 26 January 1989. One important intervention FACE made at this stage was to emphasise that greater use should be made in the directive of the term conservation, as opposed to protection, since Article 1 of the proposal made clear that conservation was the aim, whereas the term protection implied non-use to many, according to FACE (1989). FACE therefore suggested that the term Special Conservation Areas should be used in the Habitats Directive in place of Special Protection Areas (FACE, 1989, p. 4), this latter phrase being the term used in the Birds Directive. FACE's proposal was ultimately adopted, albeit it took almost two years from this point before the draft directive was changed in this way, with the wording switched around to read Special Areas of Conservation (SAC) (see Chapter 8).

Substantive negotiations begin: February 1989

Substantive negotiations regarding the text of the directive effectively began in earnest at a meeting of the Working Party on the Environment on 3 February 1989. This is not to say that the Habitats Directive proposal was by this time free from general attacks; it was not, as evidenced at the meeting on 3 February by Denmark, Ireland and the UK arguing again that it would be better to focus on the proper application of existing instruments such as the Bern Convention and the Birds Directive (CEC, 1989c), and by France, apparently keen to accommodate its hunters, which proposed proceeding in two stages, starting with the protection of habitats, with species to form the subject of a separate Directive (which might of course never materialise) (CEC, 1989b). However, from around February 1989, countries began to focus less on issues of principle and instead began to engage with the substantive provisions of the proposal. That said, progress was slow during the Spanish Presidency (House of Lords, 1989, paragraph 66), and would only really take off when the French took over the Presidency in July 1989, as detailed in Chapter 8.

Designation of Natura 2000 sites

The Commission's original proposal, as forwarded to the Member States on 16 August 1988, envisaged a step-by-step designation process of special protection areas: at the first stage, the Member States would have two years to classify as special protection areas those areas that corresponded to criteria set out in Annex V(a) (European Commission, 1988c, Article 5(2)). Annex V(a) stated that at least those areas should be classified that met one or both of the following conditions: (a) they are among the ten most important areas in the Community for the conservation of each of the species specified in Annex I or the habitats specified in Annex IV; (b) they are among the two most important areas in a region for the conservation of the species specified in accordance with Annex I or the habitats in Annex IV insofar as that region contained areas of regional significance for those species or that habitat type. If, after the expiry of the initial two-year period, the ten most important areas in the Community for each listed species/habitat had not been classified by Member States, the draft directive proposed giving the Commission the power to identify these areas, with the assistance of a Committee comprising representatives of the Member States and chaired by a Commission representative; these areas would then be classified by the Member States within one year following the decision of the Commission (European Commission, 1988c, Article 5(2)).

As a second stage, the Member States would have eight years to classify as special protection areas territories sufficient in number and size to ensure: (a) the maintenance of the species specified in Annex I at a satisfactory level in all regions where they occur; and (b) the protection of the habitats specified in Annex IV with their associated fauna and flora in all regions where they occur (European Commission, 1988c, Article 5(3)). In classifying these areas, Member States were to apply the criteria set out in Annex V(b), which included – subject to some complex exceptions – at least those areas meeting one or both of the following conditions: (a) that the area was among the one hundred most important areas in the Community for the conservation of each of the species specified in Annex I or the habitats specified in Annex IV; and (b) that the area was among the five most important areas in a region for the conservation of the species specified in accordance with Annex I or the habitat types in Annex IV insofar as that region contained areas of regional significance for those species or that habitat type (European Commission, 1988c, Article 5(3) and Annex V). In respect of these areas to be classified as part of the second stage, the Commission's proposal envisaged a derogation being available from the site selection criteria set out in Annex V(b), with the result that a site would not have to be classified for protection, "if on the basis of scientific evidence and environmental impact assessment the [national] authority competent for nature protection is satisfied that . . . the coherence of [the Natura 2000 network] and its conservation potential for the habitats and species concerned is fully maintained" (European Commission, 1988c, Article 5(5)). The reasons for the derogation would need to be made public and be communicated to the Commission. While a derogation provision along these lines did not make it in to the directive as adopted, the position finally negotiated allows for a more flexible application of the site selection criteria in certain limited circumstances (see Article 4(2) of the Habitats Directive (1992) and Chapter 9).

Areas designated under this first and second stage, together with sites classified under the Birds Directive and the Ramsar Convention,[19] would together form the NATURA 2000 network (European Commission, 1988c, Article 6(1)). Within two years of the Directive's coming into force, the Commission would adopt a Joint Programme for the development and strengthening of the NATURA 2000 network, with assistance from the Committee of Member State representatives mentioned above (European Commission, 1988c, Article 6(2)). The Joint Programme would specify the measures to be taken by the Community and the Member States to back up implementation of the Directive, and would be reviewed at the end of the eight-year period for classifying special protection areas at the second stage (European Commission, 1988c, Article 6(2)).

At the Working Group meeting on 3 February 1989, the Member States made their views on these proposals known (CEC, 1989c). Several delegations were unhappy about the "disproportionate role assigned to the Commission"; in particular, the French and German delegations thought that the list of areas to be classified ought to be determined by the Member States and that the Commission should not be able to "fill in the gaps" during the first stage. Several delegations, in particular Greece, France, Portugal and the UK, argued that the figure of ten areas per annexed species/habitat to be classified during the first two years was arbitrary;[20] or, as the UK's Secretary of State for the Environment Nicholas Ridley put it, "The proposed designation procedure bears all the hallmarks of the 'Gentleman in Brussels knows best' . . . the draft Directive [is] a Daft Directive" (Ridley, 1989). By the end of April 1989, Spain, Portugal and the UK were arguing that they would prefer to select areas according to objective criteria, without specifying a particular number in advance (CEC, 1989a), and Belgium and France proposed that each Member State should draw up a nominal list of areas to be protected on its territory that would then be submitted to the Commission (CEC, 1989a) – both ideas would later take hold.

A turning point: the RSPB's discussion paper of March 1989

On 15 March 1989, at the request of the UK government's Department of Environment, which "seemed to cautiously like the solution",[21] the RSPB produced what turned out to be a very important and influential discussion paper on possible site selection and protection criteria for the draft directive (RSPB, 1989).[22] The paper highlighted problems experienced with the Birds Directive and other conservation instruments:

> in the 1980s it has become clear that these early attempts at habitat conservation contained one major flaw. That was that they did not contain any clear criteria upon which to base action. Thus conservationists habitually asked for more action, whilst Governments generally took less. Lawyers profited! . . . [T]his inexactitude over habitat protection requirements has led at times to bitter disputes between Convention secretariats, Governments and NGOs . . . At the same time it was clear that yet more needed to be done since more and more of Europe's wildlife and wildlife habitats were declining and becoming steadily more threatened. It was in this climate that the Commission came to draft the habitats directive.
>
> *(RSPB, 1989)*

Given the resistance of certain Member States to the Commission's site designation proposals, the RSPB proposed a solution to break the impasse. They did this by posing five questions which, they said, arose in the RSPB's experience in seeking to protect habitats: which habitats/sites should be protected? How many should be protected? What size should they be? What degree of protection should be given? Under what circumstances may exceptions be envisaged? (RSPB, 1989). The draft Habitats Directive only addressed the first two questions, the RSPB said, albeit that the draft "did attempt to allow exceptions (which the Birds Directive for example did not)" (RSPB, 1989). As a central organising principle to answer all five questions, the RSPB insisted on the importance of defining the objective of the envisaged conservation measures: "It is important to define the objective, since it is only in relationship to a defined objective that the efficacy of any action may be judged. If [an] objective is agreed, then the measures needed (either for habitat conservation or other measures) are those that will achieve the objective" (RSPB, 1989). As a suggested precedent for such an objective, the RPSB's paper attached in annex an amended version of the concept of favourable conservation status, as defined in Article I(1)(c) and (d) of the Bonn Convention on the Conservation of Migratory Species of Wild Animals (RSPB, 1989). Favourable conservation status should be established as an objective in the Habitats Directive, the RSPB argued, and all actions should be judged against how they would meet this objective (RSPB, 1989).

Thus, while the UK press continued to report the government's "uncompromisingly hostile" attitude to the proposed Habitats Directive, which was reportedly being opposed "as another unwanted extension of EEC law into British life" (McCarthy, 1989), the Department of the Environment's request to the RSPB to prepare this important discussion paper suggests that attitudes were slowly beginning to change. This was likely attributable to a changing political climate, with green issues coming very much to the fore: e.g., just a few days after the RSPB's paper was written, the *New York Times* could be found reporting that "Across Western Europe, mainstream politicians are scrambling to embrace the cause of the endangered environment as small ecological parties register startling electoral triumphs", causing Margaret Thatcher and French President François Mitterrand to perform "sharp U-turns . . . to put environmental concerns at the top of their policy agendas" (Markham, 1989). Similarly, support for the environment was on the rise in Germany, with the Greens growing in support to comprise, by April 1989, the country's third political force, thus holding the balance between the two big parties (De Bresson, 1989).

Conservation NGOs continued to apply pressure in the run up to the Environment Council meeting of 5 June 1989, with the WWF's national organisations in Belgium, Denmark, Germany, Italy, the Netherlands, Spain, the UK and France writing a joint letter to heads of state on 12 May 1989, expressing their unanimous support for the adoption of a strong Habitats Directive, which "was of the highest priority for nature conservation in the European Community" (WWF, 1989). Such efforts coincided with significant political changes in the UK, with Labour sweeping to victory in the European elections in mid-June 1989, capturing ten seats from the Tories; the Greens also made a "spectacular advance" in the UK, picking up nearly two million votes (Oakley, 1989), or 15% of the poll, the highest Green vote, proportionately, in Europe (*Financial Times*, 1989). Amongst the recommendations made to Thatcher to halt the slide were the following: "Signal a more constructive and accommodating approach to Europe" and "Find re-employment for . . . Environment Secretary, Nicholas Ridley" (Jenkins, 1989). The UK's Wildlife Link NGOs immediately capitalised on these developments by writing to Lord Caithness highlighting that it would be "unacceptable to the NGOs if the Government approached negotiations on the [Habitats] Directive on the basis that it would resist any EC measures which would require the UK to strengthen its habitat protection legislation" (Wildlife Link, 1989).

Protection of Natura 2000 sites

Under Article 7 of the Commission's original proposal to the Council of August 1988, a protective provision that mirrored Article 4(4) of the Birds Directive would apply to sites classified under the Habitats Directive: "[In respect of special protection areas,] Member States shall take appropriate steps to avoid pollution or deterioration of habitats or any disturbances affecting fauna and flora, insofar as these would have significant effects, having regard to the objectives of this Directive, on the species or types of habitats for which the area has been classified" (European Commission, 1988c, Article 7(1)). In addition, Member States "shall consider in particular the appropriate protection status to be given to [special protection areas], and the setting up of integrated management plans in accordance with the ecological needs of the species and types of habitat concerned" (European Commission, 1988c, Article 7(2)). Whilst not very clearly expressed in the draft directive, the Commission's idea was that special protection areas under the directive would not all receive identical levels of protection. The Explanatory Memorandum accompanying the directive proposal makes the point explicit:

> Biotopes of major importance for the conservation of nature in the Community cover a surface area which varies between approximately 10% and 20% of the total surface area of the Member States . . . [but] only between 1 and 2% of the total surface area of the Member States will need strict measures of protection.[23] The major part of the surface area of the biotopes will need management plans reconciling appropriate measures of protection and extensive agricultural and forestry exploitation based on adapted techniques.
>
> *(European Commission, 1988c, p. 3)*

Hence the references in Article 7(2) of the proposal to Member States considering the "appropriate" protection status to be given to sites and the setting up of management plans.

Interestingly, the RSPB's March 1989 discussion paper – which had proposed the objective of favourable conservation status as the organising principle for the Habitats Directive – drew the following conclusions regarding site protection, which very much resonated with the Commission's approach in this regard:[24]

> That the protection appropriate to a habitat will be a case by case decision and must be related to the conservation of the interest of the site and hence the Objective . . . Thus in a theoretical sense we can envisage a range from small strict nature reserves; through larger reserves where certain activities continue; through the ESI [Environmental Sensitivity Index] concept where the objective is to maintain certain features of interest; to the landscape as a whole where the retention of certain features is desirable 'landscape features'. All have a role to play. Small strict reserves may be appropriate to protect certain specialist plants, but not otters; larger reserves may protect certain wetland birds, but again not otters; ESIs may protect even larger populations of birds but still not otters and finally landscape features to protect those features essential for wide-ranging species such as Golden eagle or otter as well as to provide stepping stones to prevent isolation of 'island' reserves. These have nothing in common save they all share the same objective, to restore and maintain a satisfactory conservation status for the species concerned.
>
> *(RSPB, 1989)*

In respect of the possibility of derogations from habitat conservation measures, the RSPB had an equally clear position:

> If the objective is explicit, i.e. to restore and maintain a satisfactory conservation status for the species concerned, then exceptions may be allowed which do not compromise this objective . . . [and thus] That derogations can be foreseen in principle, but only when other satisfactory alternatives are impossible and when it is in the 'Overriding national interest', only when alternative substitution is not possible or where the exception does not compromise the Objective.
>
> *(RSPB, 1989)*[25]

Here we effectively find the forerunner to the Article 6(4) derogation that would later be inserted into the text of the Habitats Directive.

The Member States were not so keen on the idea of differential standards of protection, however, which idea was further clarified as negotiations continued. By April 1989 it was clear that what the Commission had in mind were two categories of sites, linked to the proposed two-stage designation process: "The Commission proposal provides for the gradual establishment of a network of protected areas made up in the initial phase of a core of 10 areas which would be subject to very strict protective measures. In a second phase, this network would be supplemented by a larger number of areas where there would be less stringent measures, giving due weight to both socio-economic and environmental considerations" (CEC, 1989a, p. 3). Germany, Spain and Italy were in favour of this type of division, but others – Belgium, France, Ireland and the UK – expressed doubts (CEC, 1989a). France proposed a single network of protected areas, which it thought would be easier to manage and more consistent (CEC, 1989a) – and with France about to take over the Presidency this would prove crucial. Similarly, Ireland and the UK proposed adopting the same approach for all areas (most likely fearing the "very strict" protection proposed, and aiming for a single, less stringent system), based on socio-economic considerations and environmental protection objectives (CEC, 1989a).

In addition to the protective provisions just discussed, the Commission's directive proposal envisaged amending the Environmental Impact Assessment (EIA) Directive (1985) to provide that "all projects which are located in or are likely to affect the conservation potential of a special protection area forming part of the [NATURA 2000 network] shall be made subject to an assessment in accordance with Articles 5 to 10 [of the EIA Directive].[26] All development plans or programmes likely to affect special protection areas shall be made subject to an equivalent assessment" (European Commission, 1988c, Article 11).[27] The EIA Directive was to be further amended to read that "Member States shall ensure that the authorities responsible for the protection and management of special protection areas, and other relevant bodies likely to be concerned by the project, development plan or programme, are closely associated in the assessment procedures and that their final opinion is made public" (European Commission, 1988c, Article 11). In other words, at this stage in the directive's progress there was no standalone impact assessment regime in the draft Habitats Directive – the idea was to rely on the provisions of the EIA Directive. The idea of a dedicated impact assessment process for Natura 2000 came later, under the French Presidency – see Chapter 8.

Discussing this issue at the Working Party on the Environment's meeting on 3 February 1989, some delegations – in particular Belgium, Germany, Spain and Ireland – were not opposed to compulsory impact studies, but did not consider the proposed amendment of the EIA Directive to be desirable (CEC, 1989c). Ireland argued that since Article 7 of the draft contained an obligation for Member States to take appropriate steps to avoid deterioration of habitats, etc., the provisions on EIA were superfluous and should be deleted (CEC, 1989c). In March 1989 the Commission attempted to find a compromise by proposing that the EIA procedure would not apply if the national competent authorities felt on the basis of a preliminary examination that the project concerned would have no harmful effect on the environment (CEC, 1989b). However, by the end of April 1989 the number of delegations that did not regard an amendment to the EIA Directive to be desirable had grown to include: Belgium, Germany, Spain, Ireland, the Netherlands and Portugal (CEC, 1989a). When the French took over the Presidency of the Council in July 1989, one of the many major changes they made to the draft directive was to remove the idea of amending the EIA Directive altogether (DG XI, 1989a).

Notes

1 The published version of the Commission proposal is dated 31 August 1988 (European Commission, 1988a), but the proposal appears to have been submitted to the Council on 16 August (European Commission, 1988c).

2 Participants in the negotiations describe Claus Stuffmann variously as: "very clever" and "very important during the negotiations" (Goldberg, 2012); a "wonderful person" and "very helpful" to certain countries that had only just joined the EU at the time of the negotiations (Morillo, 2012); a "very genial" man and "quite wily", who could "charm you with his chat" and "win points very subtly by being charming" (Neal, 2012); someone with "a strong political sense" (Cashman, 2012); "very personable and intelligent . . . [he] could reach a compromise and [was] a man you could do business with, I found, all the way through" (Bunce, 2012). One interviewee reflected that Stuffmann should be "awarded a prize for European diplomacy" (Jouve, 2013); another said Stuffmann was "the right man" for the Habitats Directive, and that he "deserves a MEDAL" from Europe (Megret, 2013a).

3 Yourassowsky (2015) comments "From 1989 to the adoption of the Habitats Directive, three persons working at the DG XI's Nature Unit were involved on the Directive preparation: Claus Stuffmann, Henriette Bastrup-Birk and myself (as scientific expert). I was the "right-hand man" of Henriette Bastrup-Birk. I was in charge of the scientific aspects of the Directive but I also worked with her deeply on the text modifications during this very long negotiation phase. I participated to all Council meetings with the Member States and I reviewed with Henriette different articles taking into account the Member States' remarks, in close coordination with Claus Stuffmann. I did the same work on the annexes of the Directive, redrafting them in relationship with the comments of the scientific working groups and the delegations, while trying to find the best compromises. I contributed in particular to find a method and criteria for the Annex III. Geert Raeymaekers helped us, providing scientific expertise and coordination on plant species and habitats lists of the annexes".

4 These were the draft Annexes that the Commission as a whole had failed to approve, following the intervention of Commission President Delors' *cabinet* (see Chapter 6).

5 Johnson's affiliation is given as "Vice Chair of Wildlife Link/EC".

6 Hepburn describes this as being two days per week, to provide some replacement cover for a key member of the Nature Unit who was on maternity leave (Hepburn, 2012) from March to September 1990.

7 For example, Keulartz (2009, p. 446) states that, "The Habitats Directive was adopted in 1992, while the EU was under Dutch Presidency, and has been very much perceived as a 'Dutch Directive' by many officials in the Netherlands and abroad".

8 These were difficult times for UK Eurosceptics: just four months earlier, Commission President Delors had made his famous remark that, "Ten years hence, 80% of our economic legislation, and perhaps even our fiscal and social legislation as well, will be of Community origin" (Delors, 1988).

9 Alistair Gammell, then of the RSPB, confirms that he played a role in the drafting process, though he reports that the directive was "primarily" drafted by Stanley Johnson (Gammell, 2012).

10 The UK's silence during early negotiations was attributable to Secretary of State Nicholas Ridley's attitude, coupled with the fact that the UK had at the time "the most high profile and politically campaigning NGOs", leading the UK to be characterised as the main obstacle to progressing the Directive, Bunce (2013) explains. He continues, "Knowing this, other member states tended to keep their heads down and wait for us to speak against the Commission's proposals, knowing that our remarks would be leaked, and simply say 'We agree the UK's position' leaving us to take the flak. To counter this I was advised by [the UK's Permanent Representation in Brussels] to keep quiet for as long as possible and force the others to voice their own objections. Hence our relatively low profile during the early part of the negotiations. Fostering a situation where no one

was prepared to take a strong lead fitted well with our strategy of keeping the draft directive in the long grass" (Bunce, 2013).

11 Contrast this with the UK government's description of the directive receiving "a generally hostile reception when it was discussed in Brussels" (Department of the Environment, 1989, paragraph 7), which chimes with the senior French negotiator's recollection: "the newly proposed text was not very welcomed by most of the delegations of the member states" (Megret, 2013b).

12 It is worth noting that this approach did not have the support of the Secretariat of the Bern Convention, which favoured the cause of nature conservation over politics and inter-institutional competition (Fernández-Galiano, 2013).

13 A brief history of this Resolution is recorded by the Nature Conservancy Council (1989, paragraph 19).

14 The Secretariat of the Bern Convention was in no hurry to progress the task of developing the list of "endangered natural habitats requiring specific conservation measures" and the list of "species requiring specific habitat conservation measures" (Standing Committee, 1989a), on the basis that the cause of nature conservation was more important than politics and inter-institutional competition, and in any event because it would be better to await the outcome of developments regarding the Habitats Directive (Fernández-Galiano, 2013). In the end, the identification of those endangered natural habitats and species happened in 1996 and 1998, when Resolution 1 (1989) was used to extend a Natura 2000-type protected area system (the Emerald Network) to the whole of Europe (Fernández-Galiano, 2013; Standing Committee, 1989d).

15 It appeared to the Secretariat to the Convention that some countries used the occasion to seek to dilute their responsibilities in terms of habitat protection under the Bern Convention (Fernández-Galiano, 2013). In other words, at the meeting in June 1989, certain countries were not only trying to stop an EU legislative instrument on habitats, they were also seeking to "decaffeinate" the Bern Convention by adopting a restrictive interpretation of Article 4, which would thus only apply in practical terms to sites "identified by the Standing Committee" (i.e., the governments) as worthy of such protection: i.e., sites later included in the Emerald Network (Standing Committee, 1989d). Article 4 of the Convention is of course much broader than this.

16 Switzerland, Belarus and Ukraine have officially adopted Emerald Network sites (Standing Committee, 2016a); in addition, Albania, Armenia, Azerbaijan, Bosnia and Herzegovina, the former Yugoslav Republic of Macedonia, Georgia, Montenegro, Norway, Republic of Moldova, Russian Federation and Serbia have officially nominated candidate Emerald Network sites (Standing Committee, 2016b). While the EU's Natura 2000 sites have not formally become part of the Emerald Network in this way, the Bern Convention's website (undated) states that "Natura 2000 sites are ... considered as the contribution from the EU member states to the Emerald Network". This appears to be based on a provision from the preamble of the Rules for the Emerald Network, which states "Considering that for Contracting Parties which are Member States of the European Union Emerald Network sites are those of the Natura 2000 Network. Thus the procedures established by European Council Directives 79/409/EEC and 92/43/EEC will be the only rules to apply" (Standing Committee, 1998). After the Brexit vote in the UK, it remains an open question whether the UK's Natura 2000 sites might become (or perhaps remain) part of the Emerald Network. Two parliamentary questions have been put down in the UK on this subject – by Caroline Lucas (Green Party) on 11 July 2016 and by Madeleine Moon (Labour) on 1 November 2016 – but the government gave non-committal responses (Lucas and Stewart, 2016; Moon and Coffey, 2016).

17 In November 1989, just a few months after these Resolutions and Recommendations were adopted by the Standing Committee of the Bern Convention, the Berlin wall fell, and the whole mission of the Council of Europe and the Bern Convention changed (Fernández-Galiano, 2013). The Secretariat to the Bern Convention could then defend without objections that its priority was to work with the new democracies of Central and Eastern Europe, and not (at that time) on habitat conservation in Bern Convention states (Fernández-Galiano, 2013).

18 In February 1989 DG XI put out a press release setting out the new Commissioner's vision for EU environmental policy: "Particular attention will be given to urban environmental problems", the document stated, citing "the rapid deterioration of living conditions together with the dangers of a steady deterioration in the natural and architectural heritage of our cities" (European Commission, 1989).

19 Unless notice was given to the contrary in respect of sites under the Ramsar Convention (European Commission, 1988c, Article 6(1)).

20 The Commission would later note that "the figures indicated in [our] original proposal . . . were taken from the criteria guiding the selection of biotopes of Community importance in the Corine/biotope data base" (DG XI, 1989c).

21 Noted by the RSPB on the covering note sent to DG XI's Nature Unit, attaching the discussion paper.

22 The RSPB (1989) also met with the French government at this time, which was due to take over the Presidency of the Council in July 1989, and the RSPB was due to be "making some proposals to them in the next few weeks".

23 It is important to highlight that the Explanatory Memorandum did not envisage 1–2% representing the totality of the site protection network; rather, 1–2% would be in the form of strict nature reserves, while other sites would need management plans to reconcile land uses and conservation needs.

24 The RSPB's discussion paper was dated March 1989 while the draft directive was earlier, having been proposed to the Council in August 1988. However, Gammell (2012) had assisted Stanley Johnson in producing the draft directive, such that a meeting of minds on the subject of site protection is perhaps unsurprising.

25 As Gammell (2012) explains, "I never shied away from clauses like that, because it would seem to me to be, let's say, extremism to assume that nature conservation, or indeed anything else in the world, should absolutely 100% trump everything else . . . [Make] those people who wanted to exercise those clauses demonstrate the real importance and lack of alternatives, but if they could demonstrate the importance and lack of alternatives, then as I say it would seem an unreasonable expectation that society as a whole would still be denied those extremely important things that couldn't go anywhere else".

26 This would have meant that even projects listed in Annex II to the EIA Directive – which projects are not subject to mandatory assessment under the EIA Directive – would have been subject to mandatory assessment where the project was located in or likely to affect the conservation potential of a special protection area. The Commission revisited an idea along similar lines in 1994, in proposing amendments to the EIA Directive, suggesting that Annex II EIA Directive projects should be subject to mandatory assessment under the EIA Directive where "they are liable to have a significant effect on the special protection areas designated by Member States pursuant to Community law" (European Commission, 1994, Article 1(2)). However, the Member States rejected this idea, recording their conclusion as follows: "Whereas the existence of a location criterion referring to special protection areas designated by Member States pursuant to Council Directive 79/409/EEC of 2 April 1979 on the conservation of wild birds (6) and 92/43/EEC of 21 May 1992 on the conservation of natural habitats and of wild fauna and flora (7) does not imply necessarily that projects in those areas are to be automatically subject to an assessment under this Directive" (CEU, 1997, recital 10).

27 Thus requiring the assessment of projects, plans and programmes, which would have represented a significant extension of the position under the Environmental Impact Assessment (EIA) Directive (1985), which related only to projects.

References

ACE Regulation 1984. Council Regulation (EEC) No 1872/84 of 28 June 1984 on action by the Community relating to the environment. *OJ L 176*, 3 July 1984, p. 1.

Ardill, J. and Linton, M. 1988. Cautious greeting to Thatcher's 'greening'. *The Guardian*, 29 September.

Bern Convention undated. *Emerald Network of Areas of Special Conservation Interest* [Online]. Available: www.coe.int/en/web/bern-convention/emerald-network [accessed 21 June 2017].

Brooke, M. 1989. A birthday for the birds. *New Scientist* (4 February), 54–57.

Bunce, R. 2012. Interview with Roy Bunce, senior negotiator for the UK government in respect of the Habitats Directive, conducted by Andrew L. R. Jackson on 13 August 2012.

Bunce, R. 2013. Personal communication: e-mail dated 13 April 2013 from Roy Bunce, senior negotiator for the UK government in respect of the Habitats Directive, to Andrew L. R. Jackson.

Caithness, L. 1988. Extract from speech given on 18 November 1988, cited by Lord Moran in Habitat and Species Protection: ECC Report. House of Lords Debate, 14 November 1989, volume 512, cc1248-79.

Cashman, L. 2012. Interview with Liam Cashman, DG Environment lawyer, conducted by Andrew L. R. Jackson on 17 August 2012.

CEC 1988a. Outcome of proceedings from Working Party on the Environment on 23 September 1988. Subject: Proposal for a Council Directive on the protection of natural and semi-natural habitats and of wild fauna and flora. 8331/88, Brussels, 30 September 1988. Council of the European Communities.

CEC 1988b. Outcome of proceedings of Environment Council on 24 November 1988. Subject: Proposal for a Council Directive on the protection of natural and semi-natural habitats and of wild fauna and flora. Brussels, 14 December 1988, 9948/88, ENV 212. Council of the European Communities.

CEC 1988c. Report from General Secretariat to Permanent Representatives Committee. Subject: Preparation of the Environment Council meeting of 24 November 1988. Brussels, 7 November 1988. 9099/88. ENV 179. Council of the European Communities.

CEC 1988d. Report from Permanent Representatives Committee to Council. Subject: Proposal for a Council Directive on the protection of natural and semi-natural habitats and of wild fauna and flora. Brussels, 14 November 1988. 9380/88. ENV 189. Council of the European Communities.

CEC 1988e. Resultats des travaux du Groupe "Environnement" en date du 8 Septembre 1988. Objet: Proposition de directive du Conseil concernant la protection des habitats naturels et semi-naturels ainsi que de la faune et de la flore sauvages. 8078/88, ENV 140, Bruxelles, le 13 Septembre 1988. Council of the European Communities.

CEC 1989a. Introductory note from General Secretariat to Permanent Representatives Committee. Subject: Preparation for the Environment Council meeting on 8 June 1989. Proposal for a Council Directive on the protection of natural and semi-natural habitats and of wild fauna and flora. Brussels, 28 April 1989, 6284/89, ENV 74. Council of the European Communities.

CEC 1989b. Outcome of proceedings of Working Party on the Environment on 16 March 1989. Subject: Proposal for a Council Directive on the protection of natural and semi-natural habitats and of wild fauna and flora. Brussels, 6 April 1989, 5798/89, ENV 58. Council of the European Communities.

CEC 1989c. Outcome of the Working Party on the Environment on 3 February 1989. Brussels, 28 February 1989, 4708/89, ENV 25. Council of the European Communities.

CEU 1997. Council Directive 97/11/EC of 3 March 1997 amending Directive 85/337/EEC on the assessment of the effects of certain public and private projects on the environment. *OJ L 73*, 14 March 1997, p. 5.

Clinton Davis, S. 1988. Extract from the European Parliament's Intergroup on Animal Welfare bulletin, 52nd session, 13 December 1988.

Coffey, C. and Richartz, S. 2003. *The EU Habitats Directive: generating strong responses*. Berlin: Institute for European Environmental Policy.

Council of Europe 1988. Report of the 7th meeting of the Standing Committee to the Bern Convention, held in Strasbourg from 6–9 December 1988. Strasbourg, 15 December 1988, T-PVS (88) 45.

David, H. 2012. Interview with Hubert David, founder and first Secretary General of the European Environmental Bureau, conducted by Andrew L. R. Jackson on 3 August 2012.

De Bresson, H. 1989. Differing shades of green. *Guardian Weekly*, 2 April.

Delors, J. 1988. Speech by Jacques Delors before the European Parliament. Debates of the European Parliament, No. 2-367/137, 6 July 1988.

Department of the Environment 1989. Memorandum by the Department of the Environment dated 23 January 1989, headed "House of Lords European Communities Committee: Sub Committee F (Environment). Enquiry into habitat and species protection".

DG XI 1988a. DG XI file note dated 9 November 1988. Headed "Note de dossier. Object: Mission à Madrid le 7/8 novembre 1988. Discussion aves les autorités espagnoles sur la proposition de directive FFH".

DG XI 1988b. DG XI file note dated 9 September 1988. Headed "Présentation au sein du group ENVIRONNEMENT du conseil du projet de directive "Habitats" – 8.9.88".

DG XI 1988c. DG XI file note dated 21 November 1988. Headed "Mission report, London, 17 November, 1988. Discussions with the Department of the Environment on the impacts of the 'habitats' directive".

DG XI 1988d. DG XI file note headed "Reunion des hauts fonctionnaires –'Protection de la nature' – 29 September, 1988" – with handwritten amendments.

DG XI 1988e. Draft headed "Speaking notes a l'attention de M. Brinkhorst. Object: reunion de hauts fonctionnaires 'protection de la nature' le 29 Septembre 1988 de 11h30 a 13h00".

DG XI 1988f. File note dated 28 September 1988 headed "Examen 'Directive Habitats' au COREPER – 23/9/1988".

DG XI 1988g. File note headed "Principales conclusions des discussions au sein du Groupe ENV. – 6/10/88. Directive Habitat".

DG XI 1988h. Internal DG XI letter dated 25 October 1988. Headed "Rapport de mission: Object: Mission à la Haye le 24/10/88. Discussion sur la proposition de directive habitats naturels (FFH)".

DG XI 1989a. DG XI (unit B3) file note dated 7 July 1989, attaching a draft version of the habitats directive dated March 1989, with the handwritten annotation "FFH: amendements français".

DG XI 1989b. DG XI (unit B3) file note dated 23 January 1989, headed "Object: Entretien du Commissaire Ripa di Meana avec le Président du Conseil Environnement le 24.1.89".

DG XI 1989c. Document headed "Speaking note" – for responding to Hemmo Muntingh's report for the European Parliament on the Habitats Directive proposal. With handwritten annotation at the top: "13.10.89 Parlement".

Environmental Impact Assessment (EIA) Directive 1985. Council Directive 85/337/EEC of 27 June 1985 on the assessment of the effects of certain public and private projects on the environment, *OJ L 175*, 5 July 1985, p. 40, since codified as Directive 2011/92/EU of the European Parliament and of the Council of 13 December 2011 on the assessment of the effects of certain public and private projects on the environment (codification), *OJ L 26*, 28 January 2012, p. 1. Codified version since amended by Directive 2014/52/EU, *OJ L 124*, 25 April 2014, p. 1.

European Commission 1979. *State of the environment: second report.* Brussels – Luxembourg, ECSC – EEC – EAEC.

European Commission 1980. Progress made in connection with the environment action programme and assessment of the work done to implement it. Communication from the Commission to the Council. COM(80) 222 final, Brussels, 7 May 1980.

European Commission 1988a. Commission of the European Communities proposal for a Council Directive on the protection of natural and semi-natural habitats and of wild fauna and flora. COM(88) 381 final, Brussels, 31 August 1988.

European Commission 1988b. Responsibilities of the members of the new Commission. Press release IP/88/825, 16 December 1988.

European Commission 1988c. Translation of letter dated 16 August 1988 from the Commission to the President of the Council, attaching Habitats Directive proposal COM(88) 381 final. Council document 8149/88, ENV 143, Brussels, 16 September 1988.

European Commission 1989. The general outlines of a Community policy for protecting the environment. IP/89/54, 7 February 1989.

European Commission 1994. Proposal for a Council Directive amending Directive 85/337/EEC on the assessment of the effects of certain public and private projects on the environment. COM(93) 575 final. Submitted by the Commission on 21 April 1994. *OJ C 130*, 12 May 1994, p. 8.

European Parliament 1988a. Report, dated 26 September 1988, on the implementation in the EC of the Community Regulation on the Washington Convention (Convention on International Trade in Endangered Species of Wild Fauna and Flora). Drawn up on behalf of the Committee on the Environment, Public Health and Consumer Protection. Rapporteur: Mr H. J. Muntingh. Series A, document A 2-0180/88.

European Parliament 1988b. Report, dated 26 September 1988, on the implementation of the Berne Convention (on the conservation of European wildlife and natural habitats) and the Bonn Convention (on the conservation of migratory species of wild animals) in the European Community. Drawn up on behalf of the Committee on the Environment, Public Health and Consumer Protection. Rapporteur: Mr H. J. Muntingh. Series A, document A 2-0179/88.

European Parliament 1988c. Report, dated 26 September 1988, on the implementation of the Birds Directive in the EEC. Drawn up on behalf of the Committee on the Environment, Public Health and Consumer Protection. Rapporteur: Mr H. J. Muntingh. Series A, document A 2-0181/88.

European Parliament 1989. Report, dated 29 September 1989, on the proposal for a Community directive on the protection of natural and semi-natural habitats and of wild fauna and flora (COM(88) 381 final – Doc. C 3-34/89). Drawn up on behalf of the Committee on the Environment, Public Health and Consumer Protection. Rapporteur: Mr H. J. Muntingh. Series A, document A 3-0039/89.

FACE 1989. Memorandum by the Federation of Hunting Associations of the EEC (FACE), dated 26 January 1989, headed "Preliminary comments and observations on the proposal for a Council Directive 'on the protection of natural and semi-natural habitats and of wild fauna and flora' COM(88) 381 final".

Fernández-Galiano, E. 2012. Interview with Eladio Fernández-Galiano, then head of the Secretariat of the Bern Convention, conducted by Andrew L. R. Jackson on 30 August 2012.

Fernández-Galiano, E. 2013. Personal communication: e-mail dated 3 April 2013 from Eladio Fernandez-Galiano, then head of the Secretariat of the Bern Convention, to Andrew L. R. Jackson.

Financial Times 1989. Little green waves of discontent (6 July). *FT Energy Newsletters – Power Europe*.

Forsyth, V. 1988. Anti-nuke protester joins the establishment. *Sunday Mail*, 18 September.

Fourth Environmental Action Programme 1987. Resolution of the Council of the European Communities and of the Representatives of the Governments of the Member States meeting within the Council of 19 October 1987 on the continuation and implementation of a European Community policy and action programme on the environment (1987–1992). *OJ C 328*, 7 December 1987, p. 1.

Gammell, A. 2012. Interview with Alistair Gammell, formerly of the RSPB, conducted by Andrew L. R. Jackson on 31 July 2012.

Goldberg, C. 2012. Interview with Claus Goldberg, Danish government official who participated in the Habitats Directive negotiations, conducted by Andrew L. R. Jackson on 31 August 2012.

Habitats Directive 1992. Consolidated version of Council Directive 92/43/EEC of 21 May 1992 on the conservation of natural habitats and of wild fauna and flora. *OJ L 206*, 22 July 1992, p. 7. Consolidation date 1 July 2013.

Henningsen, J. 2012. Interview with Jørgen Henningsen, former Director at DG Environment, conducted by Andrew L. R. Jackson on 24 August 2012.

Hepburn, I. 2012. Interview with Ian Hepburn, English RSPB/WWF campaigner who headed an EU-wide NGO campaign in support of the Habitats Directive during the instrument's negotiation and was seconded to DG XI's Nature Unit for a period during the negotiations, conducted by Andrew L. R. Jackson on 16 August 2012.

House of Lords 1988. Letter from the European Communities Commitee of the House of Lords, dated 14 December 1988, headed "European Communities Committee Sub-Committee F (Environment). Enquiry into habitat and species protection", inviting views and evidence.

House of Lords 1989. House of Lords' Select Committee on the European Communities. 15th report: Habitat and Species Protection. Ordered to be printed 18 July 1989.

Huw Davies, G. 1988. Halting the bulldozers; EEC laws to protect the landscape. *The Times*, 29 December.

Jaffeux, H. 2013. Personal communication: e-mail dated 18 April 2013 from Henri Jaffeux, ex French government official who worked on implementing the Habitats Directive, to Andrew L. R. Jackson.

Jenkins, P. 1989. Lessons of defeat for Mrs Thatcher. *The Independent*, 20 June.

Johnson, S. 2012. Interview with Stanley Johnson, former European Commission official and former MEP, conducted by Andrew L. R. Jackson on 16 July 2012.

Jouve, M. 2013. Personal communication: e-mail dated 8 April 2013 from Marcel Jouve, ex head of the International and European Affairs division of the Nature Protection Directorate at France's Department of the Environment, to Andrew L. R. Jackson.

Keulartz, J. 2009. European nature conservation and restoration policy—problems and perspectives. *Restoration Ecology*, 17, 446–450.

Klein, R. 2012. Interview with Reinhard Klein, former member of DG XI's Nature Unit, conducted by Andrew L. R. Jackson on 31 August 2012.

Koester, V. 2004. The Bern Convention and I. *Naturopa*, 101, 5.

Koester, V. 2012. Interview with Veit Koester, senior negotiator for the Danish government in respect of the Habitats Directive, conducted by Andrew L. R. Jackson on 22 August 2012.

Lalonde, B. 2012. Personal communication: e-mail dated 31 August 2012 from Brice Lalonde, ex-Minister for the Environment of France, to Andrew L. R. Jackson.

Lasén-Díaz, C. 2010. The Bern Convention: 30 years of nature conservation in Europe. *Review of European Community and International Environmental Law*, 19, 185–196.

Leistra, G., Keulartz, J. and Engelen, E. 2008. Wintering geese in the Netherlands: legitimate policy? *In:* Keulartz, J. and Leistra, G. (eds) *Legitimacy in European nature conservation policy: case studies in multilevel governance*. The Netherlands: Springer.

Liefferink, J. D. and Zouwen, M. W. V. D. 2004. The Netherlands: the advantages of being "Mr. Average". *In:* Jordan, A. and Liefferink, D. (eds) *Environmental policy in Europe: the Europeanization of national environmental policy*. London: Routledge.

Lucas, C. and Stewart, R. 2016. Written question 42311 asked in the UK House of Commons on 11 July 2016 by Caroline Lucas MP. Answered by Rory Stewart MP on 14 July 2016.

Markham, J. M. 1989. Europe's politicians thinking green as concern grows on environment. *The New York Times*, 12 April.

McCarthy, M. 1989. Support for countryside law. British hostility to EEC wildlife plan under fire. *The Times*, 28 March.

Megret, A. 2013a. Personal communication: e-mail dated 23 March 2013 from Alain Megret, deputy director of the Nature Protection directorate of the French government's Department of the Environment at the time of the Habitats Directive negotiations, to Andrew L. R. Jackson.

Megret, A. 2013b. Personal communication: Written comments from Alain Megret, deputy director of the Nature Protection directorate of the French government's Department of the Environment at the time of the Habitats Directive negotiations, attached to e-mail dated 4 April 2013 from Alain Megret to Andrew L. R. Jackson.

Meyer, J. H. 2010. Greening Europe? Environmental interest groups and the Europeanization of a new policy field. *In:* Kaiser, W. and Meyer, J. H. (eds) *Non-State actors in European integration in the 1970s: towards a polity of transnational contestation*. Leipzig: Leipziger Univ.-Verl.

Moon, M. and Coffey, T. 2016. Written question 51347 asked in the UK House of Commons on 1 November 2016 by Madeleine Moon MP. Answered by Thérèse Coffey MP on 7 November 2016.

Morillo, C. 2012. Interview with Cosme Morillo, Spain's senior negotiator in respect of the Habitats Directive, conducted by Andrew L. R. Jackson on 23 August 2012.

Nature Conservancy Council 1989. Memorandum by the Nature Conservancy Council dated 24 January 1989, headed "House of Lords European Communities Committee: Sub Committee F (Environment). Enquiry into habitat and species protection".

Neal, H. 2012. Interview with Hilary Neal, UK government official who participated in the Habitats Directive negotiations and early implementation, conducted by Andrew L. R. Jackson on 14 August 2012.

Oakley, R. 1989. Labour sweeps to Euro win; European election results. *The Times*, 19 June.

Phillips, J. 1988. Prime minister says new elections likely soon. *United Press International*, 13 May.

PR Newswire Europe 1988. Environmental protection at home an urgent requirement. *PR Newswire Europe* 28 September.

Raeymaekers, G. 2012. Interview with Geert Raeymaekers, environmental consultant working with DG XI during the Habitats Directive negotiations, conducted by Andrew L. R. Jackson on 25 September 2012.

Ridley, N. 1989. Extract from a speech made by Nicholas Ridley, Secretary of State for the Environment, at the opening of the RSPB Centenary Reserve on 10 May 1989. Attached to a Wildlife Link note to its EC Habitats Working Group dated 29 June 1989.

Rose, C. 1988. Europe leads the way. *Natural World*, Winter edition.

RSPB 1989. Note dated 15 March 1989, prepared by the RSPB at the request of the UK government's Department of Environment, with a handwritten, undated covering note to DG XI.

Standing Committee 1989a. Standing Committee of the Bern Convention on the conservation of European wildlife and natural habitats. Recommendation no. 14 (1989) of the standing committee on species habitat conservation and on the conservation of endangered natural habitats. Adopted by the Standing Committee on 9 June 1989.

Standing Committee 1989b. Standing Committee of the Bern Convention on the conservation of European wildlife and natural habitats. Recommendation no. 15 (1989) of the standing committee on the conservation of endangered natural habitats types. Adopted by the Standing Committee on 9 June 1989.

Standing Committee 1989c. Standing Committee of the Bern Convention on the conservation of European wildlife and natural habitats. Recommendation no. 16 (1989) of the standing committee on areas of special conservation interest. Adopted by the Standing Committee on 9 June 1989.

Standing Committee 1989d. Standing Committee of the Bern Convention on the conservation of European wildlife and natural habitats. Resolution no. 1 (1989) of the standing committee on the provisions relating to the conservation of habitats. Adopted by the Standing Committee on 9 June 1989.

Standing Committee 1998. Standing Committee of the Bern Convention on the conservation of European wildlife and natural habitats. Resolution no. 5 (1998) concerning the rules for the Network of Areas of Special Conservation Interest (Emerald Network). Adopted by the Standing Committee on 4 December 1998.

Standing Committee 2016a. Standing Committee of the Bern Convention on the conservation of European wildlife and natural habitats – 36th meeting of the Standing Committee, Strasbourg, 15–18 November 2016. Updated list of officially adopted Emerald sites (October 2016), T-PVS/PA (2016) 11. T-PVS/PA (2016) 12.

Standing Committee 2016b. Standing Committee of the Bern Convention on the conservation of European wildlife and natural habitats – 36th meeting of the Standing Committee, Strasbourg, 15–18 November 2016. Updated list of officially nominated candidate Emerald sites (October 2016), T-PVS/PA (2016) 11.

Thatcher, M. 1988. Speech to the Royal Society, Fishmongers' Hall, City of London, 27 September 1988.

Travis, A. 1989. 'Miserable' Ridley attacked over green record in Europe. *The Guardian*, 13 July.

Troya, A. 2012. Interview with Antonio Troya, Spanish government official seconded to the European Commission for the Habitats Directive negotiations, conducted by Andrew L. R. Jackson on 5 October 2012.

Verschuuren, J. 2015. Connectivity: is Natura 2000 only an ecological network on paper? *In:* Born, C.-H., Cliquet, A., Schoukens, H., Misonne, D. and Van Hoorick, G. (eds) *The Habitats Directive in its EU environmental law context*. Abingdon: Routledge.

Wildlife Link 1988a. Memorandum dated 9 November 1988. Subject: FFH Draft Directive.

Wildlife Link 1988b. Minutes of a meeting of the Habitats Directive group held at 2.30pm on Thursday 15 December 1988 at the Vincent Wildlife Trust, 21 Bury Street, London EC3.

Wildlife Link 1988c. Note of the meeting of the EC Habitats Directive Working Group held at 2pm on Wednesday 21 September 1988 at the Institute of Biology, 20 Queensberry Place, London, SW8.

Wildlife Link 1989. Letter dated 23 June 1989 from Wildlife Link to the Earl of Caithness regarding the proposed Habitats Directive.

Wurzel, R. 2008. European Union environmental policy and Natura 2000. *In:* Keulartz, J. and Leistra, G. (eds) *Legitimacy in European nature conservation policy: case studies in multilevel governance.* The Netherlands: Springer.

WWF 1989. Joint letter dated 12 May 1989 from WWF European national organisations to the Prime Minister of France regarding support for the draft Habitats Directive.

Yourassowsky, C. 2015. Personal communications: e-mails dated 22 January 2015 and 5 August 2015 from Catherine Yourassowsky, former member of DG XI's Nature Unit, to Andrew L. R. Jackson.

8

NEGOTIATING THE HABITATS DIRECTIVE II

The French, Irish and Italian presidencies (July 1989 to December 1990)

French Presidency: July–December 1989

Introduction

If the Spanish Presidency of the Council from January–June 1989 had achieved "little significant progress" on the Habitats Directive proposal (Wildlife Link, 1989, paragraph 3), the same could not be said of the French Presidency under Environment Minister Brice Lalonde, which held office for six months from July 1989. The incoming Presidency, led by its senior negotiator, Alain Megret, was keen to revise the draft directive, and sought to embed a totally different approach to the adversarial one Member States had experienced in respect of the Birds Directive's implementation (Wildlife Link, 1989; Megret, 2013a, 2013b). Megret, then deputy director of the Nature Protection directorate at the Department of the Environment, became responsible in 1985 for the Birds Directive's difficult implementation in France (Megret, 2013b). Prior to this he had worked for ten years for the French government on free movement within the EU of medical specialists, dentists and pharmacists, following mutual recognition of their qualifications: "a subject as difficult, at the time, [as] the protection of birds in Europe", per Megret (2013b). However, the Commission's representatives were, says Megret (2013b), exceptional throughout long negotiations regarding these free movement rules, leading to agreements on three directives, and helping to bring along with them professionals, academics and students (Megret, 2013b). This was exactly the sort of collaborative, consensual approach Megret felt was essential to imprint on the Habitats Directive negotiations under the French Presidency, "respectful of the respective competences of the Member States and the Commission for the implementation of the new provisions" (Megret, 2013b). Moreover, the French Presidency felt it was essential to develop a directive that European citizens could take ownership of, having been fully informed and engaged during its development, particularly given the fact that many citizens would be involved during the implementation phase. The French thus reportedly maintained a "very open dialogue" with civil society throughout, with drafts of important documents being "sent to NGOs and associations, representatives of the agricultural [industry] and hunters, fishermen", both at national level and in Brussels (Megret, 2013b).[1]

Political developments

No doubt with a view to the impending Presidency, the French had produced a radically revised version of the draft Directive as early as March 1989, during the Spanish Presidency (DG XI, 1989b). The French informed the Commission that they would be engaging in bilateral discussions with other Member States regarding the Directive during July 1989, and that France hoped the Commission would put on the table, for the purposes of the next meeting of the Environment Council, an amended directive proposal that would follow closely the ideas of the French Presidency (DG XI, 1989b). With the French Presidency thus working away on its revised draft proposal, the UK's Nicholas Ridley found himself under ever-increasing pressure. Contradicting Ridley's claim that Britain led Europe in environmental protection, former EU Environment Commissioner Stanley Clinton Davis said in July 1989 that Ridley had instead "for years obstructed or frustrated progress" (Travis, 1989). "The fact is", Clinton Davis continued, "Mr Ridley is not so much a Secretary of State for the Environment as an ecological hooligan. The best news that the Prime Minister could offer to the cause of the environment would be to shunt him into the political sidings right now" (Travis, 1989).

Against this backdrop, the UK House of Lords' Select Committee report on habitat and species protection – published on 18 July 1989 – could not have come at a worse time for the embattled Ridley. Having weighed all the evidence provided to it, the Select Committee recorded that they found it

> disappointing that the British Government should have adopted an unenthusiastic approach to the prospect of a Community Directive in the nature conservation sphere. In particular, they regret the somewhat parochial attitudes which appear to have characterised the Government's evaluation of the initiative from the outset . . . Either [the Member States] can elect to equip themselves, through their own domestic administrations, with a genuinely European perspective . . . [or] they will have to accept that, increasingly, on matters like nature protection, public pressures and popular support will encourage the Community to move in to fill the gap.
>
> *(House of Lords, 1989b, paragraph 76)*

Thus, when the long-expected UK cabinet reshuffle finally took place on 24 July 1989, it was little surprise to find Nicholas Ridley moved sideways by Thatcher into the Department of Trade and Industry, to be replaced as Environment Secretary by the up-and-coming pro-European Chris Patten (White, 1989), thus equipping the Department of the Environment with the European perspective called for by the House of Lords only days before.[2] Briefing the Director General of DG XI the following day, the Nature Unit reflected that perhaps this reshuffle would mark a change in attitude on the part of the UK government (DG XI, 1989a). And indeed it did: as Roy Bunce, then head of Wildlife Policy at the UK's Department for Environment, explains:

> my brief changed . . . from being . . . keep this in the long grass for as long as possible, basically, to be seen as, if possible, one of the positive advocates of the directive, supporters of the directive, but – and this was a big but – turning it into a shape that the government could actually live with at the end of the day . . . So the idea was to show that we had very good environmental credentials, we were not afraid of the directive, we would support it, but it had to be in terms that were practicable.
>
> *(Bunce, 2012)*[3]

There were even greater political upheavals elsewhere in Europe in 1989, of course, with 9 November marking the fall of the Berlin wall. In tandem with momentous political change came more mundane challenges: the reunification of East and West Germany gave renewed political impetus to the idea of developing a high speed rail link between Berlin and Hanover, envisaged since September 1988 (PBDE, 1998). The agreement to build the link was signed in June 1990 but the project would later run into serious difficulties owing the projected impacts of the railway's route on a special protection area for the Great bustard (*Otis tarda*) under the Birds Directive, earning the species "an ambiguous fame in some media outlets in the 1990s as 'the most expensive bird in the world'" (Troya, 2012; PBDE, 1998; Flyvbjerg et al., 2003; LUGV, 2012). This dispute in respect of the Birds Directive was another issue to the fore of certain Member States' minds during negotiations regarding the Habitats Directive (Troya, 2012).

Legal developments

Although Nicholas Ridley's departure from the UK's Department of the Environment, combined with a green French Presidency under Brice Lalonde (*The Times*, 1989), certainly gave the Habitats Directive proposal renewed vigour in July 1989, DG XI's Nature Unit under Claus Stuffmann faced the tricky political task of persuading the Member States to embrace the idea of a Habitats Directive whilst those same Member States continued to see high profile infringement proceedings brought against some of their number under the Birds Directive. Two important examples arose in mid-1989. First, on 27 June 1989, in the dying days of the Spanish Presidency, the Commission issued a Reasoned Opinion against Spain in respect of the latter's failure to protect the Santoña marshes from various developments, giving Spain one month to comply (CJEU, 1993). The following month, a particularly important application was lodged with the CJEU by Ludwig Krämer's Legal Unit, relating to German plans to build a coastal protection dyke in Leybucht Bay, a project that had been opposed by conservation organisations for several years on the grounds that the project would impact negatively on important bird habitats, which had been classified as a special protection area under the Birds Directive in September 1988 (CJEU, 1989; Clough, 1989).[4] In February 1989 the Commission had applied to bring the German government before the CJEU in respect of the overall dispute, but then, in an unprecedented move in the environmental context, the Commission sought an emergency injunction on 14 July 1989 to prevent any further construction work going ahead pending the outcome of the case (CJEU, 1989).

The application was heard in Luxembourg on 9 August and the CJEU's order issued just a week later, refusing the Commission's request on the ground that the required element of urgency had not been established (CJEU, 1989). This on the basis that the Commission,

> did not apply for interim measures until a large part of the work had already been completed. In effect, the Commission is requesting the Court to stop work which has already been partially completed . . . In those circumstances, the application for interim measures can only be allowed if it is precisely the next stage in the construction work, that is to say the stage due to be carried out in 1990, which will cause serious harm to the protection of birds in the Leybucht. There is nothing in the file relating to the case or in the arguments presented before the Court to show that that is in fact true.
>
> *(CJEU, 1989, paragraphs 17–18)*

While the judgment in the broader *Leybucht Dykes* case would later emerge to play a decisive role in the history of the Habitats Directive, one wonders what might have happened had the CJEU granted the Commission its requested injunction in August 1989. That is, one wonders how the Member States might have reacted with regard to the fledgling Habitats Directive negotiations had the CJEU ordered Germany to stop works in Leybucht Bay as a result of the Birds Directive. As an indication of the political sensitivities at stake, following the negative outcome of the Commission's request for an injunction in the *Leybucht Dykes* case, it would take almost 20 years before the Commission would make another such application to the Court in an environmental case (see CJEU, 2006).

France's tour of capitals and Denmark's change of heart

On 13 September 1989 the Commission met with the French government to discuss the latter's progress with the directive proposal. The French representatives reported that they had succeeded in consulting almost all Member States, with the exception of Greece, Luxembourg, and Denmark, although the latter would that week be asked for written comments on France's proposals.[5] As Marcel Jouve, then head of the International and European Affairs division of the Nature Protection Directorate at France's Department of the Environment, comments, "[I] think that the European tour of capitals that I made at the beginning of our presidency was probably very important in the negotiation process, as it was a strong signal of our determination to make progress, and of our openness to discuss our and [others'] ideas and find compromise formulas" (Jouve, 2013). The French Presidency built on these consultations with Member States to produce an overhauled draft of the Directive (Jouve, 2013), having assembled a broad consensus around key points: namely, a site designation process that would be more consistent with the principle of subsidiarity; the integration into the Natura 2000 network of SPAs under the Birds Directive; the principle of management plans for protected areas, supportive of ongoing human activities compatible with areas' conservation objectives; and the establishment of a Management Committee under the Directive (Megret, 2013b).

By this time, Denmark's position had undergone something of a transformation, as Veit Koester, Denmark's senior negotiator, explains:

> at a certain time during the negotiations . . . the head of department, Ministry of Environment, he said to me, 'We have to give in, we have to change our position and we have to go to be positive to the negotiations, we cannot stick to our idea that the EEC shall not interfere with our legislation in that field.' And then we turned around, not to say that we were promoting or very eager with the Habitat Directive, but we were not opposed to the Habitat Directive as we had been when the negotiations started. And I think it was a mere political standpoint the head of department took. I don't know whether he had been in contact with some high officials within the EU or whether they had indicated to him that they felt that the Danish approach was unnecessarily negative, I don't know, or simply that he felt that the time had now come to change our position . . . I don't know whether he had actually discussed it with the Minister, maybe it came from the Minister.
>
> *(Koester, 2012)*[6]

Claus Goldberg, who worked under Veit Koester and acted as Denmark's main negotiator once substantive negotiations over the text began, reflects: "I do think that the Danish position was that we were more hesitant in the beginning, and when we found out, no, we can't escape, we can't use existing Danish legislation, we need to go in through this directive, we were pretty positive" (Goldberg, 2012).

The European Parliament's first report

On 29 September 1989, the European Parliament issued what turned out to be the first of two reports on the Commission's Habitats Directive proposal, following a draft of this first report, which had issued in May 1989 (European Parliament, 1989a). These reports were drawn up by the Dutch MEP Hemmo Muntingh, and were developed in consultation with nature conservation organisations, coordinated by the EEB and the RSPB;[7] account was also taken of criticisms and recommendations made by hunting organisations such as FACE (European Parliament, 1989b). The RSPB had evidently had an important role to play in the report, since Muntingh included in his suggested amendments the RSPB's March 1989 proposal of "satisfactory ("favourable" in the RSPB's draft) conservation status" as the central organising objective for the directive (European Parliament, 1989b; RSPB, 1989). More generally, Muntingh's proposed amendments sought, *inter alia*, to shorten the time limit for establishing the Natura 2000 network – which he proposed renaming "NATURA SEMPER"[8] – and also sought to delete the proposed numbers of sites to be established at each stage (i.e., 10 most important areas in the EU for each listed species/habitat, two most important areas in a region during the first stage; 100 most important areas in the EU for each listed species/habitat, five most important areas in a region during the second stage). Removing these numerical limits would of course mean there would be no ceiling on the maximum number of sites requiring designation. As Muntingh noted, "there is no need to provide arbitrary figures with regard to the areas in question; ecological necessity and potential should be the guiding principles" (European Parliament, 1989b, p. 31). Thus, both Muntingh and several Member States regarded the requirement to designate a specific number of sites as arbitrary, though one suspects that certain Member States hoped the removal of these figures would result in fewer sites being designated, whilst Muntingh most likely expected the opposite; both could not be right.

By early October 1989 the French Presidency had indicated to the Commission its intention to table its extensively revised version of the draft directive, which had been drawn up in close consultation with the Commission's services, for approval in principle at the Environment Council meeting on 28 November 1989 (DG XI, 1989g). The Commission was happy to adopt suggestions from the Muntingh report which the Commission felt would improve the proposal, though the Commission was mindful that negotiations in the Council had proven "fairly difficult, with a number of Member States still maintaining a general reservation on the whole proposal (notably Greece, Portugal, and Spain)" (DG XI, 1989g). As such, the Commission faced a tactical question regarding how best to secure the inclusion of amendments tabled by Muntingh: "One crucial consideration in that context", noted the Commission, "is to avoid any step which would seriously hamper the prospects of a swift approval of a proposal in which the 3 key elements to which the Commission attaches the greatest priority, i.e. a system of double protection of threatened species and habitat types, the setting up of a coherent network of duly protected special protection areas and flexible procedures for amending the annexes, are retained" (DG XI, 1989g).

Launch of the French Presidency's revised draft: October 1989

Designation of Natura 2000 sites

On 13 October 1989 the French Presidency formally presented its compromise text of the draft directive to the other Member States (CEC, 1989b). There had been a complete overhaul in respect of the process for site selection and designation, and the number of annexes had been reduced from eleven to eight, and soon after to six (DG XI, 1989f).[9] Under a new Article 7 (emphasis as per the original), "On the basis of the most reliable scientific information the Commission, together with the Member States, shall draw up a list of habitats [covering listed habitat types and habitats of listed species] to be classified under [the Directive] on the basis of the criteria laid down in Annex V[, within two (or three) years of its notification]" (CEC, 1989b, Articles 5, 7 and 25).[10] This list would then be submitted to a Committee comprising representatives of the Member States and chaired by a representative of the Commission (CEC, 1989b, Article 7(1)). The Committee would deliver its opinion on the list on a qualified majority basis, though the Commission as Chair would not have a vote (CEC, 1989b, Article 7 and 22). In the event that the envisaged measures were not in accordance with the opinion of the Committee, the Commission would be empowered to submit a proposal to the Council, which would act by qualified majority; and if the Council did not take action within three months of referral, the proposed measures would be adopted by the Commission (CEC, 1989b, Article 22). Pending adoption via this Committee procedure, the listed sites would be protected by the new impact assessment obligation described below (CEC, 1989b). Once its list of sites was adopted, a Member State would have 1 to 5 years within which to classify as special protection areas the habitats included on its list (CEC, 1989b). The list itself would be revised every five years, and the Member States would send the Commission all relevant information on each special protection area (CEC, 1989b).

The missing annexes and UK support for the directive

Discussing Hemmo Muntingh's report and draft opinion on the Habitats Directive proposal during the plenary from 9–13 October 1989, the European Parliament decided that it was not in a position to adopt an opinion because the proposal still lacked detailed annexes, and the Parliament's Environment Committee then asked the Commission, on 18 October, formally to complete the proposal by including the missing annexes (DG XI, 1989n; DG XI, 1989f). As a contemporaneous media report records,

> during the last 18 months representatives of [DG XI] have allowed national experts in the Member States' Committee of Permanent Representatives (COREPER) to work on the annexes, a fact which caused Parliament fury. The Parliament accused the Commission of producing 2 proposals, one for the EEC's Council of Ministers and one for Euro-MPs and withheld its Opinion, demanding that the Commission formally propose the annexes it had discussed at COREPER meetings.
>
> *(European Report, 1990)*

The Economic and Social Committee quickly followed suit and reached a similar decision to the Parliament's (DG XI, 1989f). The Director General of DG XI concluded that "the best, and fairest option open to us at this juncture is simply to respond to the [European Parliament's] demand for completion of the proposal by formal insertion of the technical annexes" (DG XI, 1989n). Whilst this would require consequential amendments of the proposal, these were minor distractions compared with the major political development reported in the Director General's letter: "the Council's deliberations have been aided by the decision of the United Kingdom's government to support the proposal for a directive" (DG XI, 1989n).

Designation of Natura 2000 sites, continued

Meanwhile, the French Presidency's compromise draft Directive fell to be considered for the first time at a Working Group on the Environment meeting on 19 October 1989 (CEC, 1989d). Several delegations were in favour of a co-operative procedure for drawing up the lists of sites to be designated, but Belgium, Denmark and Germany felt that it should be initially for the Member States to propose the areas, with any issues then being solved according to the procedure described in the Presidency's proposal (CEC, 1989d). Spain struck out on its own, arguing that the Member States should be solely responsible for designating areas for classification on the basis of criteria set out in the Directive (CEC, 1989d).

On 8 November 1989 the French Presidency presented as working documents versions of Annexes I (species requiring the designation of protected areas), II (species requiring special protection measures wherever they are found), IV (habitat types requiring the designation of protected areas) and V (comprising revised site selection criteria) (CEC, 1989c). Gone was the Commission's specified total number of protected areas;[11] in its place were mathematical formulae for calculating the surface area to be protected in respect of each habitat type/species (CEC, 1989c). Most delegations reserved judgment, but the UK expressed doubts about the proposed mathematical approach, considering that the approach did not take account of all relevant factors, and suggested that the responsibility for designating protected areas should simply be the competence of each Member State (CEC, 1989c).

The Working Party on the Environment met on 9 November 1989 for a detailed discussion regarding the French compromise draft (CEC, 1989f). Several delegations – Belgium, Spain, Greece, Ireland, the Netherlands, Portugal and the UK – foresaw difficulties with the collaborative procedure proposed for establishing the list of protected areas, especially with regard to the role given to the Commission in the event of disagreement in the course of the procedure (CEC, 1989f). The Commission responded that the collaborative procedure provided for a certain weighing up of factors, given Article 2 (general principles), Annex V (site selection criteria) and the Committee procedure (CEC, 1989f). When, following these discussions, the French Presidency presented a revised draft of the directive on 13 November 1989, the drafter had clearly picked up on the Commission's reference to Article 2: instead of merely referring to the criteria in Annex V in selecting sites, the draft now proposed allowing Member States to take account, in addition, of the matters set out in Article 2(1), which included economic, social, cultural and recreational requirements, as well as regional peculiarities (CEC, 1989a).[12] This reference to Article 2 was ultimately dropped from the provision of the directive setting out the site selection process, and the question of whether such socio-economic factors could be taken into account in site and boundary selection fell much later to be adjudicated by the CJEU, which held that they could not (CJEU, 2000).

By now, the whole approach and language of the Directive had begun to shift subtly, its title having been amended from Council Directive on "the protection of natural habitats" to "the conservation of natural habitats", etc. (DG XI, 1989l), echoing the point raised at the beginning of the year by the hunting coalition FACE (1989). Moreover, the revised draft of 13 November had moved away somewhat from the collaborative approach to the creation of an EU list of sites – following "sharp criticism" from a number of delegations, notably Spain (DG XI, 1989l) – with Member States now to take the first step alone, before submitting their lists of sites to the Commission (CEC, 1989a, Article 5(1)). As the head of the Nature Unit commented, "The latest version hereby very much picks up the model developed under the Birds Directive, whilst the former version was an attempt to innovate and introduce a more cooperative mechanism straight from the outset" (DG XI, 1989l). Given the experience with the Birds Directive, the expected benefits of such a cooperative mechanism would likely have included fewer visits to the CJEU in Luxembourg.

However, the Commission had not given up all hope of retaining some degree of control over the site designation procedure without the necessity of going to court. In the revised draft directive of 13 November, a new Article 6 proposed that where the Commission believed that areas other than those proposed by a Member State ought to be included on the EU list, the Commission could make an appropriate proposal to the Council, which would act by a qualified majority in responding to the Commission's proposal (CEC, 1989a). The provision was introduced by the French Presidency in response to the Commission's wish that the site selection procedure should not give Member States a right of veto over the sites to be designated (DG XI, 1989d). As the head of the Nature Unit commented,

> Maintenance of this provision is a *sine qua non* condition for the Commission's acceptance [of the new site designation system]. This item should, if necessary, be brought up to Ministerial level. During the discussions in Coreper, the Commission should, however, play down a bit this provision describing it as a means likely only to be topical in marginal cases with a view to ensure the coherence of the Community network of special protection areas.
>
> *(DG XI, 1989l, p. 5)*[13]

Having achieved a great deal in respect of the Habitats Directive proposal during its Presidency, the French government made it known that it intended to ask the EU's Environment Ministers to give their opinion on Articles 1 to 9 of the compromise text at the Environment Council meeting on 28 November 1989

(DG XI, 1989l).[14] According to the head of DG XI's Nature Unit, "Article 5 and 6 specifying the mechanism for designating SPA's are likely to become the main items for discussion . . . As for Annex V [site selection criteria], the initial version of which was heavily criticized by a majority of delegations, a new draft will be discussed within a technical sub-group under the Council, hopefully before the end of the year" (DG XI, 1989l, p. 1).

Explaining the UK's change of heart

Addressing the House of Lords on behalf of the UK government on 14 November 1989, Lord Reay sought to paper over any suggestion of a Damascene conversion by the Tories with regard to the Habitats Directive proposal:

> When the Commission first published its text, it is certainly true to say that we approached it with caution rather than enthusiasm . . . However, claims that we were the villain of the piece, pulling against the wishes of the majority of member states, are exaggerated. I believe that our initial lack of enthusiasm was soundly based . . . [T]he Community's previous involvement in nature conservation (the 1979 Birds Directive) has sometimes created tensions between the Commission and member state governments . . . [and in] the Government's view . . . the Commission's draft on this occasion was presented prematurely – without any prior consultation, without eight of the 11 annexes, and containing a number of features which had not been properly thought through . . . I believe for three reasons that there is light at the end of the tunnel. In the first place, the Commission has been showing greater flexibility – more over the past couple of months than in the whole of the previous year . . . Secondly, the current French Presidency has tabled an alternative text as a basis for further discussions . . . Our initial reaction, like that of most other member states, is to regard the French text as an improvement on the Commission's draft . . . The purely mechanical approach to site designation, for example, seems to have been dropped . . . [but] Commission involvement in site selection and what we regard as unacceptable powers for the proposed committee might lead . . . to site designation from Brussels against the wishes of the member state concerned. So there is some way to go on this front yet.
>
> *(House of Lords, 1989a)*

Designation of Natura 2000 sites, continued

Reporting to the Council on 21 November 1989, COREPER noted that Spain and Portugal had maintained their general reservations regarding the directive proposal, given the absence of the annexes from the proposal, and that the Council would therefore not be in a position to take a decision regarding the directive at its meeting on 28 November (CEC, 1989h). Nevertheless, progress had been made pending the receipt of detailed annexes, and only two main problems emerged during COREPER discussions on 15 November: first, the procedure for designating sites, in respect of which numerous Member States (Spain, Ireland, Portugal, the UK, Greece) had issues, feeling that Member States were the best judges of the need to protect habitats and that protected areas should not be imposed upon countries (CEC, 1989h). Denmark, Spain and Italy stressed the need to see the detailed site selection criteria before reaching a final position (CEC, 1989h), though Denmark was said to be favourably disposed towards the proposed process (DG XI, 1989f). The second major outstanding issue was the question of financing, with Greece, Spain, Ireland, Italy, the Netherlands and Portugal saying that the mobilisation of EU financial resources would need to be guaranteed (CEC, 1989h).

On 13 December 1989 the Director General of DG XI circulated draft annexes to the rest of the Commission and asked for their agreement by 20 December 1989 in order to allow the Commission to adopt them formally before 15 January 1990, for (re)insertion into the proposal (DG XI, 1989j). In the meantime, various people were contributing ideas to the Commission regarding the best way to deal with site selection criteria for the directive. One was Geert Raeymaekers, an environmental consultant working with DG XI, who proposed in December 1989 the idea of setting up a series of biogeographical working groups: that is, Atlantic, Continental, Mediterranean, and Alpine biogeographical working groups in which national scientific delegates would be represented, under the chair of the Commission (DG XI, 1989k; Raeymaekers, 2012). These working groups, which were intended to strengthen the international and transboundary aspect of the protected area network, would then be given the task of establishing site selection criteria (DG XI, 1989k). The Council met the following day – 15 December – to discuss the Annex V site selection criteria, and the discussion revealed almost as many different ideas as delegates (DG XI, 1989e; CEC, 1989i).

Protection of Natura 2000 sites: "appropriate assessment" coined

On 13 September 1989 France discussed with the Commission progress during the French Presidency regarding the directive proposal (DG XI, 1989c). The French representatives reported that a majority of Member States were not in favour of operating a system of two categories of protected areas with differing standards of protection applying to the categories, as had been proposed in the Commission's August 1988 directive proposal (DG XI, 1989c). Thus, the Member States were moving slowly towards a single category system, with the potential for derogations being available in respect of all protected areas. Further, while a majority of Member States maintained their rejection of the Commission's proposal to amend the EIA Directive to apply that directive to sites designated under the Habitats Directive, France's idea of including a new standalone provision on impact assessment within the text of the draft Habitats Directive had been met with a favourable response (DG XI, 1989c).

France launched its impact assessment proposals at the meeting of the Working Party on the Environment on 28 and 29 September 1989. The note of the meeting records:

> To take account of the concern voiced by a number of delegations . . . [France] proposed that, instead of [provisions applying the EIA Directive to special protection areas under the Habitats Directive] . . . special rules for protection areas should be included in [the Habitats Directive itself] providing for an assessment of the effects which development projects not directly connected to the management of these areas would have on the conservation potential of the areas concerned, on the understanding that the Member States would be making this assessment and would inform the Commission thereof. The rules should also be applicable to areas awaiting designation.
>
> *(CEC, 1989e, p. 21)*

Several delegations were "favourably inclined" to this proposal, although Belgium expressed doubts, and Denmark stated that any form of intervention likely to have an impact on protection areas should be ruled out (CEC, 1989e).

On 13 October 1989, the French Presidency formally submitted to the other Member States its compromise proposal regarding the text of the Habitats Directive. As expected, the French had removed all references to the EIA Directive, and in its place inserted the following tailored provision as a new Article 8:[15]

1. In respect of the special protection areas referred to in Article 7 Member States shall take appropriate steps to avoid pollution or deterioration of habitats or any disturbances affecting fauna and flora, insofar as these would have significant effects, having regard to the objectives of this Directive, on the species or types of habitat for which the area has been classified.
2. Member States shall in particular decide on the most appropriate protection status to be given to the areas referred to in paragraph 1 and consider setting up integrated management plans in accordance with the ecological needs of the species and types of habitat concerned. A common typology for the classification of protected areas at European level shall be adopted in accordance with the procedure laid down in Article 22.
3. All development projects affecting or likely to affect a special protection area and which are not directly connected with or necessary to the management of the area shall be subject to detailed assessment of their impact in relation to the area conservation objectives. Where development is carried out, the Member State shall send the Commission the results of the study and inform it of any compensatory measures adopted, showing that the integrity of the area has not been significantly affected.
4. Where necessary to ensure a satisfactory conservation status of a species Member States shall envisage the re-establishment of destroyed or degraded biotopes or the development of new ones.

(CEC, 1989b)

Discussing the proposed text at a meeting of the Working Party on the Environment on 9 November 1989, Spain argued that account should be taken of Article 2 – i.e., economic, social, cultural requirements, etc. – in applying the above Article (CEC, 1989f). In respect of Article 8(3), the UK and Ireland argued that the introductory wording should be amended to read "All development projects affecting or likely to significantly affect"; Germany opposed this on the basis that a series of non-significant effects could result in serious consequences (CEC, 1989f). When the French Presidency presented its revised draft on 13 November, the result was a compromise: "significantly" had been added, but to address Germany's point regarding cumulative effects the draft now required assessment of projects likely to have significant effects, *either alone or in combination with other projects* (CEC, 1989a). Finally, the UK challenged the term "detailed assessment" in paragraph 3, and said that it preferred "appropriate assessment", thus giving rise to the now well-known term in the context of the Habitats Directive (CEC, 1989f; CEC, 1989a).

Commenting internally, the head of DG XI's Nature Unit wrote, "The Commission should express strong support for [Article 8(3)] and be ready to carry it up to Ministerial level if so required" (DG XI, 1989l). The UK government was also positively disposed towards the new provision, noting that "the new proposal on environmental impact assessment is much less ambitious" than the earlier proposal to amend the EIA Directive (House of Lords, 1989a), an opinion likely based on the fact that the proposal to amend the EIA Directive had covered projects, plans and programmes, whereas the new standalone impact assessment provision for the Habitats Directive covered only projects in the version first proposed. Nevertheless, the provision would ultimately be extended during negotiations to cover projects *and* plans, and would also be amended to set out the circumstances in which a plan or project could be approved at national level following an impact assessment. Thus, the UK government's view of a "less ambitious" provision has been disproven by history. The original proposal to apply the EIA Directive's provisions to sites under the Habitats Directive provided only procedural protection; i.e., impact assessment would be needed in many cases but the provision did not dictate the circumstances under which projects/plans/programmes could be permitted at national level in light of the assessment. In contrast, the insertion of a standalone impact assessment procedure into the draft Habitats Directive resulted ultimately in something quite different: an impact assessment procedure that dictated whether national authorities could or could not approve a plan or project.

The UK as a forerunner: November 1989

By November 1989 the UK's position regarding the directive had changed fundamentally, from obstructionist at the proposal's outset to forerunner:

> I can say that the Government feel that there is now a real basis on which to move forward and that we are looking to play a leading role in this process . . . We do not think that the fate of each nation's threatened species,

> and of the habitats necessary for their survival, is and should be the concern only of the nation within whose frontiers they are to be found . . . we acknowledge also the right of others to know what we are doing in our own backyard. We accept that we can learn [from] the experience of others. We hope therefore that it will soon be possible to reach an agreement on this important directive.
>
> *(House of Lords, 1989a)*

However, it was by this time clear that this would not be possible during the French Presidency, despite the enormous progress made. A report from COREPER to the Council on 21 November 1989 noted that "The absence of technical annexes has seriously inhibited progress on this proposal. Delegations have indicated on several occasions that they cannot take final decisions on the enacting terms without knowing the content of the annexes. In these circumstances, [Spain and Portugal] have upheld a general reservation" (CEC, 1989g).

Financing

Financing for nature conservation in light of the proposed directive grew in stature to become a key concern during the French Presidency. On 13 September 1989 the French Presidency reported to the Commission that financial support represented a prerequisite for certain delegations in respect of the directive's entry into force (DG XI, 1989c). This was reinforced at a meeting of the Working Party on the Environment on 28 and 29 September 1989, where several delegations expressed concern at the financial implications of the directive; in particular, Spain, Greece, Ireland, Italy and Portugal (CEC, 1989e). And when, in October 1989, it became known that the UK government would be supporting the directive, the biggest remaining stumbling block, in the view of the Director General of DG XI, was the "absence of a financial-support mechanism to assist the less developed Member States to preserve the wealth of important habitats which they have – largely because of their under-development!" (DG XI, 1989n).

The Commission had a well-established position for the purposes of the Environment Council meeting on 28 November 1989:

> A directive addressed at Member States is not the place to include a section on financial resources. The Commissioner is, however, ready to make a formal statement at the Council meeting committing himself to present to the Commission in the near future a proposal for the extension of ACE/biotopes Regulation to cover other species than birds and foreseeing a substantial increase of financial means under the Regulation and also, in association with the Commissioner of Agriculture, to introduce a proposal aimed at increasing support under the Agricultural fund for the conservation of the special protection areas included in the Natura 2000 network.
>
> *(DG XI, 1989l)*

The stakes were already high by this stage: as the UK government noted in November 1989, "some member states are saying that they will not agree to the text of this directive unless the financial provisions are clearly established in advance" (House of Lords, 1989a). But the Director General of DG XI had a plan: in a briefing note sent to the Environment Commissioner on 20 November 1989 he noted that the availability of EU financial support has been "a major stumbling-block for [Spain, Portugal, Ireland, Italy and Greece] from the beginning. You should be able, at the Council [of 28 November 1989] to allay their fears . . . My services are working on a proposal which would involve the replacement of the biotopes element of the ACE Fund by a non-finite financial regulation which would reflect the decisions eventually taken by the Council on the Habitats Directive" (DG XI, 1989i).

The French Presidency: a "remarkable effort"

Overall, the Nature Unit clearly felt that the French Presidency had done an excellent job in respect of the Habitats Directive proposal. An internal DG XI note recorded that the Commission, "warmly congratulates the Presidency with its remarkable effort of synthesis aimed at reconciling very different views and resulting in a simplified and clear text. The Commission can broadly support the compromise text" (DG XI, 1989l, p. 2).

Irish Presidency: January–June 1990

Designation of Natura 2000 sites

Ireland took over the Presidency of the Council in January 1990 and made very little progress regarding the Habitats Directive during its six months in charge. In respect of site designations, the Member States picked up where they had left off, meeting on 16 January 1990 to discuss the Annex V site selection criteria. For the purposes of the meeting, the Commission had produced a working draft version of Annex V which had moved slightly away from the previous mechanistic approach (DG XI, 1990b). The Annex was now in two parts: a qualitative "general objectives" part as section A; and the quantitative, mechanistic approach now shifted to a section B (DG XI, 1990b). In part A, the Commission had introduced to the criteria the general objective of maintaining or re-establishing "a favourable state of conservation" for each listed species and habitat, picking up on the RSPB's (1989) suggestion, which had been included in Hemmo Muntingh's report for the European Parliament (DG XI, 1990b). Moreover, the Commission had also included the idea of dividing the EU into biogeographical regions for the purposes of site selection: Macaronesia having been added to Geert Raeymaeker's earlier suggested list of Mediterranean, Atlantic, Continental and Alpine regions (DG XI, 1990b).[16]

The suggestion of including favourable conservation status as an objective did not prove controversial, unlike the suggestion that each region, as defined in the

Directive, would be obliged to "allocate a minimum of X% of its area to the network of special protection areas with a view to conserving the common natural heritage made up of a coherent ecological infrastructure throughout the whole Community. Beyond this minimum threshold, the contribution of each region [would] vary according to the size of the habitats [as defined in the Directive] it contains at the time the Directive is adopted" (DG XI, 1990b, paragraph 3). This idea was the subject of a long discussion at the meeting on 16 January, with the Irish representative resisting the idea particularly strongly, pointing to the "sacrosanct" nature of property rights in Ireland, and suggesting that if a percentage needed to be included, this should be on a voluntary basis, and that the percentage should be capable of being revised subsequently, an idea supported by the UK and Germany (DG XI, 1990c). Germany pointed out that its earlier idea of protecting 10–15% of territory did not necessarily envisage this percentage comprising only special protection areas, but could also be understood as including buffer zones and ecological corridors (DG XI, 1990c). Spain suggested a compromise proposal, which received broad support, whereby each region would go about the business of designating sites for each listed species and habitats, and, where the region ultimately fell short of the minimum percentage stated, Member States could then add additional areas of their choice to reach the minimum (DG XI, 1990c). Finally, Ireland, the UK and Germany were very critical of the quantitative method contained in section B of the Annex V site selection criteria, favouring an approach that would rely on the judgment of experts at the regional and national levels (DG XI, 1990c).[17]

Protection of Natura 2000 sites

When the Member States met to discuss the Annex V site selection criteria at the beginning of the Irish Presidency, on 16 January 1990, the Commission's revised working version of the criteria included interesting language on site protection, including a definition of special protection area that effectively reintroduced the idea of differing standards of protection, as envisaged by the Commission's original directive proposal and the RSPB's March 1989 discussion paper, but earlier rejected by Member States:

> By 'special protection area' is meant an area where the necessary measures are applied for the maintenance or re-establishment of a favourable state of conservation for the species and types of habitat for which the area was designated. These measures may, if necessary, imply the creation of full nature reserves in cases where the species or types of habitat in question cannot resist disturbance by man; in practice, however, the measures will most often be limited to defining a certain number of rules on the human activities which can take place there and guaranteeing the maintenance or re-establishment of the factors on which depends the conservation of the species and types of habitat for which the area has been designated.
>
> *(DG XI, 1990b)*

Oddly perhaps, given the Member States' earlier resistance to the idea of differential standards of protection for sites, there were no objections to this definition, and the Member States instead asked that it be transferred to Article 3 of the main text of the Directive (DG XI, 1990c). However, such a definition was not ultimately retained in the Directive as adopted, and the Directive instead codified a single protection regime for special areas of conservation, albeit one that was flexible enough (on paper) to provide strict protection (i.e., no human disturbance) for certain habitat types and to allow continuing human land uses where this would be beneficial or not damaging in nature conservation terms.

Delors' cabinet *intervenes again*

With the Commission now required by the European Parliament to table annexes formally as part of its proposal, in December 1989 the Director General of DG XI had circulated the relevant draft annexes to the rest of the Commission for their agreement, commenting that "the draft annexes I, II, III and IV . . . have been drastically shortened as compared with the initial ones . . . As your Directorate General has already agreed to the far broader lists of species and habitat types presented in the initial annexes I, II, III and IV . . . we assume that you will have no problem in agreeing to the attached set of annexes" (DG XI, 1989j). Two points are worthy of note: first, it is not clear that the rest of the Commission had in fact agreed to the original annexes, given that they were dropped before the formal proposal was made to the Council;[18] second, Annex IV (habitats requiring special protection areas) had not been drastically shortened, and had instead grown considerably in length.[19]

In any event, the *cabinet* of Commission President Delors intervened (again) in February 1990, its earlier intervention having led to the annexes being excluded from the directive proposal in the first place. Briefing notes were dispatched to Environment Commissioner Ripa di Meana for the purposes of a meeting with Delors, explaining why the principle of subsidiarity was not a sound basis to block the proposal (DG XI, 1990k, 1990f). A contemporaneous media report records that

> With hands tied [by the European Parliament's demands to see the annexes], Environment Commissioner, Carlo Ripa di Meana, suggested that the Commission re-enact the annexes drafted in 1988 and put them forward by written procedure. But, Mr Ripa di Meana was not the only one with a keen memory, as Delors reproduced the same arguments he had used in 1988 for not proposing the annexes, that it would be bureaucratic and that there should be subsidiarity, and blocked the move. Commission sources report that Delors has never been very keen on conservation legislation anyway mainly because of pressure from France's powerful hunting lobby.
>
> *(European Report, 1990)*

Stanley Johnson, who had not been heavily involved with the directive proposal once it got off the ground, nevertheless re-engaged when the proposal came under threat once more, writing to the Director General of DG XI that it was "impossible to imagine [that the directive] could be blocked not by the Council or the Parliament . . . but BY THE COMMISSION ITSELF! And notably by the President and his cabinet!" (DG XI, 1990k). However, as reported at the time, "since the Commission now has little option but to put forward proposals on the annexes, Delors is not expected to block the measures" (*European Report*, 1990), and the Commission's modified proposal, including the annexes, was ultimately agreed and communicated to the Council on 14 March 1990 (CEC, 1990b).

Designation of Natura 2000 sites, continued

Despite this hiccup within the Commission, things continued to look positive in the UK, with *The Observer* reporting on 4 February 1990 that "Britain, which originally opposed the [Habitats Directive] proposals, is now understood to be strongly in favour . . . [under Environment Secretary Chris Patten]" (Ghazi, 1990). Notwithstanding this renewed political impetus, progress during the Irish Presidency was very slow, with the work that was done focusing mainly on the Annex V site selection criteria.

The Working Group on the Environment met on 23 February 1990 to discuss the best way forward regarding Annex V (DG XI, 1990j). At the meeting, the UK expressed its support for the idea of site selection based on biogeographical regions, and those present agreed that the EU list of protected areas should be established by the Commission *in agreement with* each Member State (DG XI, 1990j). The meeting mandated the Commission, with the aid of the Irish Presidency, to draw up a revised version of Annex V (DG XI, 1990j). This the Commission did, and its working document version of Annex V of 8 May 1990 revealed a continuing gradual shift away from a quantitative assessment of the sites requiring protection towards a more qualitative method, albeit that the proposal still envisaged there being a minimum percentage area of habitat to be designated as special protection area in respect of different classes of species/habitat (e.g., class I (exceptional sites), class II (sites of major importance, etc.)) (DG XI, 1990e). This type of approach had already been criticised by the UK government on the basis that it might be more logical "to aim to apply a reducing degree of protection to sites in each class rather than a declining percentage of area to be protected . . . [I]f the proposal is to designate as SPAs 80% of Class 1 sites and 20% of Class 4 sites the end result will be that the 20% in Class 4 will have full protection while the 20% not selected for Class 1 will have none. This might seem odd given the comparative value of the sites in conservation terms" (Department of the Environment, 1990).

Progress during the Irish Presidency was in fact so poor that the proposal was included on the agenda for the Environment Council meeting of 7 June 1990 "solely to try to motivate the [incoming] Italian Presidency to restart active work on the dossier" (DG XI, 1990i). Ireland sought to explain the lack of progress during its Presidency on the basis that the Commission had formally transmitted the annexes too late, and DG XI acknowledged that there was some truth in this, given that Delors' intervention had delayed matters somewhat (DG XI, 1990i). The UK was not impressed, however. Junior Environment Minister David Trippier commented that the situation "is bordering on a farce. We are all the more appalled that so little progress has been made in recent months" (Brown, 1990).

The Member States' desire to pin down precise site selection criteria, and to provide a degree of flexibility in the protection afforded to sites, was no doubt given added impetus by Environment Commissioner Ripa di Meana's continuing focus on enforcement, and his envisaged role for the new European Environment Agency, which the Member States had only recently agreed to set up. He said:

> Community legislation must be strictly applied throughout all the Member States. At the very start of my mandate I made it clear that this would be my major priority. I have stuck to that . . . I consider that my determination to have Community Law applied has led to a change in the attitude of Member States to environmental legislation in all its stages. At present the task of following the application of Community law in the Member States is undertaken by the Commission alone . . . The Council agreed that in two years it would examine again the question of how the [European Environment] Agency might assist the Commission in its task of following the application of Community law and, in particular, the task of inspection.
>
> *(Ripa di Meana, 1990)*[20]

Financing

On 2 May 1990 the Commission announced a proposal for a Council Regulation on action by the Community relating to nature conservation (ACNAT), a new instrument for the protection of habitats of EU importance and for the conservation of endangered species, with increased funding (European Commission, 1990a), to replace the biotopes part of the earlier ACE Regulation (1984) (European Commission, 1990b).[21] The target was clear: "Although such action may concern the whole of the Community, the Commission is particularly interested in helping less favoured regions to promote the development of nature conservation measures" (European Commission, 1990a). Several Member States were very critical of the failure to include details of the amounts of financing that would be available during each year of operation of the Regulation, and some queried the link between the ACNAT proposal and the Habitats Directive proposal (DG XI, 1990h). The Commission indicated that it would not be possible for the ACNAT Regulation[22] to be adopted by the Council before the Habitats Directive was adopted (DG XI, 1990h), but, nevertheless, looking to the Italian Presidency, the Environment Council mandated COREPER to press ahead with work on the Habitats Directive proposal, and to start examining the new ACNAT proposal with a view to reaching agreement on both as soon as possible (CEC, 1990a).

The European Parliament's second report

Having worked rapidly since the Commission formally added the annexes to its proposal, Hemmo Muntingh produced his second draft report on the Habitats Directive proposal on 27 June 1990 (European Parliament, 1990a). The draft built on his first draft, but now included comments and proposed changes to the detailed annexes (European Parliament, 1990a). The final report was issued on 18 October 1990 (European Parliament, 1990b), by which time the Commission had had a good opportunity to digest its contents. The Parliament debated the report and delivered its positive Opinion with proposed amendments on 19 November 1990 (European Parliament, 1990c).[23]

Italian Presidency: July–December 1990

Designation of Natura 2000 sites

During the early months of the Italian Presidency, work on the Habitats Directive continued in the footsteps of the Irish Presidency's focus on the Annex V site selection criteria. Commenting on the latest draft on 17 September 1990, the Commission suggested that protected areas under the Directive should be named Special Areas of Conservation rather than Special Protection Areas, since the former connoted the idea of management rather than outright banning of activities (DG XI, 1990g), finally reflecting a point raised almost two years before by the hunting coalition FACE (1989). Around the same time, the Commission and the Italian Presidency introduced the idea of identifying "Areas of Urgent Intervention" for species and habitats in danger of disappearance, the latter to be indicated with an asterisk on Annexes I (species) and IV (habitats), an idea that would ultimately lead to the concept of priority species and habitats (DG XI, 1990l). Spain and France, in particular, were keen on this idea, on the one hand to introduce an order of priority in respect of their conservation work, and on the other to cast light on the differences between Member States in terms of their respective contributions to the Natura 2000 network (presumably Spain and France expected to be contributing more priority sites to the network than certain northern European countries, for example) (DG XI, 1990a).

Experience with the Birds Directive

In October 1990, the Commission stressed (internally) that the Habitats Directive proposal risked reaching a stalemate, in part owing to certain Member States' ongoing issues regarding implementation of the Birds Directive, which risked seeping across and impacting on discussions regarding the Habitats Directive (DG XI, 1990a). As at 15 March 1990, the Commission had commenced 37 cases against

Member States for failure to comply with their site designation and site protection obligations under Article 4 of the Birds Directive (CJEU, 1990c), and court judgments kept coming: e.g., the CJEU held the Netherlands in breach of the Birds Directive in March 1990 (CJEU, 1990b), Germany in breach in July 1990 (CJEU, 1990a), and various other cases were in the pipeline, including against Italy in respect of bird hunting (CJEU, 1990d), and against Spain in respect of its failure to protect the Doñana wetlands (European Commission, 1990c), leading Spain to characterise the Birds Directive as a "complaints machine" (COREPER, 1991).

Antonio Troya, who was seconded to DG XI as a Spanish national expert from 1990–1994, working mainly on facilitating negotiations between the Commission and Spain concerning the Habitats Directive proposal, explains that when he joined the Commission in 1990, the directive proposal was facing a "strong blockage", led by Spain but supported by other Mediterranean countries and France to a certain extent, owing to those countries' experience with the implementation of the Birds Directive (Troya, 2012). These countries hoped to address, via the Habitats Directive, the "dysfunctionality" they perceived in the implementation of the Birds Directive, which was impacting on projects funded under the Structural and Regional Funds, for example (Troya, 2012). Spain's attitude would likely have hardened in that regard in November 1990 when the Commission lodged an application to bring the Spanish government before the CJEU in respect of Spain's failure to apply the Birds Directive to protect the Santoña marshes from various damaging developments (CJEU, 1993).

In parallel with these developments, DG XI's Legal Unit and Commissioner Ripa di Meana were beginning to gear up for a major battle with the UK government regarding an alleged failure to assess various projects under the EIA Directive, a complaint having been lodged with the Commission in November 1990 (McCarthy, 1990). However, before that conflict could fully develop, there would be even greater turmoil within the UK government, with the resignation of Prime Minister Margaret Thatcher on 22 November 1990, having failed to decisively fend off a leadership challenge from Michael Heseltine in the first round of voting by Conservative MPs (*Associated Press*, 1990). In the event, Heseltine lost out to John Major in the subsequent leadership election, and Major then appointed Heseltine as Environment Secretary in place of Chris Patten (Oakley, 1990).[24]

Another six months lost

With the Italian Presidency having, per the Commission, made "very little progress" regarding the Habitats Directive proposal, the Commission met with Luxembourg's Minister for the Environment on 28 November 1990 and with a senior official on 13 December to discuss the situation regarding the draft Habitats Directive, in view of Luxembourg's taking over the Presidency in January 1991 (DG XI, 1990d). The Commission stressed that the European Parliament had expressed strong support for the proposal in a vote on 19 November 1990[25] and that NGOs in various Member States were increasing their pressure on their respective governments to work more actively towards a political agreement in the first half of 1991 (DG XI, 1990d). The financial aspect of the proposal was now central to the Commission's concerns: "It becomes increasingly clear that the discussions in the next half year on the 'Habitats' draft directive and on the Acnat Regulation as well as on other relevant sources of funding have to take place at the same time at the various levels of work within the Council . . . [otherwise] the prospects of it getting approved quickly by the Council are very poor indeed" (DG XI, 1990d). Nevertheless, the Commission maintained "reasonable hope" for an agreement during the Luxembourg Presidency, in part owing to the motivation displayed by several Member States, notably Germany, the Netherlands, Denmark and France (DG XI, 1990d).

Germany's appetite for the Habitats Directive might perhaps have been dampened a little by legal developments in December 1990, when Advocate General Van Gerven gave his Opinion in the *Leybucht Dykes* proceedings, which had earlier been the subject of an unsuccessful application for an injunction by the Commission (discussed earlier in this chapter). Germany's arguments in the case were supported by the UK,[26] and neither can have been comforted by the Advocate General's conclusion (albeit this was only advisory and remained subject to the CJEU's judgment):

> I am in agreement with the Commission that the reduction in extent of a designated special protection area or a diminution in the quality of living conditions in such an area is only permitted where the work carried out can be justified by compelling reasons arising from a general interest which is more important than the environmental interest protected by the directive. I am also in agreement with the Commission that the protection of human life is (of course) such a higher interest, [and] that (in the absence of special or exception circumstances which are difficult to foresee) economic or recreational interests do not constitute such an interest.
>
> *(CJEU, 1990c, paragraph 39)*

Sites under the Birds Directive therefore looked set to receive strict protection, but the prospect of there being *any* sites under a Habitats Directive seemed far off, the Irish and Italian Presidencies having "made virtually no progress with the draft Directive [in 1989]", as Ian Hepburn, the joint RSPB/WWF campaigner would write in his campaign newsletter in January 1991 (RSPB/WWF, 1991). "Ireland points to the lack of a formal Commission proposal for the complete set of annexes and lack of a funding proposal (which were eventually produced in March and May, respectively)", he added (RSPB/WWF, 1991), continuing:

> Italy had no excuse for failing to make progress, but seemed unwilling to give the Habitats Directive proposal any priority in working group discussions with the Commission and equally unwilling to take any initiative to negotiate the crucial points of difficulty – Annex V [site selection criteria]

and the articles relating to establishment of Special Protection Areas and Commission powers of intervention. As a result, the Italian Environment Council in late December did not even include a progress report on the Habitats Directive: there was no progress to report!

(RSPB/WWF, 1991)

Notes

1 One can juxtapose this account with Keulartz's (2009, p. 446) statement that the Habitats Directive "was adopted in a very top-down fashion with relatively little consultation involving stakeholders and policy actors on the ground who had to implement the Natura 2000 measures . . . [T]hese low levels of consultation and representation triggered high levels of resistance". Consultation with civil society during the drafting and negotiation of the Habitats Directive would make for an interesting subject for further research. In terms of consultation post-adoption, the relative absence of public participation from the text of the directive is notable; note the conclusion of Blondet et al. (2017), in the context of Natura 2000, that the relationship between participation, acceptability and effectiveness is not straightforward.

2 In June 1990 Ridley would be forced to resign from the cabinet altogether, having given an interview to *The Spectator* comprising an "impassioned denunciation" of Germany, in which he called the European Commission "seventeen unelected reject politicians", and accused Germany of trying to take over Europe (Lawson, 1990; Dowling, 1990). "I'm not against giving up sovereignty in principle, but not to this lot", he said, adding, "You might just as well give it to Adolf Hitler, frankly" (Lawson, 1990). He then came very close to saying he would rather see Hitler in charge than Helmut Kohl: "I'm not sure I wouldn't rather have . . . er . . . the shelters and the chance to fight back" (Lawson, 1990).

3 As Bunce further notes, "under that part of the Thatcher government and under the Major government which followed it, the cabinet was, well the Tory party in fact was hopelessly split . . . [E]very time you got a new Minister, and you got a new Minister quite frequently, it was like a change of government. You know the whole policy would simply change" (Bunce, 2012).

4 For a more detailed account of the facts of the *Leybucht Dykes* case, see Freestone (1996).

5 Marcel Jouve recalls visiting all Member States bar Denmark and Greece (Jouve, 2013).

6 As Koester reflects, it was a "little bit surprising" that no-one in Denmark reacted to this change of position, given the debate that took place over retaining competence at the national level for nature conservation and land use planning prior to Denmark's accession to the EU in 1973 (Koester, 2012). It is possible, of course, that the adoption of the Birds Directive had in any event prepared the ground somewhat for this development.

7 Ian Hepburn, who was coordinating an EU-wide Habitats Directive campaign on behalf of the RSPB and the WWF, explains that he developed close links in this capacity with Hemmo Muntingh (Hepburn, 2012).

8 'Semper' being Latin for 'always'. Muntingh explained his thinking thus: "[Natura 2000 is] a somewhat unfortunate name, because it creates the impression that nature can be frozen in time. This is not the case nor, of course, may it be so. Nature reserves are intended as a permanent resource for the continued existence of living and dynamic nature, and the name of the programme should reflect this" (European Parliament, 1989b, p. 28). The Environment Commissioner responded: "The reason why the Commission chose the term 'Natura 2000' was that it thereby wished to underline that the objective is to set up, by year 2000, the foundation of a network which, obviously, by its sheer nature, is bound to remain dynamic. I also have some doubts as to whether a Latin term would strike the public opinion as much as the threshold of the 3rd millennium" (DG XI, 1989h).

9 The annexes removed at this time related to: the information to be specified about the transmission of data about the classification of special protection areas, site notices,

research, management plans (European Commission, 1988, to CEC, 1989b); and a list of regions, and the protection of natural features enabling species to live normally in their area of distribution (i.e., outside the Natura 2000 network) (CEC, 1989b, to CEC, 1989f).

10 Revised Annex V site selection criteria were not included with the proposal, having themselves been discussed at a meeting of experts on 11 October 1989 (DG XI, 1989m).

11 I.e., first phase: 10 most important sites in the EU for each listed species/habitat, 2 most important sites regionally; second phase: 100 most important sites in the EU for each listed species/habitat, 5 most important sites regionally.

12 As a matter of historical interest, Ireland and Portugal requested that the term "classify" sites be replaced with "designate", explaining the difference in terminology between the Birds Directive, which uses "classify" and the Habitats Directive, which uses "designate" (CEC, 1989f). The UK evidently agreed, reminding the Commission on 15 February 1991, "You did, I think, agree to replace 'classify' by 'designate' in the English text, to reflect the difference between these two words in normal English usage" (Department of the Environment, 1991).

13 As it turned out, this process – ultimately restricted to sites for priority species/habitats, as set out in Article 5 of the adopted version of the Directive – has never been used by the Commission (Ó Bríain, 2013; Cashman, 2013), though Cyprus argued unsuccessfully before the CJEU that the process ought to have been used by the Commission in a dispute regarding site protection for the Cypriot grass snake (*Natrix natrix*) (CJEU, 2012).

14 With the exception of Article 3, then the definitional provision, which had been submitted to COREPER for discussion.

15 Soon renumbered as Article 7 (CEC, 1989a) and later still becoming Article 6 in the Directive as finally adopted.

16 Germany would later suggest adding the Baltic region (CEC, 1991) and the Commission was prepared to accept this (DG XI, 1991) but this did not ultimately come to pass during negotiations. With later EU expansions, other biogeographical regions have since been added (European Commission, 2016); there are now also marine regions, including a marine Baltic region (European Commission, 2011).

17 Roy Bunce, then of the UK government, explains that "from the outset the UK's position was that the selection of both species and habitats for inclusion in the Directive should be based on sound scientific criteria. This informed all our objections to arbitrary numbers and percentages" (Bunce, 2013).

18 As noted in Chapter 6, as at 27 June 1988 all DGs with the exception of DG VI (agriculture) had indicated their agreement to DG XI's proposal, which at that time included the detailed annexes of species and habitats (DG XI, 1988). However, it is not clear that agreement was subsequently reached with DG VI around that time, and it is clear from the history in Chapter 6 that Commission President Delors was not prepared to sign off on the directive including the annexes in question at that time.

19 Though as Devillers (2013) explains, a shorter list of habitats might in fact capture more than a longer list: "it was possible to produce short lists of habitats, but they were in fact more encompassing . . . [than longer lists might have been]", because listed habitat types on a short list would protect all sub-habitat types, and these sub-habitats were extensive and were being listed as part of the CORINE biotopes project. The risk of producing a very detailed list of habitats for the draft Directive, Devillers (2013) continues, would be that "in making the sentence more precise, in a legal text like that, you of course restrict its application".

20 Surprisingly, perhaps, this idea is said to have had the support of UK Conservative MEPs (*Europe Environment*, 1991). The establishment of the European Environment Agency was held up for some time by the French government, which was accused by Environment Commissioner Ripa di Meana of "blackmail" and "hostage-taking" for blocking the Agency's establishment until Strasbourg was confirmed as a permanent home of the European Parliament (*IPS*, 1991).

21 For the proposal, see European Commission (1990d).
22 ACNAT Regulation (1991).
23 The Economic and Social Committee delivered its Opinion on 18 October 1990 (ESC, 1990).
24 Speaking in March 1991, Heseltine reaffirmed the UK's support for the Habitats Directive, thus continuing the positive attitude displayed towards the proposal by his predecessor: "the UK is fully committed and indeed enthusiastic about [the directive's] concepts ... We will be taking an active and constructive role in the Environment Council over the next few months to bring the Directive to fruition" (Heseltine, 1991).
25 103 votes for, 17 against the proposal as amended by Parliament, and 3 abstentions (RSPB/WWF, 1991).
26 In 1987 the UK government had advised planning authorities that planning permission could be granted in an SPA where, *inter alia*, there were economic or recreational requirements that outweighed the conservation interest (Freestone, 1996).

References

ACE Regulation 1984. Council Regulation (EEC) No 1872/84 of 28 June 1984 on action by the Community relating to the environment. *OJ L 176*, 3 July 1984, p. 1.

ACNAT Regulation 1991. Council Regulation (EEC) No 3907/91 of 19 December 1991 on action by the Community relating to nature conservation (ACNAT). *OJ L 370*, 31 December 1991, p. 17.

Associated Press 1990. Thatcher resigns as Prime Minister. *Associated Press*, 22 November.

Blondet, M., De Koning, J., Borrass, L., Ferranti, F., Geitzenauer, M., Weiss, G., Turnhout, E. and Winkel, G. 2017. Participation in the implementation of Natura 2000: A comparative study of six EU member states. *Land Use Policy*, 66, 346–355.

Brown, A. 1990. Rare species threatened in EC red tape row. *Press Association*, 8 June.

Bunce, R. 2012. Interview with Roy Bunce, senior negotiator for the UK government in respect of the Habitats Directive, conducted by Andrew L. R. Jackson on 13 August 2012.

Bunce, R. 2013. Personal communication: e-mail dated 11 April 2013 from Roy Bunce, senior negotiator for the UK government in respect of the Habitats Directive, to Andrew L. R. Jackson.

Cashman, L. 2013. Personal communication: e-mail dated 18 March 2013 from Liam Cashman, DG Environment lawyer, to Andrew L. R. Jackson.

CEC 1989a. Background note from the Secretariat to the Committee of Permanent Representatives dated 13 November 1989, attaching revised draft directive. Brussels, 9847/89, ENV 205. Council of the European Communities.

CEC 1989b. Cover note from General Secretariat to Delegations. Subject: Proposal for a Council Directive on the protection of natural and semi-natural habitats and of wild fauna and flora. Attaching French Presidency compromise text of Directive. Brussels, 13 October 1989, 9204/89, ENV 186. Council of the European Communities.

CEC 1989c. Document de travail ENV/89/37. Object: Proposition de directive de Conseil concernant la protection des habitats naturels et semi-naturels ainsi que de la faune et de la flore sauvages – Annexes. Bruxelles, le 8 novembre 1989. Council of the European Communities.

CEC 1989d. Outcome of proceedings from Working Party on the Environment on 19 October 1989. Brussels, 25 October 1989, 9464/89, ENV 194. Council of the European Communities.

CEC 1989e. Outcome of proceedings of the Working Party on the Environment's meeting on 28 and 29 September 1989. Subject: Proposal for a Council Directive on the protection of natural and semi-natural habitats and of wild fauna and flora. Brussels, 9 October 1989. 8973/89, ENV 179. Council of the European Communities.

CEC 1989f. Outcome of the Working Party on the Environment on 9 November 1989. Brussels, 13 November 1989, 9862/89, ENV 206. Council of the European Communities.

CEC 1989g. Report from Permanent Representatives Committee to Council. Brussels, 21 November 1989, 10039/89, ENV 216. Council of the European Communities.

CEC 1989h. Report from the Permanent Representatives Committee to Council. Subject: Proposal for a Council Directive on the protection of natural and semi-natural habitats and of wild fauna and flora. Brussels, 21 November 1989, 10039/89, ENV 216. Council of the European Communities.

CEC 1989i. Working document ENV/89/44, SN 3887/89, Brussels, 21 December 1989. Annex V. Council of the European Communities.

CEC 1990a. Outcome of proceedings of Environment Council on 7 June 1990. Brussels, 22 June 1990, 7082/90, ENV 144. Council of the European Communities.

CEC 1990b. Translation of letter from Commission dated 14 March 1990 to President of the Council of the European Communities. Brussels, 6 April 1990, 5807/90, ENV 82. Attaching supplementary Annexes to the Habitats Directive proposal, COM(90) 59 final, Brussels, 30 March 1990. Council of the European Communities.

CEC 1991. Outcome of proceedings of Permanent Representatives Committee on 25 September 1991. Brussels, 25 September 1991. 7992/2/91 REV 2, ENV 250. Council of the European Communities.

CJEU 1989. Order of the President of the Court of Justice of the EU of 16 August 1989, *Commission* v. *Germany*, Case 57/89 R, EU:C:1989:334.

CJEU 1990a. Judgment of the Court of Justice of the EU of 3 July 1990, *Commission* v. *Germany*, C-288/88, EU:C:1990:273.

CJEU 1990b. Judgment of the Court of Justice of the EU of 15 March 1990, *Commission* v. *Netherlands*, Case C-339/87, EU:C:1990:119.

CJEU 1990c. Opinion of Advocate General Van Gerven of 5 December 1990, *Commission* v. *Germany*, C-57/89, EU:C:1990:432.

CJEU 1990d. Opinion of Advocate General Van Gerven of 8 November 1990, *Commission* v. *Italy*, C-157/89, EU:C:1990:385.

CJEU 1993. Opinion of Advocate General Van Gerven of 9 June 1993, *Commission* v. *Spain*, C-355/90, EU:C:1993:229.

CJEU 2000. Judgment of the Court of Justice of the EU of 7 November 2000, First Corporate Shipping, C-371/98, EU:C:2000:600.

CJEU 2006. Order of the President of the Court of Justice of the EU of 19 December 2006, *Commission* v. *Italy*, C-503/06 R, EU:C:2006:800.

CJEU 2012. Judgment of the Court of Justice of the EU of 15 March 2012, *Commission* v. *Cyprus*, C-340/10, EU:C:2012:143.

Clough, P. 1989. Endangered avocets get a hearing in court: Patricia Clough in Greetsiel examines moves to secure the future of geese and avocets, which need the north German salt marshes to survive. *The Independent*, 5 August.

COREPER 1991. Note from COREPER dated 14 March 1991, addressed to the members of the Commission. Headed "Preparation Conseil 'ENVIRONNEMENT', 18 mars 1991", SI(91) 143.

Department of the Environment 1990. Letter dated 6 April 1990 from the UK government's Department of the Environment to DG XI, headed "Habitats Directive: Annex V".

Department of the Environment 1991. Letter dated 15 February 1991 from the UK government's Department of the Environment to the head of the Nature Unit of DG XI.

Devillers, P. 2013. Interview with Pierre Devillers of Belgium's IRSNB conducted by Andrew L. R. Jackson on 25 April 2013.

DG XI 1988. Draft directive on the protection of natural and semi-natural habitats and of wild fauna and flora. Communication from Mr Clinton Davis to other members of the Commission. Brussels, 27 June 1988, COM(88) 381. On the agenda of the 926th meeting of the Commission, on Wednesday 6 July 1988.

DG XI 1989a. Briefing note dated 25 July 1989 from the Nature Unit to the Director General of DG XI.

DG XI 1989b. DG XI (unit B3) file note dated 7 July 1989, attaching a draft version of the habitats directive dated March 1989, with the handwritten annotation "FFH: amendements français".

DG XI 1989c. DG XI (unit B3) file note dated 19 September 1989, regarding a meeting between DG XI and French representatives to discuss the draft habitats directive, held on 13 September 1989 in Brussels.

DG XI 1989d. DG XI document entitled "Conseil Environnement du 28 novembre 1989. Point 5 'Habitats'. Points défensifs (Commentaire article par article)".

DG XI 1989e. DG XI document headed "Réunion du Conseil du 15.12.89 relatif à annexe V".

DG XI 1989f. DG XI note headed "Conseil Environnement, le 28 novembre 1989, Point 'Habitats', Summary Background Note".

DG XI 1989g. Document headed "IV. Background note", attached to document headed "Speaking note" – for responding to Hemmo Muntingh's report for the European Parliament on the Habitats Directive proposal – with handwritten annotation at the top: "13.10.89 Parlement".

DG XI 1989h. Document headed "Speaking note" – for responding to Hemmo Muntingh's report for the European Parliament on the Habitats Directive proposal. With handwritten annotation at the top: "13.10.89 Parlement".

DG XI 1989i. Internal DG XI briefing note dated 20 November 1989 from the Director General to the Environment Commissioner, Re: Environment Council of 28 November 1989 – General brief.

DG XI 1989j. Letter dated 13 December 1989 from the Director General of DG XI to the rest of the Commission. Subject: Annexes to the Draft Directive on the protection of natural and semi-natural habitats and of wild fauna and flora.

DG XI 1989k. Note dated 14 December 1989 from Environmental Consultant to DG XI, headed "A strategy to designate special protection areas for the Council Directive proposal 88/C 247/03. (Draft version – For internal use)".

DG XI 1989l. Note dated 14 November 1989 from the head of the Nature Unit (B3) of DG XI to the Director of Directorate B, headed "Subject: Item: 'Draft Directive FFH' on the Coreper agenda of 15–16/11/1989 – Preparation of the Environment Council of 28/11/1989".

DG XI 1989m. Sign-in sheet for meeting of experts on 11 October 1989 relating to Annex V criteria.

DG XI 1989n. Undated draft briefing note from the Director General of DG XI to the Environment Commissioner, referring to a meeting of the European Parliament on 18 October 1989.

DG XI 1990a. Annex headed "Explications des points defensifs out questions supplementaires", annexed to a background note, in turn attached to an internal DG XI briefing package dated 16 October 1990, headed "Discussion du rapport de H. Muntingh sur la proposition de directive concernant la protection des habitats naturels et semi-naturels ainsi que de la faune et de la flore sauvages".

DG XI 1990b. Commission working document XI/801/89-EN Add. 1. Annex V of the Habitats Directive proposal. Behind tab "REUNION du 16.190. Annexe V".

DG XI 1990c. DG XI file note dated 9 February 1990. Object: Bref sommaire de la réunion du 16.1.90 concernant l'annexe V.

DG XI 1990d. DG XI internal note dated 26 November 1990, headed "Aide-mémoire. Subject: Meeting between Commissioner Ripa di Meana and the Luxemburg's Minister of Environment – 28.11.1990 – Habitat draft directive".

DG XI 1990e. DG XI working document dated 8 May 1990. Annex V – Criteria for the classification of special protection areas.

DG XI 1990f. Internal DG XI briefing note dated 20 February 1990, addressed to a member of the cabinet of Commissioner Ripa di Meana.

DG XI 1990g. Internal DG XI note dated 17 September 1990 headed "Commentaires sur le document relatif a l'annexe V du 10.9.90".

DG XI 1990h. Internal DG XI note dated 20 July 1990, headed "Deuxième discussion de la proposition ACNAT au group environnement du Conseil".

DG XI 1990i. Internal DG XI note dated 22 May 1990 from the head of the Nature Unit (B3) to the head of Directorate B.

DG XI 1990j. Note of meeting headed "Directive Habitat. Principales conclusions des discussions au sein du Groupe Environnement du Conseil du 23 février 1990". Brussels, 29 August 1990.

DG XI 1990k. Undated DG XI briefing note from adviser to the Director General headed "Eléments pour l'entretien Delors/Ripa".

DG XI 1990l. Working version of Annex V dated 3 October 1990, headed "Criteria for the selection of sites to be classified as special areas for conservation". Attached to DG XI internal note dated 9 October 1990, headed "Aide-memoire".

DG XI 1991. DG XI document headed "Points defensifs, Conseil Environnement, 1er octobre 1991".

Dowling, J. 1990. Reluctant Ridley opts to do the decent thing. *Australian Financial Review*, 16 July.

ESC 1990. Economic and Social Committee Opinion on the proposal for a Council Directive on the protection of natural and semi-natural habitats and of wild fauna and flora and on the Supplementary Annexes. *OJ C 31*, 6 February 1991, p. 1.

Europe Environment 1991. European Parliament: A call for a European environment police force. *Europe Environment*, 14 November.

European Commission 1988. Translation of letter dated 16 August 1988 from the Commission to the President of the Council, attaching Habitats Directive proposal COM(88) 381 final. Council document 8149/88, ENV 143, Brussels, 16 September 1988.

European Commission 1990a. The Commission proposes fresh action by the Community on nature conservation. Press release IP/90/353, 2 May 1990.

European Commission 1990b. Commission report on the implementation of Council Regulation (EEC) No 2242/87 on action by the Community relating to the environment. COM(90) 342 final, Brussels, 19 July 1990.

European Commission 1990c. Doñana National Park in danger: Commission decides to initiate infringement proceedings against Spain. Press release IP/90/571, 12 July 1990.

European Commission 1990d. Proposal for a Council Regulation (EEC) on action by the Community relating to nature conservation (ACNAT). COM(1990) 125 final, Brussels, 11 May 1990.

European Commission 2011. Commission Implementing Decision of 11 July 2011 concerning a site information format for Natura 2000 sites (notified under document C(2011) 4892) (2011/484/EU). *OJ L 198*, 30 July 2011, p. 39.

European Commission 2016. *The Natura 2000 Biogeographical Regions* [Online]. Available: http://ec.europa.eu/environment/nature/natura2000/biogeog_regions/ [accessed 23 June 2017].

European Parliament 1989a. Draft report, dated 18 May 1989, on the proposal for a Community directive on the protection of natural and semi-natural habitats and of wild fauna and flora (COM(88) 381 final – Doc. C 2-152/88). Committee on the Environment, Public Health and Consumer Protection. Rapporteur: Mr H. J. Muntingh.

European Parliament 1989b. Report, dated 29 September 1989, on the proposal for a Community directive on the protection of natural and semi-natural habitats and of wild fauna and flora (COM(88) 381 final – Doc. C 3-34/89). Drawn up on behalf of the Committee on the Environment, Public Health and Consumer Protection. Rapporteur: Mr H. J. Muntingh. Series A, document A 3-0039/89.

European Parliament 1990a. Committee on the Environment, Public Health and Consumer Protection. Second draft report on the proposal from the Commission to the Council for a directive on the protection of natural and semi-natural habitats and of wild fauna and flora (COM(88) 381 final – C 3-34/89 + COM(90) 59 final – Doc. C 3-34/89/Ann.). Rapporteur: Mr H. J. Muntingh. 27 June 1990.

European Parliament 1990b. Second report by the Committee on the Environment, Public Health and Consumer Protection on the proposal from the Commission to the Council for a directive on the protection of natural and semi-natural habitats and of wild fauna and flora (COM(88) 381 final + COM(90) 59 final – Doc. C 3-34/89). Rapporteur: Mr H. J. Muntingh. A 3-0254/90.

European Parliament 1990c. Texts adopted by the European Parliament on 19 November 1989. Protection of habitats. Proposal for a directive COM(81) 381 final and COM(90) 59 final. Proposal for a Council directive on the protection of natural and semi-natural habitats and of wild fauna and flora. Approved with the following amendments. *OJ C 324*, 24 December 1990, p. 22.

European Report 1990. Natural habitats: institutional row over draft directive annexes. *European Report*, No. 1569, 7 March.

FACE 1989. Memorandum by the Federation of Hunting Associations of the EEC (FACE), dated 26 January 1989, headed "Preliminary comments and observations on the proposal for a Council Directive 'on the protection of natural and semi-natural habitats and of wild fauna and flora' COM(88) 381 final".

Flyvbjerg, B., Bruzelius, N. and Rothengatter, W. 2003. *Megaprojects and risk: an anatomy of ambition*. Cambridge: Cambridge University Press.

Freestone, D. 1996. The enforcement of the Wild Birds Directive: a case study. *In:* Somsen, H. (ed.) *Protecting the European environment: enforcing EC environmental law*. London: Blackstone Press.

Ghazi, P. 1990. Major rescue scheme for endangered nature sites. *The Observer*, 4 February.

Goldberg, C. 2012. Interview with Claus Goldberg, Danish government official who participated in the Habitats Directive negotiations, conducted by Andrew L. R. Jackson on 31 August 2012.

Hepburn, I. 2012. Interview with Ian Hepburn, English RSPB/WWF campaigner who headed an EU-wide NGO campaign in support of the Habitats Directive during the instrument's negotiation and was seconded to DG XI's Nature Unit for a period during the negotiations, conducted by Andrew L. R. Jackson on 16 August 2012.

Heseltine, M. 1991. Speech made by Secretary of State Michael Heseltine some time during the week beginning 11 March 1991, on the subject of wildlife protection.

House of Lords 1989a. House of Lords' debate on 14 November 1989 regarding "Habitats and Species Protection: ECC Report", columns 1248–1279.

House of Lords 1989b. House of Lords' Select Committee on the European Communities. 15th report: Habitat and Species Protection. Ordered to be printed 18 July 1989.

IPS 1991. Maastricht: EC Executive and Presidency conflict on environment. *IPS-Inter Press Service*, 10 December.

Jouve, M. 2013. Personal communication: e-mail dated 8 April 2013 from Marcel Jouve, ex head of the International and European Affairs division of the Nature Protection Directorate at France's Department of the Environment, to Andrew L. R. Jackson.

Keulartz, J. 2009. European nature conservation and restoration policy—problems and perspectives. *Restoration Ecology*, 17, 446–450.

Koester, V. 2012. Interview with Veit Koester, senior negotiator for the Danish government in respect of the Habitats Directive, conducted by Andrew L. R. Jackson on 22 August 2012.

Lawson, D. 1990. Saying the unsayable about the Germans. *The Spectator*, 14 July.

LUGV 2012. *The Great Bustard – the Brandenburg ostrich*, 2nd edn. Landesamt für Umwelt, Gesundheit und Verbraucherschutz. Potsdam: State Office of Environment, Health and Consumer Protection of the Federal State of Brandenburg.

McCarthy, M. 1990. EC may step in to save ancient London wood. *The Times*, 16 November.

Megret, A. 2013a. Personal communication: e-mail dated 23 March 2013 from Alain Megret, deputy director of the Nature Protection directorate of the French government's Department of the Environment at the time of the Habitats Directive negotiations, to Andrew L. R. Jackson.

Megret, A. 2013b. Personal communication: written comments from Alain Megret, deputy director of the Nature Protection directorate of the French government's Department of the Environment at the time of the Habitats Directive negotiations, attached to e-mail dated 4 April 2013 from Alain Megret to Andrew L. R. Jackson.

Ó Bríain, M. 2013. Personal communication: e-mail dated 18 March 2013 from Micheál Ó Bríain of DG Environment's Nature Unit to Andrew L. R. Jackson.

Oakley, R. 1990. Party unity takes top priority in reshuffle. *The Times*, 29 November.

PBDE 1998. Die schnellbahnverbindung Hannover – Berlin. Planungsgesellschaft Bahnbau Deutsche Einheit mbH / Deutsche Bahn Gruppe.

Raeymaekers, G. 2012. Interview with Geert Raeymaekers, environmental consultant working with DG XI during the Habitats Directive negotiations, conducted by Andrew L. R. Jackson on 25 September 2012.

Ripa Di Meana, C. 1990. Speech by Mr Carlo Ripa di Meana to the "Club de Bruxelles" on 31 May 1990. SPEECH/90/47.

RSPB 1989. Note dated 15 March 1989, prepared by the RSPB at the request of the UK government's Department of Environment, with a handwritten, undated covering note to DG XI.

RSPB/WWF 1991. European habitats campaign: campaign circular. January 1991.

The Times 1989. Green credentials. *The Times*, 9 August.

Travis, A. 1989. 'Miserable' Ridley attacked over green record in Europe. *The Guardian*, 13 July.

Troya, A. 2012. Interview with Antonio Troya, Spanish government official seconded to the European Commission for the Habitats Directive negotiations, conducted by Andrew L. R. Jackson on 5 October 2012.

White, M. 1989. Major leaps up cabinet ladder: Howe is made deputy PM. *The Guardian*, 25 July.

Wildlife Link 1989. Internal Wildlife Link memorandum (LINK/MEM/89/21) dating from some time before 14 July 1989.

9

NEGOTIATING THE HABITATS DIRECTIVE III

The Luxembourg and Dutch presidencies (January 1991 to December 1991) and adoption of the directive in May 1992

Luxembourg Presidency: January–June 1991

Introduction

After a lost year, things looked immediately more positive for the Habitats Directive proposal under the Luxembourg Presidency: by 15 January 1991, only two weeks in, the Commission's archive reveals a draft directive modified with numerous major handwritten changes in light of discussions with the Presidency (DG XI, 1991o). The handwritten changes here included extending the impact assessment procedure to plans as well as projects – a significant extension that was of course retained in the version of the directive finally adopted. The Commission continued working closely with the Presidency on producing a revised draft throughout January and February 1991, the Commission pausing to publish its formal modified proposal on 8 February, which took onboard certain of the European Parliament's suggested amendments, including the idea of introducing "satisfactory" conservation status as an objective of the directive, thus formally incorporating into the text the RSPB's original idea of March 1989 (European Commission, 1991b; RSPB, 1989). Indeed, the Commission and Presidency were working so rapidly on the draft that on the same day the modified proposal was published, incorporating the term "satisfactory" conservation status, the Commission's working draft had already moved on to use the term "favourable" conservation status in its place (DG XI, 1991p), which would be the terminology used in the adopted directive. Not only was the Commission working hard in conjunction with the Presidency, it was also floating trial balloons of its draft with other Member States, meeting the UK on 13 February for example, and receiving back detailed comments (Department of the Environment, 1991b).

Designation of Natura 2000 sites

On 28 February 1991 the Luxembourg Presidency presented its compromise proposal, which built on and was "largely inspired by the French Presidency's compromise text of November 1989" (CEC, 1991c). In terms of site designations, the revised draft stated that the Natura 2000 network would comprise sites designated under the Habitats Directive and SPAs under the Birds Directive (but not Ramsar sites, as the Commission's directive proposal had sought) (CEC, 1991c, Article 3(1)). Each Member State was to contribute to the network by designating a sufficient number of sites of appropriate quality and size as special areas of conservation, taking account of the objective of maintaining or restoring favourable conservation status (CEC, 1991c, Article 3(3)). In terms of site selection, the draft required the Member States, on the basis of the criteria in Annex III (formerly Annex V),[1] to draw up a list of sites hosting Annex I habitats or Annex II species[2] and to transmit this list to the Commission within three years of the directive's notification (CEC, 1991c, Article 4). The revised Annex III site selection criteria were now much more qualitative than before, relying, for example, on "overall expert judgment", amongst other things (CEC, 1991c). Having received the Member States' lists, the Commission would then establish a list of sites of Community importance drawn from the national lists, within six years of notification of the directive, in the framework of the five biogeographic regions and of the whole of the EU territory (CEC, 1991c, Article 4). This list would be submitted to a Committee foreseen by the draft directive and, in the event of a disagreement concerning one or more sites, a mechanism for resolving the dispute would be triggered, potentially leading to a qualified majority vote by the Council (CEC, 1991c, Articles 4 and 19). The draft also provided a mechanism for the Commission to refer a proposal directly to the Council where, in disagreement with a Member State, the Commission considered that a *priority* site not occurring on a national list should be added to the list of sites of Community importance; the Council would then decide the issue by a qualified majority (CEC, 1991c, Article 5).

Once a site had been retained as a site of Community importance, the Member State would have a maximum of six years to designate the site as a special area of conservation, and during this period the sites on the list would be subject to the protective provisions set out in Article 6(2) and (3) of the directive (CEC, 1991c, Article 4) – more below. The draft continued, "A Member State having designated sites retained on this list which total up to at least 15% of its European territory" may consider itself as having fulfilled its site designation obligations (CEC, 1991c, Article 4(4)). This proposed ceiling was deleted from the draft before the directive was adopted.

The Commission was now targeting the Environment Council meeting of 14 June 1991 as the date for adopting both the Habitats Directive and the ACNAT Regulation,[3] which the Commission referred to as an "indispensable complement" to the former (DG XI, 1991n). At a COREPER meeting on 6 March 1991, the Director General of DG XI spoke on the issue of financing, but also noted that experience with the Birds Directive would allow the "errors of the past" to be avoided, meaning that the system of protection envisaged by the Habitats Directive should be put in place by the Commission and the Member States acting in full cooperation (COREPER, 1991b; DG XI, 1991n). The Director General also stressed the need for a flexible process regarding future revisions of the technical annexes of species and habitats; this should not be by way of unanimity in the Council, he said, but rather by way of qualified majority, in order to ensure a credible and efficiently operating directive (COREPER, 1991b; DG XI, 1991n). In this regard, DG XI was also drawing on experience from the Birds Directive, with amendments having been agreed to that Directive the same day as the COREPER meeting, after a "cumbersome" process ending with a vote from the Council on the basis of unanimity (European Commission, 1991d).

Luxembourg Presidency's compromise draft

Sandwiched either side of the COREPER meeting were meetings of the Working Party on the Environment on 5 and 7 March 1991, which met to discuss the Luxembourg Presidency's compromise draft for the first time (CEC, 1991a). The delegations in general expressed satisfaction with the proposals, but a number of basic problems nevertheless emerged from the discussions (CEC, 1991a). For example, Article 5 – which covered the circumstance of a dispute between the Commission and a Member State regarding the non-designation of a priority site – was the subject of lengthy consideration, with Spain and Portugal maintaining that responsibility for designating sites should be the exclusive competence of the Member States, resulting in revised draft text that provided that if the Council did not take a decision within three months of a proposal by the Commission, the Commission's proposal would cease to be applicable (CEC, 1991a; COREPER, 1991c). Most delegations responded favourably to this – Spain a notable exception – subject to more detailed evaluation (CEC, 1991a; COREPER, 1991c). Interestingly, therefore, it was certain of the countries that insisted most strongly on the need for EU financing which equally wished to maintain exclusive competence over the designation of sites (COREPER, 1991c), a difficult negotiating position to maintain.

Territorial extent

The Working Party on the Environment met again on 21 March 1991 to discuss the detail of the Luxembourg Presidency's proposed text (CEC, 1991p). Amongst the issues discussed was the territorial extent of the Directive, the draft having been amended to provide explicitly that "maritime areas under the sovereignty or jurisdiction" of the Member States would be covered by the Directive (CEC, 1991p). Denmark, Spain, France, Ireland and the UK expressed reservations, arguing that this was "likely to create problems of legal interpretation and implementation" (CEC, 1991p, p. 11).[4]

Luxembourg Presidency's compromise draft, continued

Most delegations expressed reservations regarding the proposal that a country that had designated 15% of its territory would be regarded as having fulfilled its contribution to the Natura 2000 network, albeit for different reasons. Germany and the Netherlands felt that the percentage could be interpreted as a maximum, to the detriment of the environment; the other delegations were concerned on the vague grounds of the "incoherences and imbalances in implementation that a single percentage would cause" (CEC, 1991p, p. 15). Answering the first concern, the Commission suggested specifying that the provision would apply where 15% of a country's territory had been designated as *priority* sites (CEC, 1991p), but the idea did not take hold.

Interestingly, the Annex III site selection criteria (formerly Annex V), which had previously been hotly contested, were by now less of a concern, the Member States seemingly broadly happy with the proposed approach (CEC, 1991p). Geert Raeymaekers, who had earlier proposed adopting the concept of biogeographical regions for the purposes of the Directive, played an important role in settling the Annex III criteria, having redrafted the Annex to "give more weight to the understanding of the definition of the habitat types in the country, and also to the [qualitative] elements of the site within the [relevant] biogeographical region" (Raeymaekers, 2012). Indeed, the adopted version of Annex III provides for a much greater degree of qualitative assessment than had been the case in early drafts (Habitats Directive, 1992).

Further debate on territorial extent

On 3 May 1991 the Luxembourg Presidency presented a revised draft of the directive, amended to reflect discussions to date (CEC, 1991d), and this draft was the subject of a detailed debate on 16 and 23 May (CEC, 1991o). The issue of the territorial extent of the Directive remained live, with the relevant provision having now been amended to read that the Directive covered (emphasis added) "the European territory of the Member States to which the Treaty applies *including Member States' territorial waters*" (CEC, 1991d);[5] Spain, France, Ireland, Greece and the UK entered scrutiny reservations (CEC, 1991o).

Site selection criteria, and the Commission's power to intervene

At the Environment Council meeting of 13–14 June 1991, progress was made on several issues, but other points of fundamental importance remained outstanding, including financing (CEC, 1991k). One issue that was more or less settled, however, was the site designation process under Article 4 (CEC, 1991k): on the basis of the criteria set out in Annex III, Member States would propose a list of sites to the Commission within three years of notification of the Directive. On the basis of criteria in Annex III, the Commission would then establish, in agreement with each Member State, a draft list of sites of Community importance. Member States whose sites hosting priority natural habitat types and priority species represented more than 5% of their national territory could, in agreement with the Commission, request that other criteria than those listed in Annex III be applied for selecting *all* the sites of Community importance on their territory.[6] The Commission would then adopt this list of sites of Community importance within six years of the notification of the directive, and in accordance with a procedure set down in the directive. As soon as a site was adopted in this way, the protective obligations set out in Article 6(2) to (4) would apply – more below. The Member State in question would have six years from a site's adoption as a site of Community importance to designate the site as a special area of conservation.

However, the issue of the Commission's power to propose sites not included on a national list continued to be controversial at the Environment Council meeting of 13–14 June 1991 (CEC, 1991k). It was noted by several delegations that there was a close link between this provision and the question of financing (CEC, 1991k); i.e., a country seeking substantial EU financing could hardly expect to retain exclusive competence for selecting sites. Nevertheless, Spain, Ireland and Portugal expressed reservations regarding this aspect of the site designation process (CEC, 1991k).

Protection of Natura 2000 sites

Crucial changes: extending 'appropriate assessment' to plans as well as projects; and the 'integrity of the site'

On 15 January 1991, in light of discussions with the Luxembourg Presidency the previous day, the Commission made two crucial changes to the directive's impact assessment provision (DG XI, 1991o): first, the provision was amended to apply an appropriate assessment to *plans* as well as to projects likely to significantly affect a special area of conservation, either alone or in combination with other plans or projects; second, the same provision was amended to provide that, in light of such an appropriate assessment, as a general rule a Member State could not authorise the plan or project in question unless they were satisfied that the plan or project would not significantly affect the integrity of the site in question. Where such a conclusion could not be reached, a Member State could allow the plan or project to proceed via a derogation provision, but only where this was required for reasons of overriding public interest, and where all necessary measures were taken to ensure that the conservation potential of the Natura 2000 network would not be adversely affected. In other words, the provision now dictated the circumstances in which a plan or project could – or could not – proceed, in contrast to the earlier proposal of applying the EIA Directive to Natura 2000 sites, which would not have had this effect.

Taking account of economic factors in protecting sites

At this point, the dispute being played out over the Birds Directive in the *Leybucht Dykes* case – between the Commission on one side and the UK and Germany on the other – spilled over to the text of the draft Habitats Directive, when the UK was given the opportunity to comment on a working draft of the directive shortly prior to the formal presentation of the Luxembourg Presidency's compromise proposal. By 12 February 1991, the working draft's derogation provision had been further amended to provide:

> If the plan or project must nevertheless be carried out for reasons of overriding public interest related to public health and security [i.e., categories recognised as legitimate reasons for damaging a special protection area under the Birds Directive by the Advocate General's Opinion in the *Leybucht Dykes* case (CJEU, 1990)], the Member State shall take all necessary steps to ensure that the conservation potential of Natura 2000 will not be affected. The Member State shall transmit to the Commission the results of the assessment and, as appropriate, the adopted compensatory measures.
>
> *(DG XI, 1991r, Article 7(2))*

Commenting on this provision on 15 February, the UK government stated, "after 'security' insert 'and economic factors'" (Department of the Environment, 1991b). This effectively sought to weave the UK's and Germany's argument in the *Leybucht Dykes* case into the text of the Habitats Directive, in advance of the CJEU's judgment on the issue in respect of the Birds Directive.

On 28 February 1991, the Luxembourg Presidency formally presented its compromise proposal, containing the following as Article 6(4):

> Article 6
>
> 4. If, in spite of a negative impact assessment and in the absence of other feasible solutions, a plan or project must nevertheless be carried out for reasons of overriding national interest, the Member State shall take all necessary steps to ensure that the coherence of Natura 2000 will not be affected and shall inform the Commission of the adopted compensatory measures.
>
> *(CEC, 1991c)*

Thus, the derogation provision had been further amended to require "other feasible solutions" to be excluded, on top of the language about compensatory measures and reasons of overriding interest. The bare reference in paragraph 4 to "reasons of overriding national interest"[7] indicates that the Commission/Luxembourg Presidency had backed off from the idea of limiting such reasons to public health and security, favouring broader language that could potentially encompass the "economic factors" the UK sought to have added to the draft before the outcome in *Leybucht Dykes* was known.

CJEU's judgment in Leybucht Dykes

The CJEU's decision in *Leybucht Dykes* was handed down on 28 February 1991, the same day the Luxembourg Presidency's compromise proposal was presented, and "created a huge wave amongst the Member States" (Neal, 2012). The CJEU agreed with the Advocate General's Opinion, holding that (emphasis added):

> the power of the Member States to reduce the extent of a special protection area [under the Birds Directive] can be justified only on exceptional grounds . . . *Those grounds must correspond to a general interest which is superior to the general interest represented by the ecological objective of the directive. In that context the interests referred to in Article 2 of the [birds] directive, namely economic and recreational requirements, do not enter into consideration* . . . With regard to the reason put forward in this case, it must be stated that the danger of flooding and the protection of the coast constitute sufficiently serious reasons to justify the dyke works and the strengthening of coastal structures as long as those measures are confined to a strict minimum and involve only the smallest possible reduction of the special protection area.
>
> *(CJEU, 1991, paragraphs 21–23)*

Thus, Germany won the case on the facts, but Germany and the UK had lost the battle – at least for now – on the issue of taking socio-economic factors into account in deciding whether and how sites should be protected. In the Member States' view, the CJEU's decision went too far (Krämer, 2009): many governments were reportedly "alarmed" (Baldock, 1992, p. 144) by the decision's "draconian consequences" (Sharp, 1998, p. 33).

Discussing Article 6 at a meeting of the Working Party on the Environment on 21 March 1991, the Netherlands and Portugal proposed deleting the word "development" from "development plan or project" in Article 6(3), and this was successfully removed, potentially broadening the application of the impact assessment obligation (CEC, 1991p).[8] Pulling in the opposite direction, Ireland tried to narrow the impact assessment obligation by arguing that the words "or in combination with other plans or projects" should be deleted, such that cumulative effects would not need to be assessed; this was successfully opposed by the Commission and the Netherlands (CEC, 1991p). In respect of the proposed derogation provision, Article 6(4), Belgium, Ireland and the UK wanted the interpretation of "overriding reasons of national interest" to be clarified, but, surprisingly, no mention of the CJEU's recent decision in the *Leybucht Dykes* case is recorded in the minutes of this meeting (CEC, 1991p).[9]

A new version of Article 6 was presented in the Luxembourg Presidency's revised draft of 3 May 1991, and this was discussed at a meeting of the Working Party on the Environment on 16 and 23 May (CEC, 1991o). Spain and France successfully argued that the words "insofar as these would have a significant effect with regard to the objectives of this Directive" should be added to the end of

Article 6(2), and revised versions of Articles 6(3) and (4) also raised debate.[10] The proposed changes to Article 6(4) included the replacement of "reasons of overriding national interest" with "imperative reasons of overriding public interest", as well as an entirely new second paragraph:

> [Article 6(4)] If, in spite of a negative assessment of the implications for the site and in the absence of alternative solutions, a plan or project must nevertheless be carried out for imperative reasons of overriding public interest, the Member State shall take all necessary steps to ensure that the coherence of Natura 2000 will not be affected. It shall inform the Commission of the compensatory measures adopted.
>
> Where the site concerned hosts a priority natural habitat type and/or a priority species, the only considerations which may be raised are those relating to human health and public safety.
>
> *(CEC, 1991d)*

The derogation provision thus now clearly distinguished between two categories of sites: regular sites, in respect of which any imperative reason of overriding public interest could be used to justify damage to a site; and priority sites, in respect of which a more restrictive rule would apply.

The UK pays a visit to the Commission re: Article 6(4)

France, the Netherlands and the UK felt that the proposed derogation in respect of priority sites was too restrictive, and Greece argued that considerations relating to the exploitation of strategic resources should also be taken into account (CEC, 1991o). On 4 June 1991, Roy Bunce of the UK government met informally with DG XI's Claus Stuffmann to discuss the implications of the CJEU's judgment in *Leybucht Dykes*. As Bunce recalls,

> I remember the *Leybucht* case causing consternation. I can remember that almost immediately I sought a private meeting with Claus Stuffmann . . . I went to Brussels, he came out of a meeting, we found a quiet [spot] in some corridor, we didn't even go into a Committee room, and we sat and I told him in terms that in the light of *Leybucht* the British government would be unable to basically continue our support for the directive. We had to have something which enabled us to take, to put precedence for human needs over those of wildlife, and the *Leybucht* seemed to destroy that. And I think it's fair to say now that the particular concern of the British government were estuaries – very important for wildlife but also extremely important for our trade: Southampton water . . . the Felixstowe, the Humber – all very important bird habitats, but absolutely vital economic interests.[11] And we just couldn't, we couldn't have lived with the way the Commission[12] interpreted the Birds Directive.
>
> *(Bunce, 2012)*

The UK followed up with a letter on 10 June 1991, stating:

> When we met informally on 4 June I explained to you the UK's need to have a single legal regime of protection applying to both SPAs classified under the Birds Directive and Special Areas of Conservation (Natura 2000 sites) envisaged under the Habitats Directive, and also the need for this regime to give Member States the right to take economic considerations into account when deciding the future of these sites . . . You kindly indicated that the Commission were not unsympathetic[13] to the UK's concerns but that you were concerned to ensure that the protection offered to these sites protected them from all but overriding issues of national importance. You indicated that this could include economic considerations with regard to Natura 2000 sites but that for 'priority' sites hosting species or habitat types in danger of extinction the protection should be waived only for matters of public health or safety or other overriding environmental benefit. You invited the UK to propose amendments to the text of the Directive that would achieve these aims.
>
> *(Department of the Environment, 1991a)*

This the UK did, as follows, keeping "as precisely as I know how to my understanding of what would be acceptable to you [i.e., DG XI]" (Department of the Environment, 1991a) (proposed insertions in italics, proposed deletions in bold):[14]

> [Article 6(4)] If, in spite of a negative assessment of the implications for the site and in the absence of alternative solutions, a plan or project must nevertheless be carried out for **imperative** reasons of overriding public *or economic* interest, the member state shall take *compensatory measures* **all necessary steps** to ensure that the *overall* coherence of Natura 2000 will not be affected. It shall inform the Commission of the compensatory measures adopted.
>
> When the site concerned hosts a priority natural habitat type or a priority species, the only considerations which may be raised are those relating the human health or public safety *or overriding environmental benefit.*
>
> *(Department of the Environment, 1991a)*

In addition to this proposed drafting, the UK made a play for "social" factors to be taken into consideration in addition to economic interests in the first paragraph of Article 6(4):

> In discussions with my colleagues here, particularly those with responsibilities for remoter areas containing small communities facing difficult conditions, I have been pressed to include the word 'social' in line 4 of paragraph 4 of Article 6. My understanding is that this is unacceptable to you but if, on reflection, you can see merit in including the means to take account of

> matters that might be vital to the survival of small human communities as social units but which could not be described as overriding economic significance at the national scale, that would certainly help the UK.
>
> *(Department of the Environment, 1991a)*

On 12 June 1991, the Presidency proposed revised text intended to address the UK's concerns, as well as to tackle certain other major issues, including financing (CEC, 1991b). In addition to an amendment to apply Article 6(2) to (4) of the Habitats Directive to SPAs under the Birds Directive, "clarifying" the obligation under Article 4(4) of the Birds Directive (the provision that had been interpreted strictly by the CJEU in the *Leybucht Dykes* case), the Presidency proposed the following in respect of the Article 6(4) derogation (proposed changes in italics):

> [Article 6(4)] If, in spite of a negative assessment of the implications for the site and in the absence of alternative solutions, a plan or project must nevertheless be carried out for imperative reasons of overriding public interest *of a social or economic nature*, the member state shall take *such compensatory measures as are necessary* to ensure that the *overall coherence of Natura 2000 is protected*. It shall inform the Commission of the compensatory measures adopted.
>
> When the site concerned hosts a priority natural habitat type or a priority species, the only considerations which may be raised are those relating to human health or public safety *or overriding environmental benefit*.
>
> *(CEC, 1991b, pp. 3 and 6)*

At first glance this might appear to have addressed the UK's concerns, but the framing of the point regarding socio-economic concerns in fact narrowed the conditions in which the derogation would be applicable, by qualifying the imperative reasons of overriding public interest that could lawfully be taken into account; in other words, *only* interests of a social or economic nature could be used to justify damage to a regular Natura 2000 site in this revised draft. When the provision arose for discussion at the Ministerial level Environment Council meeting of 13–14 June 1991, this issue was addressed, and the following text was agreed (emphasis added):

> [Article 6(4)] If, in spite of a negative assessment of the implications for the site and in the absence of alternative solutions, a plan or project must nevertheless be carried out for imperative reasons of overriding public interest, *including* those of a social or economic nature, the Member State shall take all compensatory measures necessary to ensure that the overall coherence of Natura 2000 is protected. It shall inform the Commission of the compensatory measures adopted.

> When the site concerned hosts a priority natural habitat type and/or a priority species, the only considerations which may be raised are those relating to human health or public safety or to beneficial consequences of primary importance for the environment.
>
> *(CEC, 1991k)*

As Roy Bunce, the UK's senior negotiator, notes, "We did have some problems in that obviously the Commission is always against taking into account national interest, that's why it's all written in terms of public interest rather than national interest, and they were worried about inserting 'economic', 'of an economic nature', into it, but that had to be insisted upon and was finally – at the very end – finally agreed" (Bunce, 2012). The amendment "overriding public interest, *including* those of a social or economic nature" ensured that the interests that could be taken into account in applying a derogation would not be limited to socio-economic matters.

Interestingly, the proposal that special protection areas under the Birds Directive should be subject to Article 6(2) to (4) of the new directive rather than Article 4(4) of the Birds Directive – effectively overruling the CJEU's decision in the *Leybucht Dykes* case – did not sail through, with Spain maintaining a reservation regarding the point (CEC, 1991k). Antonio Troya, a Spanish national expert seconded to DG XI, explains that Spain's objections in respect of Article 6 were directly linked to its goal of achieving the inclusion of a provision on EU co-financing within the text of the Habitats Directive (Troya, 2012): in other words, this issue regarding a single regime for SPAs and SACs was something the UK wanted very badly, while Spain was holding out strongly for a deal on financing, as detailed below, such that objecting to Article 6 gave Spain a bargaining chip. As one of the UK's negotiators puts it, "That was very much the dynamic of the final stages of negotiations, where they [Spain] weren't going to capitulate on what they knew we really, really wanted, until we were prepared to sign up to something which was . . . tricky for us" (Neal, 2012).

Financing

On 31 January 1991 the Commission had presented a new, grand proposal: for a Council Regulation establishing a Financial Instrument for the Environment (LIFE) (European Commission, 1991c).[15] The aim was to introduce an "Environment Fund commensurate with the problems to be solved and the role which the Community must play in this area", which fund would ultimately subsume other funding initiatives such as ACNAT (European Commission, 1991c, p. 5). However, LIFE's precise scope and levels of financing were unclear, and this proved problematic from the perspective of the Habitats Directive proposal.

By April 1991 the issue of financing had become of foremost political concern for the future of the Habitats Directive proposal, in particular given the position of Spain, which had made it clear that its agreement to the Directive's entry into force would be dependent on the creation of a financial instrument with guaranteed levels of funding (DG XI, 1991b). In that regard, ACNAT, which was by now said to be likely to be ready for adoption in June 1991, would be totally insufficient in Spain's view (DG XI, 1991b). The Commission's services could not resolve the issue, and recommended that everything should be done at a political level to find a solution (DG XI, 1991b).

Reporting to the Council on 7 June, COREPER reinforced this point, recording that Spain, Greece, Ireland, Italy and Portugal "questioned the ability of existing or projected financial instruments, in particular ACNAT and LIFE, to make an effective contribution to financing the Directive" (CEC, 1991r, p. 3). The Commission's representative responded that, in addition to ACNAT, other EU instruments in the framework of the environment and the agricultural and regional policies could contribute to the financing of the directive, and that the involvement of these instruments should be expressly provided for within the text of the Habitats Directive – the Commission representative undertook to submit a proposal along these lines (CEC, 1991r). The Commission worked on this idea at the beginning of June 1991, sending Spain for review a draft Article for potential inclusion within the Directive, covering Spain's two "major concerns": first, the possibility for a Member State to make clear what costs deriving from the directive it considered itself unable to cover without EU funds; second the possibility to postpone a number of measures (more below), in agreement with the Commission, in case the available EU funds proved insufficient (DG XI and Spain, 1991).

To address Spain's concerns, on 12 June 1991 the Presidency formally proposed revised text for the directive (CEC, 1991b). Article 7 of the revised text proposed obliging the Member States to inform the Commission of their programmes relating to nature conservation and activities that followed specifically from the implementation of the Habitats Directive, including the estimated costs involved (CEC, 1991b). On the basis of this information, the Commission would establish a framework of EU activities, identifying the measures essential for the survival of species or the maintenance of priority natural habitat types in the EU, and "the necessary joint financing which can be provided within the framework of the relevant budget funds and procedures" (CEC, 1991b).

In establishing this framework of activities, the Commission was to be obliged to take into account

> the specific situations of the various Member States, especially where, because of the high percentage of areas to be designated in relation to its territory, the existence of a large number of priority areas within those areas, and the socio-economic situation of the regions concerned, the financial burdens resulting from the implementation of Article 6(1), (2) and (3) [site management and protection provisions] are disproportionate for one Member State in comparison with those borne by the other Member States.
>
> *(CEC, 1991b)*

Further, the draft provision provided that implementation of Article 6(1) to (3) could be suspended if it proved impossible to provide adequate joint financing (CEC, 1991b). In such a case, the Member State would be obliged not to take any action liable to degrade the area in question (CEC, 1991b). Following further comments from Spain, the Presidency tabled a revised version of this provision for the purposes of the Environment Council meeting on 13–14 June 1991 (CEC, 1991e). However, when the proposal was discussed at that meeting, the new provision did not receive the wholehearted support Spain may have wished for, with some delegations noting that the provision might have considerable financial implications that ought to be studied closely (CEC, 1991k).

The Luxembourg Presidency: "absolutely brilliant"

According to Roy Bunce of the UK government, the Luxembourg Presidency was, in summary, "absolutely brilliant. I mean, everybody knew that they had been . . . [T]hey had no axe to grind anywhere, and they really did seek a compromise and they worked extremely hard and were extremely good, and if it wasn't for them I don't think the directive could've been finally passed" (Bunce, 2012).

Dutch Presidency: July–December 1991

Designation of Natura 2000 sites

Work on the directive continued apace under the new Dutch Presidency, with a meeting of the Working Party on the Environment held just over a week in, on 10 July 1991 (CEC, 1991m). The question of the Commission's ability to propose, and the process for proposing, priority sites for designation where they were absent from a national list remained strongly contested. As a compromise, the Dutch Presidency proposed in Article 5 that any such proposal would be subject to a unanimous decision of the Council, rather than a qualified majority, which had been the previous position (CEC, 1991m). Germany, Greece and Portugal said that they could agree to this; Spain and Ireland upheld a reservation of principle regarding the Article; Belgium, Denmark, France and Italy considered a qualified majority vote to be the most appropriate solution and pointed to the close link with the new Article 7 financing provision, effectively saying that Spain could not ask for substantial EU money to fund implementation of the directive while at the same time rejecting any possibility of the Commission stepping in to propose designation of a priority site which Spain had left off its national list (CEC, 1991m). The UK similarly preferred a qualified majority vote, but primarily wanted clarification regarding what would happen if no decision at all was taken by the Council (CEC, 1991m). DG XI's Legal Service made it clear that in such circumstances the Commission could bring a case regarding the situation before the CJEU (DG XI, 1991h).

At an Environment Council meeting on 1 October 1991 the Commission expressed itself to have a net preference for a qualified majority vote in the

Council, but, in a spirit of compromise, could accept the need for a unanimous decision (DG XI, 1991e). However, the Dutch Presidency's compromise proposal of a unanimous decision did not obtain a consensus, and the Dutch then proposed as an alternative solution that the Article be deleted altogether (CEC, 1991i). Denmark, Germany,[16] Spain, Greece, Ireland, Luxembourg and Portugal were agreeable to this solution, or could accept it as part of an overall compromise; however, Belgium, France, Italy and the UK still considered that a qualified majority was the most appropriate solution (CEC, 1991i). Indeed, the Belgian, French and UK Ministers remained strongly attached to the provision, arguing that the Article 5 procedure would help to ensure a smooth implementation of the Natura 2000 network, allowing site designation issues to be determined at the level of the Council rather than being brought before the CJEU, the latter having been the experience with the Birds Directive (CEC, 1991f).

Territorial extent of the Directive

The issue of the precise territorial extent of the Directive remained live as late as September 1991, with the Directive no longer referring explicitly to any maritime areas, instead referring simply to the "European territory of the Member States to which the Treaty applies" (CEC, 1991n). Following a statement from the Council's Legal Service to the effect that this phrase included territorial waters (CEC, 1991n), on 16 August 1991 the UK proposed a change to the site designation process – which was accepted and retained in the adopted version of the directive[17] – adding the words, "For aquatic species which range over wide areas such sites will only be proposed where there is a clearly identifiable area representing the physical or biological factors essential to their life and reproduction" (UK Government, 1991).

Twyford Down EIA dispute

All the while, in the background, Environment Commissioner Carlo Ripa di Meana was in full flow, reiterating his call for a new EU inspectorate as part of the then proposed European Environment Agency, "to ensure the Community's green laws are not flouted" (*EC Energy Monthly*, 1991).[18] A major dispute regarding the Twyford Down M3 motorway in the UK, and the UK government's alleged failure properly to assess the impacts of that and several other proposals pursuant to the EIA Directive, was beginning to heat up at this time, with sources in Brussels reporting at the end of August 1991 that a letter of formal notice would soon be sent to the UK government (McCarthy, 1991a). This dispute would later blow up into a political crisis, threatening the UK's support for the Maastricht Treaty, which would ultimately be signed in December 1991 (see further below).

Flexible site designation criteria

Spain continued to negotiate hard in respect of other aspects of the site designation process: regarding the situation where more than 5% of a country's territory comprised sites hosting priority habitats or species, the draft directive proposed allowing the country in question to request that criteria other than those set out in Annex III be applied for selecting sites of Community importance on their territory (CEC, 1991n). The UK proposed that this should be amended to instead read that the criteria listed in Annex III may be applied more flexibly in selecting such sites; Spain rejected this proposal initially, but ultimately accepted the UK position (CEC, 1991n; DG XI, 1991j).

Amidst these detailed discussions, there was a possibility that political agreement regarding the Directive might be reached by Ministers at the Environment Council meeting on 1 October 1991, and the European Parliament and NGOs kept up the pressure from outside: the Parliament, in collaboration with the WWF, hosted a conference "Europe 2000 – Let's Protect Nature" from 25–26 September 1991, where Dutch MEP Hemmo Muntingh reflected that "the Community must truly assume its responsibilities and react to the questions of nature protection if it is to be taken seriously by the rest of the world", noting that the EU's contribution of 0.2% of its budget for the environment was "ridiculous" (*Europe Environment*, 1991).

Environment Council of October 1991

Introducing the Habitats Directive proposal to the Environment Council on 1 October 1991, Commissioner Carlo Ripa di Meana noted that the public could not understand why the Council, after three years of work, was not yet in a position to take a decision regarding the Directive, and flagged that the adoption of the Directive could be considered as an important contribution looking towards the Earth Summit in Rio, due to be held in 1992 (DG XI, 1991e; CEC, 1991f).

Despite finally reaching agreement on the content of the directive's annexes (CEC, 1991g; CEC, 1991i),[19] the Council could not however reach agreement on the text of the directive at this meeting, and charged COREPER with continuing to examine the directive with a view to an agreement at the next Environment Council meeting, in December 1991 (CEC, 1991q).[20]

UK (non)compliance

Right in the heart of these final negotiations, Member States continued, of course, to be held to account for alleged failures to apply EU environmental law. Embarrassingly for the UK's Environment Secretary Michael Heseltine, who had taken a relatively positive line towards the Habitats Directive, the Commission's decision to bring legal proceedings against the UK government in respect of the Twyford Down M3 motorway was announced the day before the press reported that Heseltine would be seeking to "expose European backsliders" in respect of EU environmental law (McCarthy, 1991b; *European Report*, 1991a). Although Heseltine declined to give examples, the press reported "many flagrant failures to implement EC environmental measures, such as the Italian government's unwillingness to enforce the Birds Directive by preventing the annual slaughter of thousands of birds of prey by Italian hunters" (McCarthy, 1991b). The UK evidently felt it was above such behaviour, with Heseltine claiming that, "Many of the people who sign up to these directives don't do anything about them. We can have a position where directives are acclaimed with enthusiasm, and then shelved with dispatch, except in this country" (McCarthy, 1991b).

Protection of Natura 2000 sites

Dealing with Leybucht Dykes

At the first meeting regarding the Directive under the Dutch Presidency, on 10 July 1991, to maintain tactical leverage vis-à-vis the UK, Spain continued to query the idea that Article 6(2) to (4) of the Habitats Directive would be applied to "clarify" obligations in respect of SPAs under the Birds Directive, and requested the opinion of the Council Legal Service on the obligations stemming from the introduction of this provision in relation to the obligations that already existed under the Birds Directive (CEC, 1991m). The Council Legal Service ultimately proposed text to cover the situation, which stated that obligations under Article 6(2) to (4) of the Habitats Directive would replace (rather than clarify) obligations in respect of SPAs under the first sentence of Article 4(4) of the Birds Directive (which had been interpreted strictly in the *Leybucht Dykes* case), and a large majority of delegations agreed to the proposed text (CEC, 1991l).

At a meeting of the Working Party on the Environment on 4 September 1991, Ireland made an audacious late attempt to restrict the effect of the impact assessment provision – Article 6(3) – by proposing that only projects should require appropriate assessment and not also plans (CEC, 1991n).[21] The Commission was naturally against this, and successfully resisted Ireland's proposal (CEC, 1991n).

The second paragraph of the Article 6(4) derogation

At the same meeting, a separate attempt was made to weaken the protective regime by way of a proposed deletion of the second paragraph of the Article 6(4) derogation, which then read: "Where the site concerned hosts a priority natural habitat type and/or a priority species, the only [derogation] considerations which may be raised are those relating to human health or public safety or to beneficial consequences of primary importance for the environment" (CEC, 1991n).

Belgium, Denmark, Greece, Italy and the Commission entered scrutiny reservations regarding the proposed deletion (CEC, 1991n). Preparing for the Environment Council meeting on 1 October 1991, the Commission, internally, made its position clear regarding the proposed deletion: if a majority rallied around this, the Commission would express its strong objections to deletion and would propose that: first, the application of a derogation in respect of priority species/habitats would be made the subject of an Opinion from the Commission, to be given within three months; and second, that in respect of sites which benefited from EU co-financing under then Article 7,[22] only derogations based on human health, security or beneficial consequences of primary importance for the environment would be acceptable (DG XI, 1991d). The first idea found its way into the directive, minus the deadline, with the second paragraph of Article 6(4) amended to read, after the Environment Council on 1 October 1991: "Where the site concerned hosts a priority natural habitat type and/or a priority species, the only considerations which may be raised are those relating to human health or public safety, to beneficial consequences of primary importance for the environment[23] or, further to an opinion from the Commission, reasons of overriding public interest" (CEC, 1991h).

In other words, the Commission had accepted the principle that Member States could apply a derogation to allow damage to sites protecting priority habitats/species for reasons of overriding public interest[24] *other than* those relating to human health, public safety, or to beneficial consequences of primary importance for the environment, provided that those other reasons be made subject to an Opinion from the Commission in advance (DG XI, 1991j). That said, from the Commission's perspective, it remained to be clarified that such other "reasons of overriding public interest" should be qualified as "imperative" reasons, as per the first paragraph of Article 6(4); after all, as the Commission noted internally, it would be difficult to justify the availability of a more permissive derogation in respect of priority habitats/species than was the case for non-priority habitats/species (DG XI, 1991j). The opportunity to achieve this came when the UK government requested that the derogation provision in respect of the species protection regime – what became Article 16(1)(c) – be amended to include reference to imperative reasons of overriding public interest, to mirror the Article 6(4) provision in respect of sites;[25] one of the *quids quo pro* for this, from the Commission's perspective, was the insertion of the word "imperative" into the second paragraph of Article 6(4) (DG XI, 1991q), and the Commission set out to achieve this at the Environment Council meeting on 12 December 1991 (DG XI, 1991l).[26]

Financing

At a meeting on 10 July 1991, Germany – supported by Belgium, Denmark, France and the UK – argued that the text for then Article 7,[27] the proposed provision on EU co-financing to be included within the Habitats Directive, "should make a clearer distinction between the 'normal' situation and cases where Community co-financing was necessary, for which precise obligations and criteria would need to be defined" (CEC, 1991m, p. 23). The Presidency undertook to produce a revised version of the text based on these discussions (CEC, 1991m). Some delegations, in addition, drew attention to the link with the Article 5 site designation process (CEC, 1991m) – again, the implication being that Spain could not hold out for major EU co-financing in circumstances where it was unwilling to accept that the Commission would have power to propose priority sites for designation that Spain had left off its national list.

By the end of September 1991 the parties remained apart, with some doubting the advisability of including a financial provision in the directive itself, stressing "the exceptional nature of an approach of this type in the general context of Community environmental policy, especially in relation to the 'polluter pays' principle" (CEC, 1991s, p. 3). Some were concerned about the precedent that might be set by linking the entry into force of a directive with the provision of EU funds (COREPER, 1991a). Germany and Spain emerged as the key players in this regard, with the Dutch Presidency meeting with Spain in Madrid on 24 September 1991 to discuss new recitals desired by Germany on this issue for the Directive's preamble (DG XI, 1991c).

Having failed to reach agreement regarding the Directive proposal at the Environment Council meeting on 1 October 1991, the Council charged COREPER with continuing to work on the Habitats Directive and LIFE funding proposals, with a view to reaching agreement on both as soon as possible (CEC, 1991q). By this time financing had escalated to become *the* core political problem for the Habitats Directive:

> The technical problems raised during the [1 October Council meeting] soon gave way to substantive ones, with Spain asking for an automatic system of co-funding to be set up to implement the measures contained in the Directive so that each Community obligation would be backed up with sufficient resources. The other 11 Member States were at pains to point out that a Directive had to be applied at national level by being incorporated into national legislation. As a result, the cost of introducing these sorts of measures should be borne at national level, too . . . Thumbs down for automatic co-funding. Spain's main concern during the Council was to establish just to what extent the Community would be prepared to provide aid. The 11 other Community members were reluctant to enter into this debate as this question was thought to be a smokescreen for Spain's idea that all Community activities should be financed at EEC-level, or so some sources would have us believe. Spain wants the measures to be recognised as Community ones [within the meaning of Article 130r(4) of the EEC Treaty] so that all special protection areas established by the proposed Directive would be regarded as of relevance to the Community, thus implying that the Community should provide some funding. The other Member States felt, however, that Community funding should be reserved for exceptional cases and to give in to Spain on this point could have created a dangerous precedent for environmental policy and have repercussions for all other Community policies . . . Creating a precedent was the biggest fear lurking in the minds of the Ministers, who finally expressed their willingness to finance certain activities, but refused to recognise the activities in the natural habitat Directive as Community measures. The German Environment Minister, Klaus Töpfer, summed up the Member State's attitude when he said: "We are prepared to finance the habitat Directive through the LIFE programme but not to fund the Directive directly".
>
> *(European Report, 1991c)*

A fundamental difference in approach had opened up between two groups: the countries of the south (joined by Ireland) sought an EU co-financing mechanism that was quasi-automatic and unconditional, while certain northern European countries (Belgium, Germany, France, the UK) could accept the principle of a certain financial solidarity, but wanted any co-financing provision within the text of the Directive to be subject to strict conditions, and expressed as an exceptional case (CEC, 1991f). The ministers of Spain and Germany met in the margins of the informal Environment Council meeting of 12/13 October 1991 to see if they could break the impasse, but the Spanish attitude was reportedly "extremely hard": if the Community wanted to impose concrete obligations on Member States, Spain said, it needed to be willing to contribute the necessary financial means to allow countries to meet those obligations (DG XI, 1991g). Such a consideration would apply not only in the case of the Habitats Directive but also in other situations, and Spain thus rejected the idea of including a paragraph in the preamble citing the "exceptional" nature of a co-financing mechanism within the terms of the Habitats Directive (DG XI, 1991g). The Dutch Presidency hoped to finesse this by writing to Spain and Germany proposing that "exceptional" might be replaced with a reference to the "particular" character of nature conservation measures (DG XI, 1991g), but this did not break the impasse.

The Twyford Down EIA dispute erupts in the UK

With DG XI's Nature Unit seeking to manage this very sensitive issue, just a few days later, towards the end of October 1991, the brewing dispute between the Commission and the UK regarding the UK's alleged failure to apply the EIA Directive to several development projects, including the M3 motorway at Twyford Down, erupted, casting a shadow over the UK's support for the draft Maastricht Treaty, which was then entering a critical phase (Usborne, 1991). "I hear they are calling me Carlo the Ripper", said Ripa di Meana. "I'm not the one who wants to rip apart an 8,000-year-old English forest" (White and Wolf, 1991). By the height of the dispute, Ripa di Meana had received over 1,000 letters from the British public, including two death threats, and proudly displayed to a journalist the certificate in his office recording the fact that he had personally adopted one of the chainsaw-threatened oaks from the forest in question (Schoon, 1991). The UK's Secretary of State for Transport Malcolm Rifkind described himself as "rather irritated" by the intervention of the Commission, which had "some explaining to do", he said (Moncrieff and Archer, 1991). When told that Ripa Di Meana was seen by some as a white knight riding to the European environment's rescue, Rifkind reportedly retorted: "Knights sometimes fall off their horses" (Dynes et al., 1991). And so it would prove.

As DG XI itself noted in respect of the Habitats Directive proposal, the Maastricht discussions seemed likely to be very important for the Directive itself: that is, the solution to the Article 7 financing issue – now the main blockage – was felt likely to depend to a great degree on the ministerial discussions in Maastricht on 9–10 December 1991, where the issue of EU co-financing of measures relating to the environment, culture, education and tourism would be examined (DG XI, 1991j). In this high politics forum, the Habitats Directive proposal did not seem likely to fall under scrutiny, except that the UK was at the very same time fiercely disputing similar (and less stringent) impact assessment obligations in the EIA Directive. As the UK's then Foreign Secretary Douglas Hurd commented, when asked to give an example of the European "straitjacket" that Britain wanted to avoid, "When out of some desk in Brussels comes something telling Britain which field a road should go through" (MacIntyre, 1991).

On the brink of being withdrawn, the Habitats Directive is agreed

After more than three years of negotiations, it was very much make or break time for the Habitats Directive proposal, the Commission having by this time nearly reached the end of what was typically accepted to be the maximum number of Presidencies for securing agreement on a proposal (seven in this case) (Yourassowsky, 2012). Rumours of a potential withdrawal of the proposal were circulating: the WWF wrote to DG XI on 9 December 1991 stating that they were "very concerned to hear . . . that the Commission is considering withdrawing the Habitats Directive if it is not adopted under the Dutch Presidency", and encouraged the Commission to leave the proposal on the table throughout 1992 (WWF, 1991). However, unbeknownst to the WWF and others, Portugal had made it clear to the Commission that Portugal would not be taking the Habitats Directive dossier forward during its Presidency in the first half of 1992, in the event of a stalemate at the Environment Council in December 1991 (DG XI, 1991a).

The main sticking point on the financing issue was said by DG XI to be Germany's persistence regarding the need to make explicit the exceptional nature of a link being made between the implementation of a directive and the availability of EU funds (DG XI, 1991a). However, this framing of the issue may well have been coloured by the Dutch Presidency's decision (per the Commission) to try to isolate Germany on the point, after the Presidency's proposed drafting of Article 7 was welcomed by Spain, Portugal, Greece, Italy and Ireland at the COREPER meeting on 29 November but rejected by Germany (DG XI, 1991k; COREPER, 1991e). However, this tactic had not really worked, since France, the UK and Belgium had shown a certain solidarity with Germany (DG XI, 1991k; COREPER, 1991e). The UK then proposed changes to Article 7 for consideration at the Environment Council meeting on 12 December (CEC, 1991u). These changes were, however, considered "too brutal" by the Commission, as well as having been tabled too late in the day, which risked alienating Spain et al. (DG XI, 1991i).

Going in to the crunch Environment Council meeting on 12 December 1991, there was therefore no certainty that a deal could be done to allow the Habitats Directive to be adopted. Introducing the Directive proposal at that meeting, Environment Commissioner Ripa di Meana noted that it would be a sad first – which from the outside would risk appearing "grotesque" – if a proposed piece of EU legislation were to fall owing to a dispute over the text of the preamble (where Germany was seeking insertion of the "exceptional" language relating to financing) (DG XI, 1991m). Ripa di Meana then played to the German part of his audience a little, referring to the Article 7 financing mechanism as "exceptional" and "*sui generis*", before pausing to hear the interventions of the Member States (DG XI, 1991m).[28] The Commissioner was then presented with two scenarios by his speech-writer: in the first, he was to call upon Spain to compromise by acceding to Germany's demand to record in the directive's preamble the exceptional nature of Article 7, Spain having already obtained most of what it had asked for in respect of that Article (DG XI, 1991m). There then followed the second scenario, under a section headed "In extremis", for use should Spain refuse to give any concessions, in which case the Commissioner would propose withdrawing the directive proposal altogether, call the outcome "deplorable" in light of the efforts of the Dutch, Luxembourg and French Presidencies, and highlight the disastrous outcome for habitats and species in the EU (DG XI, 1991m).[29] This latter scenario was evidently considered a serious possibility, for in the Commission's archives is a draft press release – unused in the event – for use should Spain have refused to budge, announcing the Commission's "profound regret" at having to withdraw the proposal, name-checking Spain as the guilty party, and promising a mobilisation of all the Commission's resources to enforce strictly the Birds Directive as well as to monitor more closely and have a decisive say regarding the execution of policies, plans and programmes with significant environmental impacts (DG XI, 1991f).

Ultimately, of course, this nuclear option did not come to pass, and political agreement was reached regarding the Habitats Directive and the LIFE funding instrument at the meeting on 12 December 1991 (COREPER, 1991d; *European Report*, 1991b). Spain secured the inclusion of the Article 7 provision on financing,[30] and Germany secured a provision for the preamble stating (emphasis added) "in this *exceptional* case, a contribution by means of Community co-financing should be provided for within the limits of the resources made available under the Community's decision" (Habitats Directive, 1992). Spain, which had "wanted the Community to finance all measures provided for in environmental Regulations or Directives" (*European Report*, 1991b) had therefore compromised,[31] but sought 2 billion Ecus for LIFE, while the Dutch Presidency proposed a compromise of 500 million Ecus, which many delegations still felt was too high. Agreement was reached on a figure of 400 million Ecus for the first stage of the programme (*European Report*, 1991b).

As a parting shot, Germany's Minister Töpfer agreed to sign up to the Habitats Directive only with a declaration recorded in the Council's minutes to the effect that Germany presumed that, from 1994, the financing of measures foreseen by Article 7 of the Habitats Directive would be financed by the Cohesion Fund in eligible countries (COREPER, 1991d; CEC, 1991j). The Cohesion Fund was a new fund – "the fruit of a political trade-off on the eve of the Maastricht Treaty" – intended to direct EU funding to four specific less-developed Member States: Spain, Portugal, Greece and Ireland (Borrás, 1998, p. 211). For their part – and unsurprisingly – Spain, Portugal, Greece and Ireland made an equal but opposite declaration, to the effect that measures under the Habitats Directive should *not* be financed by the Cohesion Fund (COREPER, 1991d; CEC, 1991j). The majority of the other Member States, plus Commissioner Ripa di Meana, supported Spain et al., noting that the Cohesion Fund was intended only for use in respect of four Member States, and not to fund environmental measures in all twelve, such as might be foreseen by the Habitats Directive (COREPER, 1991d).[32] To secure the agreement, however, Ripa di Meana declared that the Commission would take into consideration, in its report on the operation of the LIFE programme, all relevant elements pertaining to environmental policy, including the manner in which the Cohesion Funds had been put in place (COREPER, 1991d). And in this spirit of compromise and goodwill, after more than twelve hours of negotiations at this final crucial session (*European Report*, 1991b), the Habitats Directive was agreed![33]

Final position: designation of Natura 2000 sites

The final position agreed in respect of site designations was that there would be a standard designation procedure via Article 4, whereby Member States would draw up national lists of sites hosting habitats listed in Annex I and species listed in Annex II; these would be proposed to the Commission within three years of the notification of the Directive; the Commission would then adopt a list of sites of Community importance on the basis of biogeographical regions; each Member State would then have six years to designate the sites in question as SACs (Habitats Directive, 1992). The special procedure set out in Article 5 was retained, allowing the Commission to propose for designation a priority site that had not been proposed by a Member State, subject ultimately to a unanimous decision by the Council in the event of a dispute (Habitats Directive, 1992). Many Member States had felt that this was an important inclusion if Spain was successful in its attempts to agree a financing provision within the text of the directive.

On the subject of modifications to the annexes, the general position was that such modifications would be subject to a qualified majority vote by the Council, with the exception of Annex IV, which listed the species requiring strict protection wherever they are found, which was made subject to a unanimous vote by the Council at France's insistence, causing Denmark and Italy to abstain from the final vote (COREPER, 1991d).[34] Thus, the Habitats Directive was adopted unanimously by 10 Member States, with Denmark and Italy abstaining. France's Environment Minister Lalonde sought to justify the French position – which was essentially a concession to France's hunting lobby – on the grounds that if the designation of a priority site on a proposal from the Commission was to be made subject to unanimity, so too should be proposals to add new species to Annex IV (COREPER, 1991d). Thus, despite the Commission stressing throughout the negotiations the need for a flexible process for amending the annexes, the mechanisms finally agreed have made making changes difficult in practice (Evans, 2013).[35]

Final position: protection of Natura 2000 sites

The final text of Article 6(3) and (4), as agreed at the Environment Council meeting on 12 December 1991, read:

> 3. Any plan or project not directly connected with or necessary to the management of the site but likely to have a significant effect thereon, either individually or in combination with other plans or projects, shall be subject to appropriate assessment of its implications for the site in view of the site's conservation objectives. In the light of the conclusions of the assessment of the implications for the site and subject to the provisions of paragraph 4, the competent national authorities shall agree to the plan or project only after having ascertained that it will not adversely affect the integrity of the site concerned and, if appropriate, after having obtained the opinion of the general public.
>
> 4. If, in spite of a negative assessment of the implications for the site and in the absence of alternative solutions, a plan or project must nevertheless be carried out for imperative reasons of overriding public interest, including those of a social or economic nature, the Member State shall take all compensatory measures necessary to ensure that the overall coherence of Natura 2000 is protected. It shall inform the Commission of the compensatory measures adopted.
>
> Where the site concerned hosts a priority natural habitat type and/or a priority species, the only considerations which may be raised are those relating to human health or public safety, to beneficial consequences of primary importance for the environment or, further to an opinion from the Commission, [to other imperative][36] reasons of overriding public interest.
>
> *(CEC, 1991j)*

Also, in order to ensure that the CJEU's decision in the *Leybucht Dykes* case was effectively overruled, an Article 6a – later renumbered Article 7 – was inserted to read:

> Obligations arising under Article 6(2), (3) and (4) of this Directive shall replace any obligations arising under the first sentence of Article 4(4) of Directive 79/409/EEC in respect of areas classified pursuant to Article 4(1) or similarly recognized under Article 4(2) of the latter Directive,[37] as from the date of implementation of this Directive or the date of classification or recognition by a Member State under Directive 79/409/EEC, where the latter date is later.
>
> *(CEC, 1991j)*

Per Scott (1998, pp. 115 and 112), this amounted to "a dramatic reassertion of Member State sovereignty over 'their' natural resources" and "a slap in the face for the European Court". But this was not the end of the story.

Notes

1 The reference to taking account, during site selection, of Article 2's "economic, social and cultural requirements and regional characteristics" had by now been dropped.
2 Formerly Annexes IV and I, respectively, which had been renumbered.
3 ACNAT Regulation (1991) – a funding instrument.
4 DG XI subsequently sought advice from the Commission Legal Service regarding the definition of maritime areas (European Commission, 1991a).
5 Had this language been retained in the adopted version of the directive, the territorial extent of the directive in the marine area may have proved significantly less extensive than the marine area covered in light of the CJEU's (2005) judgment in Case C-6/04.
6 See CEC (1991d) and CEC (1991o, p. 17) for earlier discussions of this issue.
7 The possibility of a derogation based on "overriding national interest" had been raised almost two years before by the RSPB (1989). It is unclear why "overriding public interest" in the 12 February 1991 working draft (DG XI, 1991r) became "overriding national interest" in the Presidency's formal proposal of 28 February (CEC, 1991c). In any event, the adopted version of the directive went with "overriding public interest".
8 Similarly, the Netherlands successfully suggested deleting the words "responsible in the field of the environment" from the Article 6(3) phrase "the national authorities responsible in the field of the environment", thus imposing the Article 6(3) obligation on all relevant national authorities (CEC, 1991p). While this "responsible in the field of the environment" language was still in place in the Presidency's revised draft of 3 May 1991 (CEC, 1991o), it was later deleted.
9 As Roy Bunce explains, following the *Leybucht Dykes* judgment, "we [the UK] contacted as many other Member representatives on the Committee, as many of them individually as we could . . . the Dutch were not onboard with the rest of us at first, they still were very, they were still worried that if we put in a sort of a derogation like this, that the southern European countries would exploit it and negate the whole purpose of the directive. They were very concerned that this was going to undermine what they thought was a very important step forward in nature conservation in Europe, and I had conversations with them on the telephone and we went to a meeting with them about this – one to, you know, sort of, bilateral meeting with them about this, and I pointed out – I can recall this – I pointed out that, you know, our worries about the, say the Humber and Felixstowe [ports], applied just as strongly to Rotterdam, and they hadn't seen it in that light, quite, until then, or that was my impression, and their view somewhat changed after that" (Bunce, 2012).
10 The proposed text of Article 6(2) and (3) read: "[Article 6(2)] The Member States shall take appropriate steps to avoid, in the special areas of conservation, the deterioration of natural habitats and the habitats of species as well as disturbance of the species for which the areas have been designated [insofar as these would have a significant effect with regard to the objectives of this Directive].

[Article 6(3)] Any plan or project not directly connected with or necessary to the management of the site but likely to have a significant effect on such a site, either individually or in combination with other plans or projects, shall be subject to appropriate assessment of its implications for the site in view of the site's conservation objectives. In light of the conclusions of the assessment of the implications for the site and subject to the provisions of paragraph 4, the national authorities responsible for the environment shall agree to the plan or project only after having ascertained that it will not affect the integrity of the site concerned and, if appropriate, after having obtained the opinion of the general public".
11 Scotland was said to be particularly sensitive about the inclusion of rivers, inter-tidal areas and marine areas on the Annex requiring the designation of SACs (Neal, 2012).
12 Presumably the UK could not have lived with the CJEU's interpretation in addition to the Commission's, though this was given as "the Commission" during the interview.
13 This pragmatic approach is reminiscent of Stuffmann's 1977 statement that "our most dangerous friends are those who think that conservation of the environment has to

take place through a profound change in the whole structure of our economy or society . . . It therefore seems necessary to underline that there is no conflict between the environment and job creation, nor is our approach based on hippyism" (quoted in Van de Velde, 2014, p. 390).

14 According to Fairbrass and Jordan (2002, p. 14), citing Freestone (1996, p. 248), "the UK Government supported amendments to the draft habitats directive, which strongly resembled instructions issued by the UK Government in a 1987 circular".

15 'LIFE' is an acronym taken from the French: L'Instrument financier pour l'environnement.

16 Germany's position is explained by the fact that it would be the Länder that would have the power to designate protected areas under the directive (CEC, 1991f).

17 With only a very minor change in language.

18 This ultimately led, under the UK Presidency in the second half of 1992, to the creation of the IMPEL network – a "permanent network of EC environmental enforcement agencies" – to exchange information and experience (UK Government, 1992), thus falling well short of Environment Commissioner Carlo Ripa di Meana's original plans for an EU-wide environmental inspectorate.

19 The UK and France would subsequently seek additional changes in advance of the key Environment Council meeting in December 1991 (DG XI, 1991l).

20 Which would be a joint Council of Environment/Energy ministers on 10 December 1991 (*European Report*, 1991c).

21 Interestingly, in a note regarding the Environment Council meeting of 1 October 1991, DG XI records that Italy had accepted the possibility of mentioning "programmes" in Article 6(3) (DG XI, 1991j), but this does not seem to have come to anything.

22 Article 8 in the adopted version of the directive.

23 At a COREPER meeting on 29 November 1991, Greece attempted – but failed – to have "national defence" added to the list of reasons justifying an impact on a priority site without the need for a Commission Opinion (COREPER, 1991e).

24 Portugal had earlier suggested including language relating to overriding public interest (CEC, 1991n).

25 The UK feared that unless the derogation provisions in respect of the habitats and species parts of the directive were aligned, a protected species could prevent an activity on a special area of conservation even where that activity had been sanctioned under Article 6(4) (CEC, 1991t).

26 DG XI's internal briefing note for the meeting records that the omission of the word "imperative" was no doubt a typographical error but was of serious consequence (DG XI, 1991l).

27 Article 8 in the adopted version of the directive.

28 Given the content of what followed in his speech, the importance of stopping at this point was underlined by the stage instruction "!!!! stop !!!! s'arrêter ici !!!!!!!!!!!!!!!!!".

29 It is perhaps indicative of the character of Commissioner Ripa di Meana that the Nature Unit felt it necessary to add a "NB" to their briefing note: "It might be worth reminding the Commissioner that the threat of withdrawing the directive is a weapon to be used with great prudence and, if at all, as a very last resort" (DG XI, 1991i, p. 2).

30 Later renumbered Article 8 in the adopted version of the Directive. This provision, which caused such problems during the Habitats Directive negotiations, in fact "says nothing", per Cosme Morillo, the senior Spanish civil servant involved in the negotiations (Morillo, 2012). That said, Article 8(4) (which provides for the adoption in respect of Natura 2000 sites of 'prioritized action frameworks' of measures involving co-financing) came into its own in the past few years: the Member States were expected to submit their initial 'prioritized action frameworks' by the end of 2012 (European Commission, 2012).

31 For an account that argues that Spain's input into the text of the directive was of "utmost importance" and altered the "Central-Northern bias" of the directive to be "more sympathetic toward the ecological situation of Southern Europe", see Aguilar Fernández (2003, p. 693).

32 The Greek, Spanish, Irish and Portuguese delegations later included a formal declaration to this effect in the Council's minutes for the meeting on 12 December 1991 (CEC, 1992a).

33 Minor amendments continued to be made to the text up until its adoption in May 1992. For example, a revised working version of the preamble was produced on 16 January 1992 (CEC, 1992c), and this was presented to the Member States on 27 January (CEC, 1992b).

34 Denmark, Italy and Commissioner Ripa di Meana expressed their surprise that Ministers for the environment would render decisions about protecting species more difficult at a time when the EU was seeking to simplify decision-making procedures (COREPER, 1991d).

35 And although there are relatively frequent calls for changes to be made to the annexes (e.g., see Cardoso, 2012, and Hochkirch et al., 2013), there would appear to be a reluctance on the part of the Commission to open the annexes for debate at this time.

36 These words in square brackets, which the Commission had been seeking to insert (see earlier in this chapter), were not included in the formal version of the document circulated following agreement by the Environment Council on 12 December 1991 (CEC, 1991j), but they somehow found their way into the final version of the Directive adopted in May 1992. How this happened is unclear, though the inclusion of these words must presumably have been based on an agreement reached at or prior to the meeting in 1992 when the directive was formally adopted (Habitats Directive, 1992).

37 The phrase "of the latter Directive" was later changed to "thereof" (Habitats Directive, 1992), though again it is unclear when this change was agreed.

References

ACNAT Regulation 1991. Council Regulation (EEC) No 3907/91 of 19 December 1991 on action by the Community relating to nature conservation (ACNAT). *OJ L 370*, 31 December 1991, p. 17.

Aguilar Fernández, S. 2003. Spanish coordination in the European Union. *Administration & Society*, 34, 678–699.

Baldock, D. 1992. The status of Special Protection Areas for the protection of wild birds. *Journal of Environmental Law*, 4, 139–144.

Borrás, S. 1998. EU multi-level governance patterns and the cohesion fund. *European Planning Studies*, 6, 211–225.

Bunce, R. 2012. Interview with Roy Bunce, senior negotiator for the UK government in respect of the Habitats Directive, conducted by Andrew L. R. Jackson on 13 August 2012.

Cardoso, P. 2012. Habitats Directive species lists: urgent need of revision. *Insect Conservation and Diversity*, 5, 169–174.

CEC 1991a. Addendum to 5112/91 ENV 87. Outcome of proceedings of meetings on 5 and 7 March 1991 of the Working Party on the Environment regarding the Habitats Directive proposal. Brussels, 11 March 1991, 5112/91 ADD 1, ENV 87. Council of the European Communities.

CEC 1991b. Addendum to 6755/91 ENV 180, Brussels, 12 June 1991, 6755/91 ADD 1, ENV 180. Council of the European Communities.

CEC 1991c. Cover note from General Secretariat to Delegations dated 28 February 1991. Subject: Amended proposal for a Council Directive on the conservation of natural habitats and of wild fauna and flora – Presidency compromise draft. Brussels, 4961/91, ENV 68. Council of the European Communities.

CEC 1991d. Cover note from General Secretariat to Delegations. Brussels, 3 May 1991, 5949/91, ENV 134. Council of the European Communities.
CEC 1991e. Document de séance CONS/ENV/91/4 ADD 1. Luxembourg, 14 June 1991. Council of the European Communities.
CEC 1991f. Note from General Secretariat of the Council: SI (91) 651/2 to members of the Commission, dated 10 October 1991, headed "COMPTE RENDU DETAILLE de la 1516eme session du CONSEIL DES COMMUNAUTES EUROPEENNES 'ENVIRONNEMENT', Luxembourg, le 1er octobre 1991".
CEC 1991g. Outcome of proceedings of Environment Council dated 1 October 1991. Brussels, 7 October 1991, 8568/91, ENV 291. Council of the European Communities.
CEC 1991h. Outcome of proceedings of Environment Council on 1 October 1991. Brussels, 9 October 1991, 8557/91, ENV 287. Council of the European Communities.
CEC 1991i. Outcome of proceedings of Environment Council on 1 October 1991. Brussels, 10 October 1991, 8579/91, ENV 293. Council of the European Communities.
CEC 1991j. Outcome of proceedings of Environment Council on 12 December 1991. Brussels, 20 December 1991, 10468/91, ENV 405. Council of the European Communities.
CEC 1991k. Outcome of proceedings of Environment Council on 13 and 14 June 1991. Brussels, 1 July 1991. 7042/91, ENV 199. Council of the European Communities.
CEC 1991l. Outcome of proceedings of Permanent Representatives Committee on 25 September 1991. Brussels, 25 September 1991. 7992/2/91 REV 2, ENV 250. Council of the European Communities.
CEC 1991m. Outcome of proceedings of the Working Party on the Environment on 10 July 1991. Brussels, 29 July 1991, 7700/91, ENV 237. Council of the European Communities.
CEC 1991n. Outcome of proceedings of Working Party on the Environment on 4 September 1991. Brussels, 10 September 1991, 7992/91, ENV 250. Council of the European Communities.
CEC 1991o. Outcome of proceedings of Working Party on the Environment on 16 and 23 May 1991. Brussels, 28 May 1991, 6163/91, ENV 149. Council of the European Communities.
CEC 1991p. Outcome of proceedings of Working Party on the Environment on 21 March 1991. Brussels, 8 April 1991, 5572/91, ENV 107. Council of the European Communities.
CEC 1991q. Press release 8086/91 (Presse 153), 1516th session of the Environment Council, Luxembourg, 1 October 1991. Council of the European Communities.
CEC 1991r. Report from Permanent Representatives Committee to Council. Brussels, 7 June 1991, 6755/91, ENV 180. Council of the European Communities.
CEC 1991s. Report from Permanent Representatives Committee to Council. Brussels, 26 September 1991, 8343/91, ENV 274. Council of the European Communities.
CEC 1991t. Working document ENV/91/91 of 2 December 1991. SN/4400/91 (ENV). Attaching note from the UK delegation headed "Technical issues relating to the Habitats Directive". Council of the European Communities.
CEC 1991u. Working document ENV/91/100, Brussels, 11 December 1991, SN/4513/91 (ENV). Attaching a note from the UK delegation concerning some suggested amendments to Article 7 and preambular clauses. Council of the European Communities.
CEC 1992a. Corrigendum to document 10468/91 ENV 405. Brussels, 13 January 1992, 10468/91 COR 1, ENV 405. Council of the European Communities.
CEC 1992b. Working document ENV/92/1/3 REV 1, SN/1094/1/92 REV 1, Brussels, 27 January 1992. Council of the European Communities.

CEC 1992c. Working document ENV/92/3, SN 1094/92, Brussels, 16 January 1992. Council of the European Communities.
CJEU 1990. Opinion of Advocate General Van Gerven of 5 December 1990, *Commission* v. *Germany*, C-57/89, EU:C:1990:432.
CJEU 1991. Judgment of the Court of Justice of the EU of 28 February 1991, Leybucht Dykes, *Commission* v. *Germany*, C-57/89, EU:C:1991:89.
CJEU 2005. Judgment of the Court of Justice of the EU of 20 October 2005, *Commission* v. *United Kingdom*, C-6/04, EU:C:2005:626.
COREPER 1991a. Note dated 26 September 1991 from COREPER to the members of the Commission regarding the 1490th meeting of COREPER, on 25 September 1991, relating to the preparation for the Environment Council meeting of 1 October 1991.
COREPER 1991b. Note from COREPER dated 8 March 1991, addressed to the members of the Commission. Headed "Preparation Conseil 'ENVIRONNEMENT', 18 mars 1991", SI(91) 127.
COREPER 1991c. Note from COREPER dated 14 March 1991, addressed to the members of the Commission. Headed "Preparation Conseil 'ENVIRONNEMENT', 18 mars 1991", SI(91) 143.
COREPER 1991d. Note SI (91) 935 dated 14 December 1991 from COREPER for the attention of members of the Commission. Headed "Object: 1543ème session du Conseil (Environnement) du 12 décembre 1991".
COREPER 1991e. Report from Permanent Representatives Committee to Council. Brussels, 3 December 1991, 9933/91, ENV 373.
Department of the Environment 1991a. Letter dated 10 June 1991 from the UK government's Department of the Environment to the head of the Nature Unit of DG XI.
Department of the Environment 1991b. Letter dated 15 February 1991 from the UK government's Department of the Environment to the head of the Nature Unit of DG XI.
DG XI 1991a. DG XI background note, headed "CONSEIL ENVIRONNEMENT, 12–13 DECEMBRE 1991, POINT: 'HABITATS'".
DG XI 1991b. DG XI briefing note dated 25 April 1991 for meeting between the Environment Commissioner and Spain on 30 April 1991.
DG XI 1991c. DG XI document headed "Note introductive, Conseil Environnement, 1er octobre 1991, Point: 'Habitats'".
DG XI 1991d. DG XI document headed "Points defensifs, Conseil Environnement, 1er octobre 1991".
DG XI 1991e. DG XI document headed "Speaking note, Conseil Environnement, 1er octobre 1991".
DG XI 1991f. DG XI draft press release headed "PROJECT DE COMMUNIQUE DE PRESSE. *CONSEIL ENVIRONNEMENT*, 12–13 DÉCEMBRE 1991".
DG XI 1991g. DG XI file note dated 18 October 1991, written by head of the Nature Unit, entitled "Object: FFH Entretiens hispano-allemands en marge du Conseil informel".
DG XI 1991h. DG XI file note dated 29 July 1991. "Object: article 5 de la directive 'Habitats' – conséquences d'une non-décision du Conseil".
DG XI 1991i. DG XI file note headed "Object: Briefing du Commissaire: Conseil 12–13 Décembre 1991".
DG XI 1991j. DG XI note dated 22 October 1991, headed "Rapport de mission", relating to the Environment Council meeting of 1 October 1991.
DG XI 1991k. DG XI note, headed "NOTE INTRODUCTIVE. *CONSEIL ENVIRONNEMENT*, 12–13 DÉCEMBRE 1991, POINT: 'HABITATS'".
DG XI 1991l. DG XI note, headed "POINTS DEFENSIFS. *CONSEIL ENVIRONNEMENT*, 12–13 DÉCEMBRE 1991, POINT: 'HABITATS'".

DG XI 1991m. DG XI note, headed "SPEAKING NOTE. CONSEIL ENVIRONNEMENT, POINT: 'HABITATS', 12–13 DÉCEMBRE 1991".

DG XI 1991n. DG XI speaking note for COREPER meeting of 6 March 1991.

DG XI 1991o. Draft version of Habitats Directive (in French). Date hand-corrected from 12.11.90 to 15.1.91. A handwritten comment at the top notes "Version modifiée à la lumière des discussions avec [Luxembourg Presidency] 14.1.91".

DG XI 1991p. Draft version of Habitats Directive, hand-dated 8 February 1991.

DG XI 1991q. Note dated 28 November 1991 from the Nature Unit to the Director General of DG XI.

DG XI 1991r. Working version of Habitats Directive dated 12 February 1991.

DG XI and Spain 1991. Letter dated 3 June 1991 from the Director of DG XI's Directorate B to the Spanish government, together with faxed reply from Spain dated 5 June 1991.

Dynes, M., McCarthy, M. and Ford, R. 1991. Tories try to avoid damage over EC (19 October 1991). *The Times*, 19 October.

EC Energy Monthly 1991. Flouting of EC legislation by Member States increases. *FT Energy Newsletters – EC Energy Monthly*, 16 August.

Europe Environment 1991. Nature protection: Europe 2000 conference. Europe Environment, 1 October.

European Commission 1991a. Letter dated 23 April 1991 from the Commission Legal Service to the Nature Unit of DG XI. JUR(91) 02598.

European Commission 1991b. Modified proposal for a Council Directive on the conservation of natural and semi-natural habitats and of wild fauna and flora. COM(91) 27 final. Brussels, 8 February 1991.

European Commission 1991c. Proposal for a Council Regulation (EEC) establishing a Financial Instrument for the Environment (LIFE). COM(91) 28 final, Brussels, 31 January 1991.

European Commission 1991d. Protection of wild birds: the Commission revises the rules in force. Press release P/91/18, 6 March 1991.

European Commission 2012. LIFE+ Nature and Biodiversity. Guidelines for applicants 2012. Part 1 – Content of the proposal.

European Report 1991a. Environment: commission opens infringement proceedings against UK. *European Report*, 18 October.

European Report 1991b. Environment: EEC agreement on LIFE and habitats. *European Report*, No. 1729, 14 December.

European Report 1991c. Environment: question mark over natural habitats directive. *European Report*, No. 1709, 5 October.

Evans, D. 2013. Personal communication: e-mail dated 14 April 2013 from Doug Evans of the European Topic Centre on Biological Diversity to Andrew L. R. Jackson.

Fairbrass, J. and Jordan, A. 2002. EU environmental governance: uncomplicated and predictable policy-making? CSERGE Working Paper EDM 02–03. Centre for Social and Economic Research on the Global Environment, University of East Anglia.

Freestone, D. 1996. The enforcement of the Wild Birds Directive: a case study. *In:* Somsen, H. (ed.) *Protecting the European environment: enforcing EC environmental law*. London: Blackstone Press.

Habitats Directive 1992. Consolidated version of Council Directive 92/43/EEC of 21 May 1992 on the conservation of natural habitats and of wild fauna and flora. *OJ L 206*, 22 July 1992, p. 7. Consolidation date 1 July 2013.

Hochkirch, A., Schmitt, T., Beninde, J., Hiery, M., Kinitz, T., Kirschey, J., Matenaar, D., Rohde, K., Stoefen, A., Wagner, N., Zink, A., Lötters, S., Veith, M. and Proelss, A. 2013. Europe needs a new vision for a Natura 2020 network. *Conservation Letters*, 6, 462–467.

Krämer, L. 2009. The European Commission's Opinions under Article 6(4) of the Habitats Directive. *Journal of Environmental Law*, 21, 59–85.
Macintyre, D. 1991. Tebbit warns Prime Minister not to sign treaty at Maastricht. *The Independent*, 27 October.
McCarthy, M. 1991a. Campaign to save Twyford Down wins backing from EC. *The Times*, 30 August.
McCarthy, M. 1991b. Heseltine aims to expose European backsliders. *The Times*, 18 October.
Moncrieff, C. and Archer, P. 1991. Government defiant over EC building block (18 October 1991). *Press Association*, 18 October.
Morillo, C. 2012. Interview with Cosme Morillo, Spain's senior negotiator in respect of the Habitats Directive, conducted by Andrew L. R. Jackson on 23 August 2012.
Neal, H. 2012. Interview with Hilary Neal, UK government official who participated in the Habitats Directive negotiations and early implementation, conducted by Andrew L. R. Jackson on 14 August 2012.
Raeymaekers, G. 2012. Interview with Geert Raeymaekers, environmental consultant working with DG XI during the Habitats Directive negotiations, conducted by Andrew L. R. Jackson on 25 September 2012.
RSPB 1989. Note dated 15 March 1989, prepared by the RSPB at the request of the UK government's Department of Environment, with a handwritten, undated covering note to DG XI.
Schoon, N. 1991. Interview / Woodland 'saviour' with a lonely vision; Britain has until next week to answer claims that it has breached European environmental law. As the deadline nears, Nicholas Schoon interviews Carlo Ripa di Meana, the Environment Commissioner. *The Independent*, 9 December.
Scott, J. 1998. *EC Environmental Law*. London: Longman.
Sharp, R. 1998. Responding to Europeanisation: a governmental perspective. *In:* Lowe, P. and Ward, S. (eds) *British environmental policy and Europe: politics and policy in transition*. London: Routledge.
Troya, A. 2012. Interview with Antonio Troya, Spanish government official seconded to the European Commission for the Habitats Directive negotiations, conducted by Andrew L. R. Jackson on 5 October 2012.
UK Government 1991. Fax to DG XI dated 16 August 1991 attaching handwritten UK government suggestions for amendments to Article 4 of the draft Habitats Directive.
UK Government 1992. Press release No. 124 dated 9 November 1992, headed "Announcement of permanent network of EC environmental enforcement agencies", British Embassy Brussels, Information Section.
Usborne, D. 1991. Environment row threatens EC Treaty, Major says. *The Independent*, 22 October.
Van de Velde, C. 2014. Environmental and consumer protection. *In:* Bussière, E., Dujardin, V., Dumoulin, M., Ludlow, P., Brouwer, J. W. and Tilly, P. (eds) *The European Commission, 1973–86 – history and memories of an institution*. Luxembourg: Publications Office of the European Union.
White, M. and Wolf, J. 1991. 'Bully' row threatens EC talks: Environment commissioner accuses PM of intimidation. *The Guardian*, 26 October.
WWF 1991. Letter dated 9 December 1991 from the WWF to the Nature Unit of DG XI.
Yourassowsky, C. 2012. Personal communication: e-mail dated 29 December 2012 from Catherine Yourassowsky, former member of DG XI's Nature Unit, to Andrew L. R. Jackson.

10

IMPLEMENTATION AND THE LAW OF NATURA 2000

Three phases in the history of EU environmental policy

Third phase (1992–present day): weakening environmental policy dynamics

On 31 March 1992 Brice Lalonde – France's green Minister for the Environment – announced that he would not serve in a new French cabinet (*AFP*, 1992). A few days later, the Tories were returned to government in the UK's general election of 9 April 1992, and Michael Heseltine was replaced as Secretary of State for the Environment by Michael Howard (Oakley, 1992), a prominent Eurosceptic who had played an important role in the Maastricht negotiations (Forster, 1999). Just over one month later, on 21 May 1992, the Habitats Directive was formally adopted (Habitats Directive, 1992). Less than two weeks after that, the EU was plunged into "chaos" by the Danish 'no' vote in respect of the Maastricht Treaty (Palmer and White, 1992), just as the UK took over the Presidency of the Council, with subsidiarity as its focus (*European Report*, 1992): "the [UK] Government . . . sees a diminished role for the EC in social and environment issues as lying at the heart of its interpretation of the new doctrine of subsidiarity", commented *The Guardian* (1992).

Having challenged the UK government over the controversial M3 motorway at Twyford Down in the run up to the key Maastricht Treaty negotiations in December 1991,[1] Environment Commissioner Ripa di Meana continued to ruffle feathers in 1992. On 30 May 1992, Ripa di Meana published in *The Guardian* newspaper an extraordinary account of his reasons for staying away from the Earth Summit in Rio, which was due to start on 3 June. He had decided not to attend this "parade of vanity and hypocrisy", which could be nothing more than "a luxury container with no content", he wrote, pointing the finger of blame squarely at the USA and "the Europeans" (Ripa di Meana, 1992).

Immediately following the Earth Summit, Commission President Delors was said to be actively looking at ways in which the Commission could hand back powers to EU Member States, with environmental policy seen as a likely target (Walker, 1992; Wolf, 1992a). It therefore perhaps came as little surprise when, on 29 June 1992 – just two days before the UK government took over the Presidency of the Council – it was announced that Ripa di Meana would be standing down as Environment Commissioner with immediate effect, "assisted by a little push [by Delors] from within the Commission" (Meade, 1992; Marshall, 1992).[2] "He has been a difficult man to deal with", commented a British official, "because he believes so fervently in green issues that he has lost touch with political reality and political protocol. At a time when the EC may be considering handing some environmental powers back to national governments, it is appropriate that Ripa will buzz off back to Rome", the official added (Fraser, 1992). A month into the UK's Presidency it was announced that the Commission had dropped five cases against the UK, including its controversial challenge relating to the M3 motorway at Twyford Down; Karel Van Miert, the acting Environment Commissioner, said the Commission had acted "reasonably and pragmatically" in closing the cases (Wolf, 1992b).

EU environmental policy had by now entered the "hugely defensive mode" described by Ken Collins, former Chair of the European Parliament's Environment Committee (Wurzel, 1999). Environmental policy had not been defensive, according to Collins, until Ripa di Meana pursued the UK for alleged breaches of the EIA directive in Twyford Down immediately prior to the Maastricht negotiations, but "it became so very, very quickly, in a period of just a few months" (Wurzel, 1999, p. 20). While Ripa di Meana's actions can be seen as one trigger for the renewed debate about subsidiarity that followed, it would be unfair to cite his intervention in Twyford Down as the sole, or perhaps even as a major, cause for the retrenchment in EU environmental policy that followed (cf. Jordan and Jeppesen, 2000).

By February 1993 the acting Environment Commissioner Van Miert had been replaced by a "cautious" Environment Commissioner, Ioannis Paleokrassas, formerly Greece's Finance Minister (*Greenwire*, 1992a). In May 1994, Director General of DG XI Laurens Jan Brinkhorst announced he was leaving post to stand for election as an MEP (*European Report*, 1994b) and, by the end of that year, a restructuring of DG XI had been announced that would lead to Ludwig Krämer, head of the Legal Unit, being moved sideways out of that post,[3] as well as the retirement of head of the Nature Unit, Claus Stuffmann (*European Report*, 1994a). Thus it was that the "golden era" of EU environmental policy came to a rather abrupt end, and the process of implementing the Natura 2000 network then began in a challenging political landscape.

On 6 August 1992 then UK Foreign Secretary Douglas Hurd asked all UK government departments to search for European Community laws that should be sent to "the knacker's yard" (Gunn, 1992). Golub (1996) continues, "[such] efforts were reflected in Vice-President of the Commission Leon Brittan's statement that 'environmental policy should be the first target for subsidiarity' and that powers in this area should be 'repatriated' – returned to national parliaments . . . Two hit-lists were then submitted to the Commission prior to the December 1993 Brussels Summit, one by Germany, the other by France and Britain". Such a targeting of EU environmental laws for repeal should be resisted, commented one Boris Johnson, then a 28-year old journalist (*Greenwire*, 1992b). Ultimately, the EU environmental *acquis* emerged unscathed (Wurzel, 2002) and the period from 1992 onwards did not result in "the wholesale repeal of green laws that some environmentalists feared" (Jordan, 1998, p. 20).

Nevertheless, as Jordan and Jeppesen (2000, p. 73) argue, there was a watershed moment in the early 1990s: after "25 years of spectacular [policy] growth", the Commission increasingly focused on horizontal issues such as the integration and implementation of existing laws, with EU environmental governance and the allocation of political competences undergoing "unprecedented levels of scrutiny".

Subsequent trends in EU environmental policy: an overview

Since *c.*2005, and in particular since the financial crisis, the EU has moved to the right politically (Jordan and Adelle, 2013), accompanied by a rise in far right populism (Vieten and Poynting, 2016) and a surge in support for Eurosceptic parties (Hobolt and de Vries, 2016). This political balance is significant because "environmental policy has tended to advance further and faster when parties of the centre and centre left are in control" (Jordan and Adelle, 2013, p. xix).

Within the European Commission, under presidents Barroso (2004–2014) and Juncker (2014–), power has become centralised in the President (and his *cabinet*), the Secretariat General and (under Juncker) also the new Vice-Presidents, at the expense of individual commissioners of policy Directorates General (Becker et al., 2016), such as DG Environment.[4] The Commission has also "become more directly linked to and dependent on EU national capitals in a time of acute crisis" (Peterson, 2017). Distant (for now) are the days of radical commissioners such as Spinelli and Mansholt and of strong EU environment commissioners providing the "dynamic leadership which allowed [DG XI/ Environment] to 'punch above its weight' in internal Commission battles" (Ward, 1997, p. 180). Current president Juncker's ten political priorities for the Commission – rigorously policed and delivered by the new Vice-Presidential tier (Becker et al., 2016) – reveal a "narrow and outdated jobs and growth agenda", per the Secretary General of the European Environmental Bureau (Wates, 2016), neglecting many environmental concerns.[5]

With an increase in Euroscepticism, new EU treaties and proposed treaty changes have been threatened and effectively overturned in national referenda (Jordan and Adelle, 2013), and the UK Brexit vote in June 2016 has of course been destabilising and distracting in terms of new sectoral policy formulation at EU level. Finally, and perhaps most importantly, the economic situation of course worsened considerably in recent years as a result of the financial crisis. These factors combined have arguably lent weight to the view that support for the EU as a political project is at base sustained by economic growth and prosperity, and not by strong environmental policies (Jordan and Adelle, 2013).

Indeed, soon after political agreement regarding the Habitats Directive had been reached in December 1991, a slowdown in new EU environmental policy-making began in earnest, including a sharp decline in the amount of new EU environmental legislation being proposed (Ward, 1997). This trend has been particularly marked since the onset of the financial crisis in 2007/8,[6] with Groote (2012), former Chair of the European Parliament's Environment Committee, commenting on the dwindling environmental output of the EU, citing amongst other things the economic crisis and a shift away from DG Environment initiatives towards initiatives in the monetary and economic field (cf. Becker et al., 2016). The Seventh Environmental Action Programme (2013), which covers the period to 2020, effectively accepts the lack of political impetus for new EU environmental legislation, despite the continuing existence of important gaps in coverage, and focuses instead on what then EU Environment Commissioner Potočnik (2012) called the four 'i's: better implementation of existing EU environmental legislation; better information, by improving the scientific evidence base; increased investment for the environment; and full integration of environmental issues into other policies.

The years since 1992 have also seen a move away from command and control regulation towards new policy approaches, triggered in part by the Fifth Environmental Action Programme (1993) (Knill and Liefferink, 2013). This new approach favours instruments that are "less legally demanding" (Knill and Liefferink, 2013, p. 28): e.g., economic incentives and self-regulation. This has been accompanied by the Commission "increasingly opting for less intrusive measures", in terms of its choice of legislative instruments, with the number of Regulations and Directives ('hard law') falling dramatically in this phase, accompanied by significant growth in the number of Decisions, Communications and Reports ('soft law') (Becker et al., 2016, p. 1018).

Further, statistics from DG Environment reveal a striking drop in environmental infringement actions since the financial crisis hit (European Commission, 2016f). In 2007 there were 121 open infringement actions in the nature conservation sector (European Commission, undated), but by 2015 this had fallen to 51 (a drop of 58%) (European Commission, 2016f). Notably, DG Environment's Legal Unit was significantly restructured in 2013, "to get on better terms with some of the more influential member states by weakening enforcement capacity", claimed an NGO representative (*ENDS Europe*, 2013). As Becker et al. (2016, p. 1026) comment, "A new more restrained model [of Commission] has emerged, with the expansionist and interventionist ways of the past firmly set aside".

The Commission has not abandoned the use of infringement actions altogether in this field; e.g., in late 2016 the Commission escalated proceedings on long-running issues such as protection of the ortolan bunting (France) and protection of marine mammals (the UK) (European Commission, 2016a, 2016b), and in July 2017 secured an emergency injunction to prevent logging in Białowieża Forest in Poland (CJEU, 2017c). However, the Commission's reduced reliance on infringement actions places increased emphasis on the importance of access to justice at the national level, and on the preliminary reference procedure under Article 267 TFEU, although this itself raises many difficulties (e.g., see Slepcevic, 2009).

Further, while the accession to the EU of traditionally environmentally progressive states such as Austria, Finland and Sweden in 1995 arguably made environmentally ambitious Member States more influential within the Council, the accession of thirteen central and eastern European countries since 2004 might be expected to have had the opposite effect, given the focus of those countries principally on economic as opposed to environmental policy development (Knill and Liefferink, 2013).

In sum, reflecting on the history of EU environmental policy since the early 1990s, we are presented with a paradox: "the increasing legal and institutional anchoring of EU environmental policy coincides with the stagnation of its political dynamics" (Knill and Liefferink, 2013, p. 26).

Progress to date in establishing Natura 2000

Concluding its *State of nature in the EU* report, the EEA (2015, p. 8) commented, "Although the [Natura 2000] network is considered almost complete on land, its marine component is far from complete, particularly for offshore sites (i.e. those more than 12 nautical miles from the coast)". This reveals two important facts about the establishment of Natura 2000: first, it has been a long process, beginning in principle in 1981 when the Birds Directive came into force; second, something evidently caused the difference in the state of completion of the marine and land parts of the network.

The gradual growth of Natura 2000

Data compiled by the European Commission in March 1993 reveals that, twelve years after the Birds Directive's entry into force, work on establishing the network of SPAs on the ground had only just begun (DG XI, 1993), with work up to that point having focused on hunting controversies and other species protection issues (Ó Bríain, 2012).[7] As Stuffmann (1999, p. 16) reflects, Denmark – the country that was most reluctant during negotiations regarding the Birds Directive – was the "only one to correctly implement it more or less in due time", and by 1994, "only Denmark and Belgium [had] fully complied with their obligations [under the Birds Directive], and Spain [had] made a considerable effort as 40 per cent of the areas notified in the Community to date [had] been designated by this country" (Stuffmann, 1994, p. 205).[8]

The early implementation of the network of SPAs under the Birds Directive has been described by Stuffmann (1999, p. 17) as a "sad story", resulting in "disillusion", according to Gammell (1999, p. 9): "[s]ome of us fondly thought that once the Directive was adopted, everything would be perfect. We were to be proven wrong. Member States did not set up their SPA systems within two years.

Member States did not always enact adequate legislation. What legislation was enacted, was not always enforced". One result of this was the Commission's many infringement proceedings against Member States. A second result was that, on the adoption of the Habitats Directive in 1992, the Nature Unit faced the daunting task of implementing the site designation aspects of both the Birds Directive (to a large extent) and the Habitats Directive (in its entirety). And this was a task the Unit had to complete largely without several of the key figures involved in the drafting and negotiation of the Habitats Directive: e.g., Henriette Bastrup-Birk and Catherine Yourassowsky left the Unit in 1993, followed in 1994 by the retirement of Claus Stuffmann (European Commission, 1994, 1995). Stuffmann was replaced by a Frenchman, Bruno Julien (*European Report*, 1994a), who is said to have played a central role in translating the Habitats Directive into a living reality (Fernández-Galiano, 2012).[9] Another new recruit to the Nature Unit was an Irishman, Micheál Ó Bríain, who joined the Unit in 1993 and is now Deputy Head of Unit, and has thus participated in more or less the entire process of the establishment of Natura 2000 (Ó Bríain, 2012).

As discussed in Chapters 6–9, in light of their experiences with the Birds Directive the Member States set out, in negotiating the Habitats Directive, to codify a different, more flexible, and therefore potentially less adversarial, approach to site designations and site protection. Roy Bunce, the UK's senior negotiator at the time, comments, "By the time I came on to this particular scene [in 1987] . . . we were just beginning to sort of wake up to the fact that the Birds Directive was not just about species, that this business about the provision of SPAs had very serious implications which I think the original negotiators had rather glossed over" (Bunce, 2012). Even in light of this dawning realisation, Member States did not necessarily appreciate fully the implications of the Habitats Directive. Hilary Neal, a fellow UK negotiator, reflects,

> As it turned out, I think we were a little bit too relaxed about the implications of the [Habitats] Directive, you know, kind of assuming that it would be OK and that we wouldn't have to do very much because the SSSIs [UK national site protection system] did it already . . . We knew that [marine designations] would be the difficult one, but we, I think, realised once the directive had been adopted and we were starting to think about designation that actually the implications of implementation were going to be much more far reaching than we had ever expected.
>
> *(Neal, 2012)*

The UK was not alone in this respect, as Claus Goldberg of Denmark reflects:

> When the . . . negotiations started, I don't think we realised the later impact of the [Habitats] Directive . . . on Danish nature protection . . . From the decision on the Birds Directive, I mean it had lived to some extent a rather

silent and quiet life . . . And in the beginning of the Habitats Directive, I do think we had the same feeling. We had the Bern Convention, there's a lot of similarities, so we thought this might be something like the Bern Convention, while knowing that there would be the European Court and the whole EU system. But still we felt that it might not have a very detailed, a strong impact on nature protection in Denmark, which for sure it has had later on, we have realised later on the huge impact of both directives in relation to Danish nature protection

(Goldberg, 2012)[10]

The legal implications of the Directives, both in terms of site designations and site protection obligations, unfolded over an extended period. It is clear from this history: first, that there was a general reluctance on the part of many Member States to designate and protect sufficient sites, given concerns about potential restrictions on activities (Krämer, 2011, p. 191) and the "considerable additional legal protection" provided by EU law (Stuffmann, 1994, p. 205); and second, in light of this fact, the Commission ended up bringing many legal proceedings against Member States,[11] leading to decisions of the CJEU clarifying (typically in quite stringent ways) obligations under the Birds and Habitats Directives, which arguably created a feedback loop of sorts, resulting in further reluctance to designate sites. Such factors played an important role in the drawn out establishment of the Natura 2000 network (see Figure 10.1).[12]

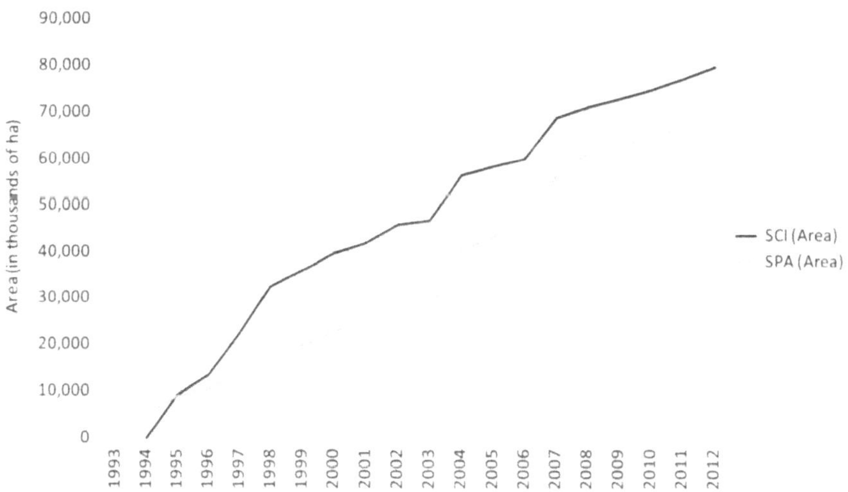

FIGURE 10.1 Development of the Natura 2000 network: cumulative surface area of sites from 1993 to 2012

(EEA, 2015, p. 121, Figure 5.1, grayscale applied)

Territorial extent: the marine area

As detailed in Chapters 6–9, the marine area proved a tricky issue during negotiations regarding the Habitats Directive, resulting in wording to effect that the Directive applies to "the European territory of the Member States to which the Treaty applies" (replicating the wording used in the Birds Directive) (Habitats Directive, 1992, Article 2(1)). When the precise marine extent of the Habitats Directive fell to be considered in 1999 in a case brought before the UK courts by Greenpeace, the result was quite unexpected from the UK's perspective. As Lasén-Díaz (2001, p. 294) records,

> [i]n the absence of a stated limit, the UK Government maintained that the Directive should be taken as referring only to territorial waters up to 12 nautical miles. However, the judge accepted Greenpeace's argument that most EU Directives with marine implications such as those relating to fisheries and oil exploration specify a 200-mile limit. The fact that the United Kingdom has applied the 1985 EIA Directive with a limit of up to 200 miles may have helped Greenpeace's case.

This interpretation of the territorial extent of the Directive was later confirmed by the CJEU in Case C-6/04, against the UK, in which the CJEU held that the Habitats Directive (and by extension also the Birds Directive) indeed applies to Member States' exclusive economic zones where they exercise sovereign rights in that area, meaning that the Directives cover a much larger marine area than the Member States had believed – out to *c*.200 nautical miles (CJEU, 2005b).

As Hilary Neal, then with the UK government, explains,

> we assumed we were signing up to a 12 mile limit from baselines and fought the case against Greenpeace on that basis, and it turns out that the court . . . eventually declares that it was 200 miles . . . For me it was absolutely clear cut that it wasn't 200 miles . . . we never in our wildest dreams thought that we were going to have to go out to 200 miles . . . I don't believe it was in anyone's mind that that was what we would have to do . . . no Member State would have been thinking that.
>
> *(Neal, 2012)*

In 2005 the CJEU thus played an important role in clarifying the territorial extent of the Habitats and Birds Directives in what could be described as an unintended direction, at least from certain Member States' perspective. This, together with a lack of knowledge regarding the marine area (EEA, 2015), explains why the marine part of the Natura 2000 network is lagging behind the terrestrial part in terms of its state of completeness.

Pushing the boundaries: CJEU rulings on Natura 2000[13]

The CJEU has through its judgments interpreting and applying the Habitats and Birds Directives undoubtedly played a vital and decisive role – indeed what can fairly be described as a leadership role – in the development of EU nature conservation law, as detailed below.

Site designation and de-designation under the Birds Directive

To recap, under the Birds Directive the EU's Member States in principle designate[14] SPAs without the European Commission's involvement.[15] Pursuant to Article 4(1) of the Directive, the Member States must, amongst other things, designate the "most suitable territories in number and size" as SPAs for the conservation of the species listed in Annex I to the Directive. Article 4(2) imposes a similar site designation obligation in respect of regularly occurring migratory species not listed in Annex I, "bearing in mind their need for protection in the geographical sea and land area where the Directive applies, as regards their breeding, moulting and wintering areas and staging posts along their migration routes", paying particular attention to wetlands. As noted in Chapter 4, in 1998 the CJEU held that Important Bird Areas (IBA) lists, although not legally binding on the Member States concerned, can, by reason of their acknowledged scientific value, be used by the Court as a basis of reference for assessing the extent to which a Member State has complied with its obligation to designate special protection areas (CJEU, 1998).[16]

While the EU's network of SPAs was to have been established by 1981 (Article 18 of the original version of the Birds Directive, 1979), new sites are still being added to the network, including in countries that were Member States at the time the Birds Directive was adopted (EEA, 2015). Reasons for this long delay in completing the network are easy to find from the Member States' perspective. First, until the CJEU's 1991 decision in the *Leybucht Dykes* case, the Member States were unsure as to the standard of legal protection to be afforded to SPAs, and were arguably wary of site designation as a result (Baldock, 1992). Second is the simple fact that once a SPA is designated it must legally be protected, imposing management costs and potential restrictions on land uses within the SPA network. Third – and connected to this second reason – is the fact that the selection of SPAs (and their boundaries) is itself a controversial issue, made particularly controversial by two decisions of the CJEU post-dating the adoption of the Habitats Directive in 1992.

First was the CJEU's 1996 decision in the *Lappel Bank* case (CJEU, 1996), in which the UK government, supported by the French government, argued that, in selecting SPAs and their boundaries, Member States should be allowed to take economic considerations into account.[17] The CJEU disagreed, building on its 1993 decision in the *Santoña marshes* case (CJEU, 1993) by holding that, in selecting SPAs and defining their boundaries, Member States must act solely on the basis of the ornithological criteria set out in Article 4(1) and (2) of the Birds Directive (CJEU, 1996). Next came a 1998 decision of the CJEU that established that the Member States' margin of discretion in selecting SPAs "does not concern the appropriateness of classifying as [SPAs] the territories which appear the most suitable according to ornithological criteria, but only the application of those criteria for identifying the most suitable territories for conservation of the species in question" (CJEU, 1998, paragraph 61). The combined effect of these judgments was to reinvigorate just the sort of ecology-trumps-economy principles that the Member States had earlier sought to legislate against, in effect, by introducing the Article 6(4) derogation for damaging projects when negotiating the terms of the Habitats Directive. In other words, while the strict protection of SPAs might have been compromised, in part, by the Member States' introduction of the Article 6(4) derogation, these CJEU judgments in 1996 and 1998 ensured that the SPA *designation process* would, at least in principle, remain the preserve of ecology, to the exclusion of socio-economic factors.

De-designation of SPAs[18]

There is no specific provision in the Birds Directive that envisages the de-designation of an SPA, albeit Article 4(1) of the Directive provides that "Trends and variations in population levels shall be taken into account as a background for evaluations" regarding the required special conservation measures for Annex I species, including SPAs. In 2006, the CJEU (2006a) held against Portugal for altering the boundaries of an SPA, thereby excluding from it almost 3,000 hectares providing habitat for species of wild birds for whose protection the SPA was designated. Portugal had altered the boundaries on the basis that the SPA in question included areas that were not important habitats for steppe-land birds. The CJEU (2006a, paragraph 12, emphasis added) countered that the areas nevertheless "did provide a habitat for *other* species of wild birds listed in Annex I to the Directive for whose protection that SPA had been designated", continuing (at paragraph 13), "In those circumstances a Member State may not reduce the surface area of an SPA or alter its boundaries unless the areas excluded from the SPA are no longer the most suitable territories for the conservation of species of wild birds within the meaning of Article 4(1) of the Directive". This would potentially leave open the possibility of a Member State de-designating an SPA on the basis that the area is no longer a most suitable territory for *any* Annex I species (or any regularly occurring migratory species not listed in Annex I), even where this situation has arisen owing to human action or inaction. While the CJEU did not consider this point directly, the Advocate General in the case evidently would not have countenanced such an interpretation: "Only if the Member State can show that a deterioration in quality in the meantime is due to objective circumstances over which it has no influence, for example volcanic eruptions, may it justify the reduction in the extent of an SPA" (CJEU, 2006d).

By analogy with the CJEU's (2007, paragraphs 82–89) judgment in C-418/04, a Member State which does not, at least, comply with its obligation under Article 6(2) of the Habitats Directive by taking "appropriate steps to avoid . . . the deterioration of natural habitats and the habitats of species as well as disturbance of the species for which the [SPA has] been designated, in so far as such disturbance could be significant" should not legally be entitled to de-designate the SPA on the grounds that it is no longer a most suitable territory for the species in question. This would be consistent with the CJEU's (2014a) judgment in *Cascina Tre Pini*, in respect of sites under the Habitats Directive, discussed below.

Site protection under the Birds Directive

Sites that have not been designated as SPAs but which should have been so designated

The CJEU has given the Member States a powerful incentive to designate their SPAs by breathing new life into its (apparently overruled) decision in the *Leybucht Dykes* case (CJEU, 1991): in the *Basses Corbières* case, the CJEU held that "areas which have not been classified as SPAs but should have been so classified continue to fall under the [old, stricter] regime governed by the first sentence of Article 4(4)" of the Birds Directive (CJEU, 2000a, paragraph 47). This was, of course, the protective provision at issue in the *Leybucht Dykes* case. The first sentence of Article 4(4) provides, simply, that "[Within SPAs,] Member States shall take appropriate steps to avoid pollution or deterioration of habitats or any disturbances affecting the [relevant] birds, in so far as these would be significant having regard to the objectives of [Article 4 of the Birds Directive]". The Article 4(4) protection regime does not provide for a derogation for damaging projects, and is buttressed by the CJEU's decision in *Leybucht Dykes*: i.e., economic considerations cannot be regarded as exceptional grounds justifying the reduction in size of an area protected by Article 4(4) of the Birds Directive (CJEU, 1991). Thus, sites that have not been designated as SPAs but which should have been so designated receive stricter legal protection than designated SPAs, since the derogation for damaging projects – Article 6(4) of the Habitats Directive – is available only in respect of designated SPAs (CJEU, 2000a). As the CJEU reasoned, a Member State should not derive an advantage from its failure to comply with EU law (CJEU, 2000a). Per the CJEU (2016c, paragraph 70) there will be a breach of the first sentence of Article 4(4) where there is a probability or risk that a project will cause deterioration to the habitats of protected species of birds or cause significant disturbance to those species.

Designated SPAs

The legal protection afforded to designated SPAs is discussed below, since protective provisions of the Habitats Directive apply to SPAs under the Birds Directive as well as to SACs under the Habitats Directive, for the reasons discussed in Chapters 6–9.

Site designation and de-designation under the Habitats Directive

Designation under the Habitats Directive

Given the difficulties experienced in establishing the SPA network, it is unsurprising that the Commission and the Council negotiated a rather different designation process for sites under the Habitats Directive – a process involving a much greater degree of Commission involvement.

To recap: first, under Article 4(1) of the Habitats Directive, each Member State must propose a list of sites – proposed Sites of Community Importance (pSCIs) – in its territory that host habitats listed on Annex I of the Directive or species listed on Annex II.[19] This list of pSCIs must then be transmitted to the European Commission.[20] The Commission examines these national lists at a biogeographical level,[21] and then, in agreement with the relevant Member States, adopts a list of Sites of Community Importance (SCIs) for each biogeographical region, by way of a formal legal Decision.[22] (In practice, several such lists have been adopted over the years for most biogeographical regions, since the Member States have needed periodic nudging by the Commission to submit sufficient sites.) Once a site has been adopted as an SCI, the relevant Member State has six years to designate the site as an SAC,[23] in principle to allow the Member State time to put in place conservation measures, such as a management plan. Under the terms of the Habitats Directive, this process was due to have been completed by June 2004 (Hamer, 2007), but, as with sites under the Birds Directive, the process remains incomplete (European Commission, 2017e). As well as new sites still being proposed on a regular basis, as at 2015 one quarter of the EU's Member States had not formally designated *any* of their SCIs as SACs (EEA, 2015).[24]

In addition to its rather different site selection process, the Habitats Directive differs from the Birds Directive in another important way: as highlighted in Chapters 6–9, of the subset of EU biodiversity protected by the Habitats Directive, the Directive singles out certain habitats and species as priority habitats/species,[25] to which certain special protective rules apply. In the context of site selection, it suffices to note that, in the case of sites hosting priority species or habitats, Article 5 of the Habitats Directive allows the Commission to initiate a bilateral consultation process with the relevant Member State where the Commission believes that a site is "essential for the maintenance of that priority natural habitat type or for the survival of that priority species", and should have been submitted for adoption as an SCI. If the dispute remains unresolved after six months of consultation, the Commission can submit a proposal to the Council relating to the selection of the site as an SCI. The Council, acting unanimously, must take a decision within three months of the Commission's proposal. As described in Chapter 9, this provision was a key subject of debate towards the end of the Habitats Directive negotiations. The Member States and the Commission had high hopes that this provision would allow disputes to be resolved at the level of the Council rather than ending up before the CJEU. In the event, however, the Article 5 process has never been used by the Commission (Ó Bríain, 2013; Cashman, 2013),[26] and many 'failure to designate' cases have ended up before the CJEU.

Mirroring the situation in respect of SPAs under the Birds Directive, the CJEU has adopted a strict ecological approach to the Member States' selection of sites under the Habitats Directive. In 2000, the CJEU held that

> on a proper construction of Article 4(1) of the Habitats Directive, a Member State may not take account of economic, social and cultural requirements or regional and local characteristics, as mentioned in Article 2(3) of [the Habitats Directive],[27] when selecting and defining the boundaries of the sites to be proposed to the Commission as eligible for identification as [SCIs].
>
> *(CJEU, 2000b, paragraph 25)*

De-designation under the Habitats Directive

Land can effectively be de-designated from Natura 2000 status under the Habitats Directive in the following circumstances.[28]

First, in the circumstances envisaged by Article 9 of the Directive: declassification of a special area of conservation[29] in the case of "natural developments". At present, this will only very rarely arise in practice; query the future impacts of climate change in this regard, though it is hard to see site impacts as a result of climate change being properly characterised as a "natural development".

Second, in the circumstances considered by the CJEU (2014a, paragraph 36) in the *Cascina Tre Pini* case, in which the CJEU held that (emphasis added)

> Member States are *required* to propose to the European Commission the declassification of a site on the list of sites of Community importance, where those authorities have received a request from the owner of land included in that site, alleging an environmental degradation of the site, *provided that that request is based on the fact that, despite compliance with the provisions of Article 6(2) to (4) of [the Habitats Directive]*, that site can definitively no longer contribute to the conservation of natural habitats and of the wild fauna and flora or the setting up of the Natura 2000 network.

The "despite compliance with" proviso is key, likely serving to give the judgment limited practical application: if a Member State has not complied with its obligations under Article 6(2) to (4), de-designation is not an option.

Note that the procedure for de-designation must follow the same procedure as that for adoption as an SCI, such that the Member State must propose an adaptation of its list of SCIs and the Commission must adopt a decision adopting a revised list (CJEU, 2014a). Per Advocate General Kokott a Commission proposal to de-designate an area of SCI pursuant to this process is reviewable by the CJEU (2017b).

Site protection under the Habitats Directive

Article 6 of the Habitats Directive is the key protective provision for SCIs and SACs under the Habitats Directive, and for SPAs under the Birds Directive. The position in respect of pSCIs under the Habitats Directive – i.e., sites that have been proposed to the European Commission but which have yet to be adopted as SCIs – is more complex, and is addressed under a separate heading below.

Legal protection afforded to SCIs, SACs and SPAs

Article 6 comprises four paragraphs that the European Commission (2010a, p. 21) summarises as follows: "Whereas Article 6(1) and (2) of the Habitats Directive concern the day-to-day management and conservation of Natura 2000 sites, Articles 6(3) and 6(4) lay down the procedure to be followed when planning new developments that might affect a Natura 2000 site".[30]

Article 6(1) of the Habitats Directive requires the EU's Member States to "establish the necessary conservation measures [for SACs] involving, if need be, appropriate management plans specifically designed for the sites or integrated into other development plans, and appropriate statutory, administrative or contractual measures which correspond to the ecological requirements of the natural habitat types in Annex I and the species in Annex II present on the sites". Note that Article 6(1) applies only to SACs and not to pSCIs, SCIs or SPAs. Pursuant to Article 4(5) of the Habitats Directive, Article 6(2), (3) and (4) (but not 6(1)) apply to SCIs, since, as noted above, the Member States are given six years from a site's adoption as an SCI to put in place the necessary conservation measures under Article 6(1), and to designate the site as an SAC.

In respect of SPAs under the Birds Directive, pursuant to the Member States' decision to (in effect) overrule the CJEU's decision in the *Leybucht Dykes* case (discussed in Chapter 9), Article 7 of the Habitats Directive substituted obligations arising under Article 6(2), (3) and (4) of the Habitats Directive for obligations arising under the first sentence of Article 4(4) of the Birds Directive (the old, stricter protective regime in respect of SPAs). Presumably Article 6(1) of the Habitats Directive was not applied to SPAs because Article 4(1) and (2) of the Birds Directive already require Member States to take conservation measures of the type described in Article 6(1).[31] In the case of SPAs, however, such conservation measures must be adopted "immediately upon [designation]", according to the European Commission (2010a, p. 20).[32] In respect of SACs, the objective of such conservation measures is to ensure that the species and habitat types are maintained or restored to a "favourable conservation status"[33] across their natural range; according to the European Commission (2010a) an analogous requirement applies in respect of SPAs under the Birds Directive.

Article 6(2) of the Habitats Directive might best be regarded as a broad, continuing "general obligation of protection" (CJEU, 2013, paragraph 33), capturing, amongst other things, activities that do not fall neatly within the impact assessment regime governed by Article 6(3) (e.g., activities that are not subject to an authorisation regime; or activities that are subject in principle to such a regime but in practice are not authorised).[34] Article 6(2) (which was based on Article 4(4) of the Birds Directive, as Chapter 6 records) provides: "Member States shall take appropriate steps to avoid, in [SPAs, SCIs, and SACs], the deterioration of natural habitats and the habitats of species as well as disturbance of the species for which the areas have been designated, in so far as such disturbance could be significant in relation to the objectives of this Directive". Interestingly, the CJEU has held that the obligation under Article 6(2), "is not to be limited to measures intended to avoid external anthropogenic impairment and disturbance but must also, according to the situation that presents itself, include positive measures to preserve or improve the state of the site" (CJEU, 2010b, paragraph 59). Such positive measures might include, for example, "measures to prevent natural developments that may cause the conservation status of species and habitats . . . to deteriorate" (CJEU, 2005b, paragraph 34).

In terms of the substantive level of legal protection provided by Article 6(2), the position is clear: "Article 6(2) and Article 6(3) are designed to ensure the same level of protection of natural habitats and habitats of species (see, to this effect, Case C-404/09 Commission v Spain . . .)" (CJEU, 2013, paragraph 32). It is therefore no surprise that the CJEU (2016d) has held that Article 6(2) requires in certain cases a review of any plan/project approved before a site was adopted as an SCI; and that such a review pursuant to Article 6(2) must be carried out in accordance with the requirements of Article 6(3).[35] Further, it is equally unsurprising to find that the CJEU has held, "taking into account the similar levels of protection required by Article 6(2) and 6(3) of the Habitats Directive", that a plan/project will comply with Article 6(2) only if it is "ensured" (CJEU, 2010a, paragraph 32) or "guaranteed" (CJEU, 2011b, 2016c) that it will not cause any disturbance likely significantly to affect the objectives of the directive, particularly its conservation objectives. It is not enough for national authorities to take enforcement steps such as commencing criminal proceedings or court cases alleging unlawful behaviour if the substantive protection required by Article 6(2) is not thereby secured; further, a failure to take "provisional measures" at the national level pending the outcome of any criminal proceedings or court cases is likely to amount to a breach of Article 6(2) (CJEU, 2016b, paragraphs 55–57).

The leading case on the general standard of protection provided by Article 6(3) of the Habitats Directive is the CJEU's 2004 decision in Case C-127/02 *Waddenzee* (CJEU, 2004). To recap, Article 6(3) determines the situations in which an appropriate assessment is needed in respect of the impacts of any plan or project on an SCI, SAC, or SPA, and the conditions under which permission for such a plan or project can be granted following an appropriate assessment. The CJEU's decision in the *Waddenzee* case thus impacts directly on the circumstances in which Member States have to rely on the Article 6(4) derogation for damaging projects: if a plan or project fails the test in Article 6(3), the only legal route available is the derogation in Article 6(4).

On the issue of the need for an appropriate assessment, Article 6(3) provides:

> Any plan or project not directly connected with or necessary to the management of the site but likely to have a significant effect thereon, either individually or in combination[36] with other plans or projects, shall be subject to appropriate assessment of its implications for the site in view of the site's conservation objectives.[37]
>
> *(Habitats Directive, 1992, Article 6(3))*

The CJEU gave a precautionary spin to this wording in the *Waddenzee* case: the wording, the Court held:

> must be interpreted as meaning that any plan or project not directly connected with or necessary to the management of the site is to be subject to an appropriate assessment of its implications for the site in view of the site's conservation objectives *if it cannot be excluded*, on the basis of objective information, that it will have a significant effect on that site, either individually or in combination[38] with other plans or projects.
>
> *(CJEU, 2004, paragraph 44, emphasis added)*

In other words, an appropriate assessment is needed where significant effects on a protected site cannot be ruled out.

Further, Article 6(3) provides: "In the light of the conclusions of the assessment of the implications for the site and subject to the provisions of [the Article 6(4) derogation], the competent national authorities shall agree[39] to the plan or project only after having ascertained that it will not adversely affect the integrity of the site concerned" (Habitats Directive, 1992, Article 6(3)).

That is the case, the CJEU (2004, paragraph 59) held, only "where no reasonable scientific doubt remains as to the absence of such effects".[40] (In this regard, an appropriate assessment under Article 6(3) "cannot have lacunae and must contain complete, precise and definitive findings and conclusions capable of removing all reasonable scientific doubt" (CJEU, 2013, paragraph 44).) This clearly sets a high threshold, which many plans and projects cannot pass. If they are to proceed, the only available route is the derogation in Article 6(4) of the Habitats Directive.

To recap, Article 6(4) allows plans or projects that have been the subject of a negative assessment to proceed, but only where three strictly-interpreted conditions are met: (1) an "absence of alternative solutions" must be demonstrated; (2) "imperative reasons of overriding public interest" for proceeding with the plan or project must be demonstrated (and these are limited in cases involving priority species or habitats);[41] and (3) the Member State in question must take "all compensatory measures necessary to ensure that the overall coherence of Natura 2000 is protected" (Habitats Directive, 1992, Article 6(4)).[42]

Per the CJEU (2016d, paragraph 73), Article 6(4) must be interpreted strictly,[43] albeit not many cases have come before the CJEU regarding the provision (Clutten and Tafur, 2012).[44] In a case regarding the routing of a motorway in Portugal, the CJEU made it clear that the burden of establishing an absence of alternatives for the purpose of Article 6(4) falls on the Member State in question (CJEU, 2006c, paragraph 38). In other words, it is not for an objector to demonstrate that an alternative solution exists for the purposes of Article 6(4); rather, the burden is on the relevant national authority to demonstrate an absence of alternatives.

In the Commission's view Article 6(4) is "only applicable under . . . strict conditions and will only be passed in exceptional circumstances" (European Commission, 2010a, p. 86). Specifically, "it falls on whoever wants to make use of [the Article 6(4)] exception to prove, as a prerequisite, that the [conditions required by the Directive] do indeed exist in each particular case" (European Commission, 2007/2012, p. 4). In the case of "imperative reasons of overriding public interest", the Commission argues that "the public interest can only be overriding if it is a long-term interest; short term economic interests or other interests which would only yield short-term benefits for . . . society would not appear to be sufficient to outweigh the long-term conservation interests protected by the Directive" (European Commission, 2010a, p. 89). Schoukens (2017) therefore seems justified in arguing that Article 6(4) will not in principle be available for many private projects given the need to demonstrate "imperative reasons of overriding public interest".

Further, in considering compensatory measures under Article 6(4), the Commission's guidance provides, in effect, for like-for-like compensation (e.g., like-for-like replacement habitat) (DECC, 2010), at ratios "generally well above 1:1" (European Commission, 2007/2012, p. 18). Such compensatory measures can be provided within the affected site or in another part of the Natura 2000 network (and might, for example, involve creating an entirely new Natura 2000 area or site) – the crucial factor is that the overall coherence of Natura 2000 is protected (CJEU, 2014b). Note that the CJEU has taken a firm line regarding the difference between mitigation (for purposes of Article 6(3) appropriate assessment) and compensatory measures (under Article 6(4)): "protective measures provided for in a project which are aimed at compensating for the negative effects of the project on a Natura 2000 site cannot be taken into account in the assessment of the implications of the project provided for in Article 6(3)" (CJEU, 2014b). In other words, a developer/national authority cannot dress up compensatory measures as "mitigation", and use this "mitigation" to argue that there will be no adverse effect on the integrity of a Natura 2000 site for Article 6(3) purposes, meaning there is no need to consider and satisfy the tests in the Article 6(4) derogation provision.[45] As Moules (2017) comments, "Both *Briels* [(CJEU, 2014b)] and *Orleans* [(CJEU, 2016e)] demonstrate that the CJEU is unwilling to consider uncertain measures as part of an appropriate assessment. They also confirm that the CJEU adopts a very narrow definition of mitigation which only extends to measures that lessen or avoid a negative effect on the site. It does not encompass measures that seek to counterbalance a negative impact and ensure no net loss".

Priority habitats: integrity of the site

The CJEU has clarified the protection to be afforded under Article 6 to Natura 2000 sites for priority habitats. The case concerned a road scheme in Ireland that would have resulted in the permanent loss of approximately 1.47 hectares of protected limestone pavement – a priority habitat – within a distinct sub-area of 85 hectares, forming part of a total area of 270 hectares of such limestone pavement in the Lough Corrib Natura 2000 site. Granting planning permission for the project, Ireland's planning appeals board (An Bord Pleanála) had decided that the project would have a locally significant negative impact on the site, but that such an impact would not "adversely affect the integrity of the site" for the purposes of Article 6(3) of the Habitats Directive, such that it would not be necessary to apply the Article 6(4) derogation. The issue of the correct interpretation of the point was referred to the CJEU by Ireland's Supreme Court, following a case brought by environmentalist Peter Sweetman. The Irish government – under a Green Party Minister for the Environment when the Supreme Court heard the case and decided to make the reference – supported Sweetman's interpretation that Article 6(4) was necessarily engaged; so too did the Commission. Opposing this view were Galway City and County Councils, Ireland's planning appeals board, and the UK. The CJEU (2013, paragraph 46) sided with Sweetman et al., holding that the integrity of a Natura 2000 site like Lough Corrib will be adversely affected where "the competent national authority concludes that [the] plan or project [in question] will lead to the lasting and irreparable loss of the whole or part of a priority natural habitat type whose conservation was the objective that justified the designation of the site concerned" as a Natura 2000 site.

The decision has far-reaching implications for priority habitats across the EU: any partial, irreplaceable loss of a priority habitat whose conservation was the objective that justified the designation of the site concerned must automatically be held to affect the integrity of the Natura 2000 site in question. The result is that projects which will permanently damage priority habitats within the Natura 2000 network will in principle be able to proceed only in very limited circumstances, via the derogation in Article 6(4).

Legal protection afforded to pSCIs

As noted above, the protection afforded to sites that have been proposed to the European Commission but which have yet to be adopted as SCIs (pSCIs) is complex.

The first relevant judgment at EU level came in the *Dragaggi* case in 2005, with the CJEU holding that, on a proper construction of Article 4(5) of the Habitats Directive, the protective measures prescribed in Article 6(2), (3) and (4) of the Directive are required only as regards sites that have been adopted as SCIs by the Commission (CJEU, 2005a). However, the CJEU continued, "This does not mean that the Member States are not to protect sites as soon as they propose them . . . [to the Commission on their national list] as [pSCIs]" (CJEU, 2005a, paragraph 26). In respect of such pSCIs, the CJEU held, Member States are required to take protective measures that are "appropriate, from the point of view of the Directive's conservation objective, for the purpose of safeguarding the relevant ecological interest which those sites have at national level" (CJEU, 2005a, paragraph 30). In 2006, in the *Bund Naturschutz* case, the CJEU attempted to clarify the content of such "appropriate" protective measures, holding that "Member States must . . . take all the measures necessary to avoid interventions which incur the risk of seriously compromising the ecological characteristics of [pSCIs]" (CJEU, 2006b, paragraph 51). Citing the criteria in Annex III (Part 1) of the Habitats Directive, the CJEU continued (emphasis added), "This is *particularly* the case when an intervention poses the risk either of significantly reducing the area of a site, or of leading to the disappearance of priority species present on the site, or, finally, of having as an outcome the destruction of the site or the destruction of its representative characteristics" (CJEU, 2006b, paragraph 46).[46] The CJEU (2012b) has since clarified that this standard of protection applies not only to sites that have been proposed to the European Commission on a national list, but also to sites that satisfy the ecological criteria in Article 4(1) of the Habitats Directive and that therefore *should have been* included in a national list sent to the Commission.

There is a strong argument that the CJEU's (2006b) judgment in *Bund Naturschutz* excludes the use of the Article 6(4) derogation in respect of sites that are eligible for adoption by, but which have not yet been adopted by, the Commission as SCIs, at least in certain cases.[47]

Horizontal issues of relevance to the Habitats and Birds Directives

Access to justice and public participation

Neither the text of the Habitats Directive nor the text of the Birds Directive covers the subject of access to justice at the national level: i.e., rules enabling individuals and their associations to exercise at the national level rights conferred on them by EU law, helping to ensure that environmental obligations are complied with.

The European Commission (2003) proposed a directive to address access to justice in environmental matters. However, the Member States blocked the proposal for over a decade and it was eventually withdrawn by the European Commission (2014). A sector by sector legislative approach succeeded in adding access to justice provisions to the EIA Directive, the Integrated Pollution Prevention and Control Directive[48] (EP and CEU, 2003) and several other instruments.[49] However, this approach has not resulted in comprehensive coverage, "and the EU legislature does not appear to be currently receptive" to addressing the gaps (European Commission, 2017a, p. 6). As such, the European Commission (2017a) has released an interpretative communication on access to justice, bringing together and drawing inferences from the CJEU's developing case law on the subject with a view to providing greater clarity in the absence of legislation.

The CJEU has handed down two important judgments regarding the relationship between the Habitats Directive and Article 9 of the Aarhus Convention, the access to justice provision of the key international law instrument addressing access to information, public participation and access to justice in environmental matters, to which the EU and in addition its Member States are parties.

In the first case of interest, Case C-240/09 *LZ I*, the CJEU (2011a) held that the CJEU has jurisdiction to interpret the provisions of Article 9(3) of the Aarhus Convention and to give a ruling on whether they have direct effect. Ultimately, the CJEU (2011a, at paragraphs 49–52) held that Article 9(3) is not directly effective, but that a strong duty of consistent interpretation applies: "50 . . . in so far as concerns a species protected by EU law, and in particular the Habitats Directive, it is for the national court, in order to ensure effective judicial protection in the fields covered by EU environmental law, to interpret its national law in a way which, to the fullest extent possible, is consistent with the objectives laid down in Article 9(3) of the Aarhus Convention".

Jans (2011) is sharply critical of the CJEU's decision, arguing that the CJEU's reasoning was unconvincing; that its assessment was incorrect; that it should have abstained from opining on whether Article 9(3) has direct effect as a matter of EU law; that the duty of consistent interpretation identified by the CJEU arguably applies across the full EU environmental law *acquis* and appears to have been conjured out of thin air; that the CJEU's judgment makes the EU's declaration on approving the Aarhus Convention (CEU, 2005) "more or less irrelevant"; and that with its judgment the CJEU "might be stepping into the EU legislature's shoes, a legislature who has, thus far, intentionally chosen not to act".

The second judgment of interest in this area is C-243/15 *LZ II*, in which the CJEU (2016a, paragraph 45) held that the public participation requirements of Article 6(1)(b) of the Aarhus Convention apply in the context of appropriate assessment under Article 6(3) of the Habitats Directive. This is highly significant because the text of Article 6(3) contains only a weak public participation requirement (in light of an appropriate assessment "the competent national authorities shall agree to the plan or project only after having ascertained that it will not adversely affect the integrity of the site concerned and, if appropriate, after having obtained the opinion of the general public").

Further, the CJEU held that because the public participation requirements of Article 6(1)(b) of the Convention apply to such situations, the access to justice requirements of Article 9(2) and Article 9(4) of the Convention apply too (CJEU, 2016a, paragraphs 57 and 62). This is also highly significant, because the Habitats Directive (unlike, say, the EIA Directive) does not contain a standalone access to justice provision. The implication is that those seeking to enforce obligations under Article 6(3) of the Habitats Directive must at the national level have access to a review procedure that provides adequate and effective remedies, including injunctive relief as appropriate, and be fair, equitable, timely and not prohibitively expensive.

Addressing unlawful harm to Natura 2000 sites under the Habitats Directive

The CJEU (2016d) has handed down an important judgment regarding the remedial action that ought to be taken under the Habitats Directive where a plan or project is carried out in circumstances that did not comply with Article 6(2) to (4). This is the same judgment discussed above in the context of reviewing plans/ projects that were approved before a site became a Natura 2000 site. Here we are interested in a slightly different angle: what EU law requires where a plan/project that is damaging to Natura 2000 has been carried out without complying with Article 6(2) to (4). In the case in point, a bridge had been constructed in a Natura 2000 site without first being permitted in accordance with Article 6(2) to (4). Essentially the CJEU required this sort of situation to be addressed remedially, such that an appropriate assessment must be carried out in respect of the plan/project in accordance with the requirements of Article 6(3), followed by a consideration, by analogy, of the tests in the Article 6(4) derogation where adverse effects on site integrity cannot be excluded. In considering alternatives under Article 6(4), the option of closing or demolishing the plan/project must be considered (CJEU, 2016d, paragraph 74). In considering such a step as an alternative, the economic cost of closure/demolition is not of equal importance to the conservation objective pursued by the Habitats Directive, and the economic cost alone cannot therefore be a determining factor in the choice of alternative solutions (CJEU, 2016d, paragraph 77). Any proposal for demolition must be regarded as a plan/project and may therefore need to be subject to Article 6(3) appropriate assessment before being carried out (CJEU, 2016d, paragraph 76).

Further, as the European Commission comments,

> it can be deduced from the judgment that compensation needed to be provided for any unlawful damage already done to the protected site . . . This is because the CJEU required the future of the bridge to be considered in the context of Article 6(4) of the Habitats Directive, 92/43/EEC. This requires compensatory measures if a damaging project is allowed to proceed on the grounds of absence of alternative solutions and overriding public interest.
>
> *(European Commission, 2017a, paragraph 168 and footnote 176)*

This approach, requiring compensation for damage done to designated Natura 2000 sites, mirrors the Commission's practice in respect of areas unlawfully excluded from Natura 2000 designation and then damaged. For example, see the habitat creation scheme at Wallasea island, England (DEFRA, 2007), and the compensatory SPA designation/extension in Cork, Ireland (DAHG, 2017), which were compensation for (respectively) a car park built in an area of the Lappel Bank that should have been an SPA (CJEU, 1996), and an area of land excluded from SPA designation and destroyed in the construction of Dublin port tunnel (CJEU, 2007).

Is Natura 2000 working?

Despite the focus here on legal conflicts, it is important to highlight some of the many positive aspects of what has proven an extraordinary achievement of EU environmental policy.

First, the Habitats and Birds Directives have together led to the establishment of a coherent, EU-wide network of protected areas – the largest coordinated network of nature conservation areas in the world (EEA, 2015) – which currently covers more than 27,500 sites, comprising about 18% of the EU's land territory plus marine areas comprising *c*.4% of the EU's seas (European Commission, 2017e; EEA, 2015). Thus, although 1992 may have marked, in a broad sense, the drawing to a close of a "golden era" in EU environmental policy, in a narrower sense this was not true of the Habitats Directive because the Directive acted as a catalyst for major new ecological survey work and mapping across Europe (Evans, 2006) and ultimately led to the creation of the enormous network just described.

Second, the Habitats Directive and the LIFE funding instruments have together resulted in a long-term flow of funding into Natura 2000 and environmental projects more generally: since 1992 there have been four complete phases of the LIFE programme, resulting in the co-financing of "3954 projects across the EU, contributing approximately €3.1 billion to the protection of the environment" (European Commission, 2017d). That said, this is a relatively small financial contribution spread across 25 years, compared with the estimated total annual cost of managing the Natura 2000 network of €5.8 billion (European Commission, 2011b), particularly when one considers that "the benefits that flow from Natura 2000 are of the order of €200 to 300 billion/year" for terrestrial sites alone (European Commission, 2013b, p. 3).

Third, because the Habitats Directive requires Member States to monitor the conservation status of listed species and habitats and report to the Commission every six years (Habitats Directive, 1992, Article 17), we now know much more about 233 habitat types and more than 1,250 species than we did before:[50] not only their distribution and populations, but also their current status and we are beginning to understand trends. Similarly, in respect of wild birds, Member States recently and for the first time provided data to the Commission on population size and trends in their national territory under Article 12 of the Birds Directive, allowing for an assessment of population status and trends at EU level for around 450 wild bird species (European Commission, 2015).

The composite report for the EU-27[51] – covering the period 2007–2012 (Habitats Directive) and 2008–2012 (Birds Directive) – states that "Whilst limited progress may appear to have been made so far, the assessment shows that EU nature legislation has been fundamental in stemming the further decline of Europe's most vulnerable species and habitat types, in spite of the ever-increasing demand for land and resources" (EU, 2015, p. 5).

Important caveats are provided: first, while the assessment in respect of the Birds Directive covers all species of naturally occurring wild birds in the EU, the assessment under the Habitats Directive is only of a subset of listed habitats/species, "most [of which] were already in a critical state when listed in the directives, which implied that considerable time and effort would be necessary to ensure their recovery" (European Commission, 2015, p. 3). It is therefore unsurprising that the percentage of species with an unfavourable conservation status is higher under the Habitats Directive than under the Birds Directive (EEA, 2015). There are also limitations in terms of detecting trends given that there have to date been only two reporting periods under the Habitats Directive (2001–2006 and 2007–2012).

In relation to wild birds, the Commission's *State of Nature* report recorded "encouraging feedback": the "status of more than half of all wild bird species assessed (52%) is secure [i.e., no foreseeable risk of extinction]. On the other hand, around 17% of the species are still threatened and another 15% are near threatened, declining or depleted", with the status of 16% unknown (EU, 2015, p. 11; EEA, 2015). The most frequently reported threats and pressures on EU birds are agriculture (including intensification and abandonment), hydrological changes (particularly in respect of wetlands) and exploitation (including hunting) (EEA, 2015).

Across the EU-27 only 16% of the habitat assessments were favourable, with more than three quarters unfavourable (EU, 2015).[52] Some habitat types were faring worse than others, of course, with dune habitats, coastal habitats, grasslands, bogs, mires and fens having the highest percentages of 'unfavourable' assessments (EU, 2015). The composite report found that for terrestrial systems, 'agriculture' and human-induced 'modifications of natural conditions' were the "greatest problems identified for all three groups (birds, other species and habitats)" (European Commission, 2015, p. 11). Focusing on conservation status trends, "a third of the habitat types are unfavourable but stable (33%). However, a further 30% are still deteriorating, which is a serious cause for concern" (EU, 2015, p. 19).

In respect of species protected under the Habitats Directive, 23% of assessments were favourable, 60% were unfavourable and in 17% of cases the status was reported as unknown (EU, 2015). In terms of conservation status trends, focusing on the 60% whose status is 'unfavourable': 4% of species are improving, 20% are stable, 22% are deteriorating and 14% are without a known trend (EU, 2015, p. 15). Information gaps were particularly marked in the marine area: "marine habitats and species remain poorly known, with 18.5% of marine habitats reported as unknown (5% for non-marine) and 50% unknown for marine species (13% for non-marine species)" (EEA, 2015, p. 62).

In terms of the influence of Natura 2000 on conservation outcomes, the EEA (2015) cautions that measuring the network's effectiveness is difficult owing to a paucity of baseline data and the difficulty in finding controls. The EEA (2015, p. 8) also highlights that while there are published studies demonstrating Natura 2000's role in improving the conservation status of birds, "it can find no similar studies for habitats or non-bird species", which is perhaps unsurprising given that the Birds Directive is 13 years older than the Habitats Directive, giving longer for trends to emerge. In respect of the former, a major study published in *Science* in 2007 revealed strong evidence that the Birds Directive has had a significant positive impact on listed species (Donald et al., 2007). In respect of species and habitats requiring SACs under the Habitats Directive, the Commission comments that

> The overall conservation status of habitats and species cannot be attributed to Natura 2000 coverage. However, for both habitats and species with unfavourable status, the trend in conservation status is closely associated with Natura 2000 coverage. The proportion of assessments with a deteriorating status is higher in situations of low coverage (0–35%) than where there is high (75–100%) coverage. By contrast, those with relatively higher Natura 2000 cover are more likely to show stable assessments. This underlines the crucial role of the network in stabilising conservation status.
>
> *(European Commission, 2015, p. 17)*

In addition, the Natura 2000 network also provides "umbrella" protection for species and habitats that are not directly targeted by the Habitats or Birds Directives, amounting to "important additional value for a range of biodiversity" (Van der Sluis et al., 2016, p. 11).

Whilst the Natura 2000 network is therefore playing a vital role in respect of nature conservation in Europe, the European Commission (2015, p. 16) recognises that the network's full potential has not yet been realised, because "the necessary conservation measures" have not yet been introduced across the board. The Commission cites, for example, the fact that only 50% of Natura 2000 sites are reported as having comprehensive management plans; investment in some Member States has been insufficient; and "the opportunities offered by, for example, the Common Agricultural Policy, the Common Fisheries Policy and the EU Regional Policy have not been fully realised" (European Commission, 2015, p. 16). As Chapters 6–9 revealed, the importance of these complementary sources of funding to the management of Natura 2000 was fully recognised during the Habitats Directive negotiations. It is therefore disappointing that, some 25 years on, the European Court of Auditors (2017) concluded, following a detailed study, that EU funds have not been well mobilised to support the management of Natura 2000.

The Birds and Habitats Directives under threat

As introduced in Chapter 1, following interventions from certain Member States and others, it was announced by the Commission in 2013 – clearly against the wishes of the then Environment Commissioner (Potočnik, 2015) – that Natura 2000 was to be the subject of a legislative "Fitness Check" as part of the European Commission's REFIT "Fit for Growth" programme to "prepare for future legislative initiatives to reduce regulatory burden" (European Commission, 2013a, p. 7). When Jean-Claude Juncker took over from Barroso as Commission President in 2014, he effectively pre-empted the outcome of the Fitness Check by mandating the new EU Environment Commissioner Vella to "carry out an in-depth evaluation of the Birds and Habitats directives and assess the potential for merging them into a more modern piece of legislation" (Juncker, 2014, p. 4). Many environmentalists feared that "more modern" was a euphemism for deregulation and weakening, and there then began a high profile campaign by environmentalists to save the directives.

This campaign mobilised an unprecedented public response to the EU public consultation held as part of the REFIT exercise: over the 12-week period of the consultation in mid-2015, more than 552,000 public responses were received, over 491,000 of which came via the Nature Alert campaign (European Commission, 2016c) organised by a pan-European coalition of NGO networks.[53] The campaign, and the massive show of public interest it galvanised, was undoubtedly influential. While Commission First Vice-President Timmermans, who had "wooed" the UK at his confirmation hearing by promising to cut red tape (Peterson, 2017, p. 354), stated in June 2015 that "we need to update [EU nature legislation]" (Crisp, 2015), in late 2015 Ministers for the Environment of nine EU Member States (Germany, Croatia, France, Italy, Luxembourg, Poland, Romania, Slovenia and Spain, collectively representing 63% of the EU population) wrote to the European Commission to support retaining the directives in their current form: "We believe that amending or merging the directives is not expedient", they concluded (Environment Ministers, 2015). Greece, Belgium, Slovakia, the Czech Republic, Hungary and Bulgaria later similarly "stressed the importance of protecting and better implementing" the Directives (Khetani-Shah, 2016), as did Lithuania (Government of Lithuania, 2015). At the Environment Council meeting in December 2015, Member States collectively emphasised their support for the Habitats and Birds Directives as "essential components" of EU nature protection, urging their full implementation and underlining the importance of maintaining the goals and of not lowering the nature protection standards of the Directives (CEC, 2015). The European Parliament (2016) added its voice, with 86% voting in favour of a resolution urging "EU leaders to listen to the half a million citizens who have called for our strong nature protection laws to be upheld and better implemented".

Even the UK got in on the act in the end: just three weeks before the Brexit referendum in June 2016, then Prime Minister David Cameron commented that "EU membership underpins many crucial environmental protections in the UK, while amplifying our voice in the world on vital issues like cutting global emissions. I will use our seat at the table to ensure . . . that the birds and habitats directives are maintained and better implemented, both in the UK and across Europe, to ensure the diversity of our countryside and wildlife" (Neslen, 2016). In light of the subsequent vote in favour of Brexit, reports suggest that "while most EU rules will be adopted into [UK] domestic law . . . the EU habitats directive is among measures set to be repealed" (Parker and Hammond, 2017). The appointment in June 2017 of Michael Gove as the UK's Secretary of State for the Environment certainly does not augur well, Gove having reportedly cited the Habitats Directive as an "absurd" EU rule which the UK could scrap post-Brexit (Taylor, 2017). If this comes to pass, it would mark the end of a remarkable history of UK involvement with the Habitats Directive, as detailed in this book.

The final crucial piece of the jigsaw in preventing renegotiation of the Directives was the evidence base. Under the REFIT exercise, the Habitats and

Birds Directives were assessed against five criteria: effectiveness, efficiency, relevance, coherence, and EU added value. This involved extensive evidence-gathering and consultation exercises, informed by a supporting study by consultants to the Commission.[54] Deadlines for publishing this study were repeatedly missed by the Commission, leading some to speculate as to what was happening behind closed doors (Baumüller, 2016). Ultimately, a draft final version of the report was leaked by the media in June 2016, revealing the study's conclusion that the Habitats and Birds Directives were fit for purpose (Crisp, 2016). Commission First Vice-President Timmermans was "grilled" by MEPs in the European Parliament's Environment Committee the same day (Baumüller, 2016).

At a meeting of the College of Commissioners in December 2016, the Commission discussed the findings of the Fitness Check and formally decided not to reopen the directives and instead to "develop an Action Plan to correct the deficiencies encountered in the implementation of the Birds and Habitats Directives" (European Commission, 2016e). This conclusion was informed by the Fitness Check evaluation, which concluded that "the Nature Directives are fit for purpose but fully achieving their objectives and realising their full potential will depend on substantial improvement in their implementation in relation to both effectiveness and efficiency, working in partnership with different stakeholder communities in the Member States and across the EU, to deliver practical results on the ground" (European Commission, 2016c, p. 8).

The Action Plan

The promised Action Plan was delivered by the Commission in April 2017, covering four priority areas with 15 concrete actions, most of which were to be launched in 2017 (European Commission, 2017b). The priority areas are: Priority A: Improving guidance and knowledge and ensuring better coherence with broader socioeconomic objectives; Priority B: Building political ownership and strengthening compliance; Priority C: Strengthening investment in Natura 2000 and improving synergies with EU funding instruments; and Priority D: Better communication and outreach, engaging citizens, stakeholders and communities. Given the importance of litigation in the history of the Birds and Habitats Directives, the following sentence under Priority B is particularly interesting: "For strengthening Member States' compliance with the Nature Directives, other measures in the broader area of environment policy, on training of national judges and prosecutors, access to justice, and assuring compliance with EU environmental law, will also be very relevant" [citing "Commission Work Programme 2017 Delivering a Europe that protects, empowers and defends (Commission Communication COM(2016), 710 final"] (European Commission, 2017b).

The Council's broadly supportive conclusions on the Action Plan, dated June 2017, did not address the above-mentioned sentence, instead focusing on the "voluntary" (the Council's addition) "bilateral dialogues between the Commission and Member States within the framework of the new Environmental Implementation

Review process" (CEU, 2017), a process intended to "address the causes of implementation gaps and find solutions before problems become urgent and reach the stage of infringements" (European Commission, 2017c). Whether such voluntary, dialogue-based solutions will be the way of the future, or whether the CJEU will continue to play a major role in this policy area, remains to be seen.

Notes

1 Causing the UK to react with "ill-concealed fury" (Hardy, 1991; Usborne, 1991).

2 There is of course every possibility that Ripa di Meana was negotiating a new job for himself while the Rio summit was ongoing. After all, he became Italian Minister for the Environment immediately on leaving his role as Environment Commissioner (Fraser, 1992).

3 As Wurzel (2002, p. 70) records, a contemporaneous report in *Environment Watch* stated that: "Krämer's effective demotion after 10 years in the job and the downgrading of his [legal] unit are widely seen as politically motivated revenge by national governments for his strict interpretation of EU environmental law and his uncompromising pursuit in the European Court of Justice of member states that fail to properly transpose or implement them". A resolution protesting the restructuring of DG XI and the treatment of Ludwig Krämer in particular was adopted in February 1995 by 60 well-known experts in EU environmental law, but this failed to alter the outcome (*Europe Environment*, 1995).

4 DG XI was renamed DG Environment after the Amsterdam Treaty (Kingston et al., 2017).

5 Climate action features among the priorities in the context of the EU's energy union. No other environmental policy area is mentioned.

6 Cf. Hey's (2005) argument that there was a renaissance in EU environmental policy-making between 1997 and 2003, as well as Wurzel's (2008) division of the post-1992 period into a "sedate phase" from 1992–2005 and a "selective activism" phase since 2005. Note also Becker et al.'s (2016, pp. 1018 and 1024) statement that between 1989 and 2009 "The field of environment, customers and health has also experienced strong growth", but that legislative activity in the area fell by 30% between 2010 and 2014.

7 That being said, by February 1988 DG XI's Legal Unit had launched 29 infringement proceedings against Member States seeking to force them to comply with their obligations to designate SPAs under the Birds Directive (DG XI, 1988).

8 However, even Denmark would later face the potential prospect of Commission-led infringement proceedings (ultimately averted), because Denmark's designation of sites remained provisional until 1994 (Cichowski, 2007), pending some potential minor changes in boundaries (Koester, 2012).

9 Without Julien's strength and resolve, commented one interviewee, the Habitats Directive would have died out (Fernández-Galiano, 2012).

10 On reflection, according to Veit Koester, the impacts on Danish nature conservation have been positive, not negative: "I think that the Habitat Directive and also the Bird Conservation Directive, both of them, they have been very, very important . . . And the situation would have been rather bad without them, especially in the previous years when we had the big development and a good economy and projects everywhere and wise ideas of building this or that and so forth, so I think that they have been very, and they are still very, important" (Koester, 2012).

11 Lasén-Díaz (2001) cites the fact that as early as November 1997, 36 infringement proceedings were ongoing against all Member States for inadequate implementation of the Habitats Directive.

12 Krämer (2011, p. 188) records details of the development of the network of SPAs over time.

13 It is beyond the scope of this book to consider the overlaps between the Habitats/Birds Directives and complementary instruments of EU law (e.g., such as the Water Framework Directive (2000): on which see De Smedt and van Rijswick (2015) and Janauer et al. (2015); the Marine Strategy Framework Directive (2008); and the

Environmental Liability Directive (2004): on which see Milieu and IUCN (2014), European Commission (2016d) and Fogleman (2015)).

14 The Birds Directive uses the term "classify" (see Article 4(1)); however, "designate" is used throughout this chapter for simplicity, since the Habitats Directive uses "designate".

15 See Article 4(1) and (2) of the Birds Directive. The Commission has frequently become involved in the process by bringing infringement proceedings where a Member State has not designated sufficient SPAs, for example. But in principle the SPA designation process under the Birds Directive should operate without the Commission's involvement.

16 For an application of this point see CJEU (2016f).

17 Note that this differs from the situation at issue in the *Leybucht Dykes* case. That case related to the reduction in size of a *designated* SPA for birds to accommodate a construction project. The *Lappel Bank* case, on the other hand, related to the selection of SPAs and their boundaries (i.e., pre-designation).

18 In addition to the scenarios discussed in this section, note that an area of SPA may be de-designated to correct an administrative error relating to the site (CJEU, 1999).

19 Per Advocate General Kokott, areas proposed for SCI status can include "potential areas for restoration measures" (CJEU, 2017b). Note also the special rule for aquatic species that range over wide areas (Article 4(1)).

20 Under Article 4(1) of the Habitats Directive.

21 Article 1 of the Habitats Directive mentions nine biogeographical regions: the Alpine region, the Atlantic region, the Black Sea region, the Boreal region, the Continental region, the Macaronesian region, the Mediterranean region, the Pannonian region, and the Steppic region. Further, there are five marine biogeographical regions (European Commission, 2011a). Examining sites at the biogeographical level "makes it easier to check species and habitat conservation trends under similar natural conditions, irrespective of national boundaries" (European Commission, 2010b, p. 19).

22 Article 4(2) of the Habitats Directive.

23 Article 4(4) of the Habitats Directive.

24 Nevertheless, Article 6(2) to (4) apply to sites as soon as they are adopted as SCIs (see Article 4(5)).

25 Identified by an asterisk on Annexes I and II to the Directive.

26 Though Cyprus recently argued, unsuccessfully, before the CJEU that the process ought to have been used by the Commission in a dispute regarding site protection for the Cypriot grass snake (*Natrix natrix*) (CJEU, 2012b).

27 Article 2(3) of the Habitats Directive provides "Measures taken pursuant to this Directive shall take account of economic, social and cultural requirements and regional and local characteristics".

28 In addition, an area of SCI/SAC may be de-designated to correct an administrative error relating to the site (CJEU, 1999). The judgment cited relates to de-designating an area of SPA under the Birds Directive but, per the Advocate General in CJEU (2017b), applies also to de-designating areas under the Habitats Directive. At the time of completing the present research the CJEU had not yet given its judgment in the case.

29 Note the specific reference to SACs here. While there is no provision of the directive that expressly provides for the de-designation of a site on the list of SCIs, the CJEU (2014a, paragraphs 25–26 and 36) specifically considered this point and interpreted the Habitats Directive as providing for SCIs to be de-designated in certain circumstances.

30 This reference to "new" developments is arguably misleading, since Article 6(3) has been held by the CJEU to cover, for example, an activity that has been carried out for many years but for which a licence is granted annually for a limited period (CJEU, 2004).

31 Cf. CJEU (2010b).

32 After all, pursuant to Article 4(1) of the Birds Directive, the species mentioned in Annex I to the Birds Directive must be the subject of "special conservation measures" concerning their habitat in order to ensure their survival and reproduction in their area of distribution, and Article 4(2) of the Directive requires the Member States to take similar measures for regularly occurring migratory species not listed in Annex I.

33 Defined in Article 1 of the Habitats Directive.
34 Note that the CJEU (2016b, paragraphs 121–122) has held that Article 6(3) "is relevant only when the competent national authorities grant authorisation for a project, as that authorisation must in such a case be preceded by an appropriate assessment of the implications of the project for the site concerned . . . Consequently, Article 6(3) . . . does not apply in respect of any action whose implementation was subject to authorisation but which was carried out without authorisation and thus unlawfully". Presumably the potential breach in such a case would be of Article 6(2).
35 For temporal aspects of the review, see CJEU (2016d).
36 Note that "national authorities are required, when assessing cumulative effects, to take into account all projects which, in combination with the project for which an authorisation is sought, are likely to have a significant effect on a protected site in the light of the objectives pursued by that directive, even where those projects precede the date of transposition of that directive" (CJEU, 2017a, paragraph 61).
37 On "conservation objectives" see Stahl (2015).
38 See Note 36.
39 See Note 34.
40 Note that a mitigation measure whose effectiveness is uncertain at the time of the consent decision cannot be relied upon to guarantee beyond all reasonable doubt that the plan/project in question will not adversely affect the integrity of the site (CJEU, 2017a).
41 Article 6(4) provides: "Where the site concerned hosts a priority natural habitat type and/or a priority species, the only [imperative reasons of overriding public interest] which may be raised are those relating to human health or public safety, to beneficial consequences of primary importance for the environment or, further to an opinion from the Commission, to other imperative reasons of overriding public interest".
42 On "compensatory measures" see McGillivray (2015).
43 For example, see the Commission's guidance documents on Article 6(4) (European Commission, 2007/2012), and on wind energy developments and Natura 2000 (European Commission, 2010a).
44 For two recent examples on the issue of imperative reasons of overriding public interest, see CJEU (2012a) and CJEU (2012c).
45 See Schoukens (2017) for a detailed consideration of these issues.
46 The CJEU (2012a, paragraph 104) has referred to these circumstances as (mere) "examples" of interventions that incur the risk of seriously compromising the ecological characteristics of such sites. In other words, this is not a definitive list. For examples of the CJEU applying these "examples" in practice, see CJEU (2010c) and CJEU (2011b).
47 For a detailed analysis of this point see Jackson (2014).
48 Then Directive 96/61/EC, which was codified in 2008 and later replaced by Directive 2010/75/EU.
49 See European Commission (2017a, p. 10, footnote 27) for a summary.
50 These figures are taken from EEA (2015, p. 10). EEA (2007) contains information on the numbers of species and habitats listed in each annex to the Birds and Habitats Directive for the EU-27 (current EU minus Croatia).
51 Excluding Croatia, which only joined the EU in July 2013, and using data from the previous period (2001–2006) for Greece, since Greece reported too late for the period 2007–2012 (EEA, 2015).
52 The text of the Commission's brochure states "more than two thirds [of habitat assessments] are unfavourable", which is technically accurate, though the percentage is 77% (EU, 2015, p. 19).
53 The campaign was led by BirdLife, the EEB, Friends of the Earth, and WWF, with many national NGOs participating within these networks.
54 The study was awarded to "a consortium of experts led by Milieu Ltd, and also comprised of the Institute for European Environmental Policy (IEEP), ICF International and Ecosystems Ltd" (European Commission, 2016c, p. 18).

References

AFP 1992. Brice Lalonde ne veut plus rester au gouvernement. *Agence France Presse*, 31 March.

Baldock, D. 1992. The status of Special Protection Areas for the protection of wild birds. *Journal of Environmental Law*, 4, 139–144.

Baumüller, A. 2016. 1,001 nights: the long saga of the EU Nature Directives. *Euractiv*, 16 June 2016.

Becker, S., Bauer, M. W., Connolly, S. and Kassim, H. 2016. The Commission: boxed in and constrained, but still an engine of integration. *West European Politics*, 39, 1011–1031.

Birds Directive 1979. Council Directive 79/409/EEC of 2 April 1979 on the conservation of wild birds, *OJ L 103*, 25 April 1979, p. 1, since codified as Directive 2009/147/EC of the European Parliament and of the Council of 30 November 2009 on the conservation of wild birds. *OJ L 20*, 26 January 2010, p. 7.

Bunce, R. 2012. Interview with Roy Bunce, senior negotiator for the UK government in respect of the Habitats Directive, conducted by Andrew L. R. Jackson on 13 August 2012.

Cashman, L. 2013. Personal communication: e-mail dated 18 March 2013 from Liam Cashman, DG Environment lawyer, to Andrew L. R. Jackson.

CEC 2015. Outcome of the 3441st Council of the EU meeting (Environment), Brussels, 16 December 2015, 15380/15 (OR. en), PRESSE 84, PR CO 76.

CEU 2005. Council Decision (2005/370/EC) of 17 February 2005 on the conclusion, on behalf of the European Community, of the Convention on access to information, public participation in decision-making and access to justice in environmental matters. *OJ L 124*, 17 May 2005, p. 1.

CEU 2017. Council conclusions on EU Action Plan for nature, people and the economy. Press release 387/17, 19 June 2017.

Cichowski, R. A. 2007. *The European Court and civil society: litigation, mobilization, and governance*. Cambridge: Cambridge University Press.

CJEU 1991. Judgment of the Court of Justice of the EU of 28 February 1991, *Leybucht Dykes, Commission* v. *Germany*, C-57/89, EU:C:1991:89.

CJEU 1993. Judgment of the Court of Justice of the EU of 2 August 1993, *Santoña marshes, Commission* v. *Spain*, C-355/90, EU:C:1993:331.

CJEU 1996. Judgment of the Court of Justice of the EU of 11 July 1996, *Lappel Bank*, C-44/95, EU:C:1996:297.

CJEU 1998. Judgment of the Court of Justice of the EU of 19 May 1998, *Commission* v. *Netherlands*, C-3/96, EU:C:1998:238.

CJEU 1999. Judgment of the Court of Justice of the EU of 25 November 1999, *Commission* v. *France*, C-96/98, EU:C:1999:580.

CJEU 2000a. Judgment of the Court of Justice of the EU of 7 December 2000, *Commission* v. *France*, C-374/98, EU:C:2000:670.

CJEU 2000b. Judgment of the Court of Justice of the EU of 7 November 2000, *First Corporate Shipping*, C-371/98, EU:C:2000:600.

CJEU 2004. Judgment of the Court of Justice of the EU of 7 September 2004, *Waddenzee*, C-127/02, EU:C:2004:482.

CJEU 2005a. Judgment of the Court of Justice of the EU of 13 January 2005, *Dragaggi*, C-117/03, EU:C:2005:16.

CJEU 2005b. Judgment of the Court of Justice of the EU of 20 October 2005, *Commission* v. *United Kingdom*, C-6/04, EU:C:2005:626.

CJEU 2006a. Judgment of the Court of Justice of the EU of 13 July 2006, *Commission* v. *Portugal*, C-191/05, EU:C:2006:472.

CJEU 2006b. Judgment of the Court of Justice of the EU of 14 September 2006, *Bund Naturschutz*, C-244/05, EU:C:2006:579.

CJEU 2006c. Judgment of the Court of Justice of the EU of 26 October 2006, *Commission* v. *Portugal*, C-239/04, EU:C:2006:665.

CJEU 2006d. Opinion of Advocate General Kokott of 23 February 2006, *Commission* v. *Portugal*, C-191/05, EU:C:2006:150.

CJEU 2007. Judgment of the Court of Justice of the EU of 13 December 2007, *Commission* v. *Ireland*, C-418/04, EU:C:2007:780.

CJEU 2010a. Judgment of the Court of Justice of the EU of 4 March 2010, *Commission* v. *France*, C-241/08, EU:C:2010:114.

CJEU 2010b. Judgment of the Court of Justice of the EU of 14 October 2010, *Commission* v. *Austria*, C-535/07, EU:C:2010:602.

CJEU 2010c. Judgment of the Court of Justice of the EU of 20 May 2010, *Commission* v. *Spain*, C-308/08, EU:C:2010:281.

CJEU 2011a. Judgment of the Court of Justice of the EU of 8 March 2011, *LZ I*, C-240/09, EU:C:2011:125.

CJEU 2011b. Judgment of the Court of Justice of the EU of 24 November 2011, *Commission* v. *Spain*, C-404/09, EU:C:2011:768.

CJEU 2012a. Judgment of the Court of Justice of the EU of 11 September 2012, *Nomarchiaki*, C-43/10, EU:C:2012:560.

CJEU 2012b. Judgment of the Court of Justice of the EU of 15 March 2012, *Commission* v. *Cyprus*, C-340/10, EU:C:2012:143.

CJEU 2012c. Judgment of the Court of Justice of the EU of 16 February 2012, *Solvay*, C-182/10, EU:C:2012:82.

CJEU 2013. Judgment of the Court of Justice of the EU of 11 April 2013, *Sweetman*, C-258/11, EU:C:2013:220.

CJEU 2014a. Judgment of the Court of Justice of the EU of 3 April 2014, *Cascina Tre Pini*, C-301/12, EU:C:2014:214.

CJEU 2014b. Judgment of the Court of Justice of the EU of 15 May 2014, *Briels*, C-521/12, EU:C:2014:330.

CJEU 2016a. Judgment of the Court of Justice of the EU of 8 November 2016, *LZ II*, C-243/15, EU:C:2016:838.

CJEU 2016b. Judgment of the Court of Justice of the EU of 10 November 2016, *Commission* v. *Greece*, C-504/14, EU:C:2016:847.

CJEU 2016c. Judgment of the Court of Justice of the EU of 14 January 2016, *Commission* v. *Bulgaria*, C-141/14, EU:C:2016:8.

CJEU 2016d. Judgment of the Court of Justice of the EU of 14 January 2016, Grüne Liga Sachsen, C-399/14, EU:C:2016:10.

CJEU 2016e. Judgment of the Court of Justice of the EU of 21 July 2016, Orleans, C-387/15, EU:C:2016:583.

CJEU 2016f. Judgment of the Court of Justice of the EU of 24 November 2016, *Commission* v. *Spain*, C-461/14, EU:C:2016:895.

CJEU 2017a. Judgment of the Court of Justice of the EU of 26 April 2017, *Commission* v. *Germany*, C-142/16, EU:C:2017:301.

CJEU 2017b. Opinion of Advocate General Kokott of 15 June 2017, VHL, C-281/16, EU:C:2017:476.

CJEU 2017c. Order of the Vice-President of the Court of Justice of the EU of 27 July 2017, C-441/17 R, *Commission* v. *Poland*, EU:C:2017:622.

Clutten, R. and Tafur, I. 2012. Are imperative reasons imperiling the Habitats Directive? An assessment of Article 6(4) and the IROPI exception. *In*: Jones, G. (ed.) *The Habitats Directive: a developer's obstacle course?* Oxford: Hart Publishing.

Crisp, J. 2015. Recording of part of Timmermans' speech included in article: 'Better Regulation won't hurt environment', vows Timmermans. *Euractiv*, 4 June.

Crisp, J. 2016. Commission research shows Nature Directives don't need 'better regulation'. *Euractiv*, 15 June.

DAHG 2017. A programme of measures by Ireland to ensure full compliance with the Judgment of the Court of Justice of the European Union in Case C-418/04 "the Birds Case" (update May 2017). Department of the Arts, Heritage and the Gaeltacht.

De Smedt, P. and Van Rijswick, M. 2015. Nature conservation and water management: one battle? *In*: Born, C.-H., Cliquet, A., Schoukens, H., Misonne, D. and Van Hoorick, G. (eds) *The Habitats Directive in its EU environmental law context*. Abingdon: Routledge.

DECC 2010. Severn tidal power: potential for compensatory measures (May 2010). UK Department of Energy and Climate Change.

DEFRA 2007. Wallasea wetland creation project, Newsletter 3, September 2007.

DG XI 1988. Internal DG XI note dated 18 February 1988 from the Legal Unit to the Director of Directorate B, the adviser to the Director General, the Nature Unit, and the Economics Unit. Subject: Draft of a directive on habitats protection. Ref: Version of FFH directive of 15.2.1988.

DG XI 1993. DG XI document dated 7 April 1993, headed "Implementation of the Birds and Habitats Directives".

Donald, P. F., Sanderson, F. J., Burfield, I. J., Bierman, S. M., Gregory, R. D. and Waliczky, Z. 2007. International conservation policy delivers benefits for birds in Europe. *Science*, 317, 810–813.

EEA 2007. *Number of habitats types and species in the annexes of the Birds and Habitats Directives. EU 27 – February 2007* [Online]. Available: https://bd.eionet.europa.eu/activities/Natura_2000/pdfs/Nb_annexes_Directives-2007.pdf [accessed 30 June 2017].

EEA 2015. *State of nature in the EU: results from reporting under the nature directives 2007–2012*. European Environment Agency. EEA Technical report No 2/2015.

ENDS Europe 2013. DG Environment faces major reorganisation, staff cuts. *ENDS Europe*, 27 February.

Environment Ministers 2015. Letter dated 26 October 2015 from Ministers of the Environment of Germany, Croatia, France, Italy, Luxembourg, Poland, Romania, Slovenia and Spain to EU Commissioner for Environment, Maritime Affairs and Fisheries [Online]. Available: www.birdlife.org/sites/default/files/attachments/fitness_check-letter_nine_ministers_to_vella_26.10.2015.pdf [accessed 23 March 2017].

Environmental Liability Directive 2004. Directive 2004/35/EC of the European Parliament and of the Council of 21 April 2004 on environmental liability with regard to the prevention and remedying of environmental damage. *OJ L 143*, 30 April 2004, p. 56.

EP and CEU 2003. Directive 2003/35/EC of the European Parliament and of the Council of 26 May 2003 providing for public participation in respect of the drawing up of certain plans and programmes relating to the environment and amending with regard to public participation and access to justice Council Directives 85/337/EEC and 96/61/EC.

EU 2015. Brochure: The State of Nature in the EU – Reporting under the EU Habitats and Birds Directives 2007–2012.

Europe Environment 1995. 60 legal experts oppose restructuring of Commission's DGXI. *Europe Environment*, 7 February.

European Commission 1994. Directory of the European Commission (16 June 1994).
European Commission 1995. Directory of the Commission of the European Communities (October 1995).
European Commission 2003. Proposal for a Directive of the European Parliament and of the Council on access to justice in environmental matters. COM/2003/0624 final, 2003/0246 (COD).
European Commission 2007/2012. Guidance document on Article 6(4) of the 'Habitats Directive' 92/43/EEC.
European Commission 2010a. Guidance document: Wind energy developments and Natura 2000 (October 2010).
European Commission 2010b. Guidance document: Wind energy developments and Natura 2000 (October 2010) [Online]. Available: http://ec.europa.eu/environment/nature/natura2000/management/docs/Wind_farms.pdf.
European Commission 2011a. Commission Implementing Decision of 11 July 2011 concerning a site information format for Natura 2000 sites (notified under document C(2011) 4892)(2011/484/EU). *OJ L 198*, 30 July 2011, p. 39.
European Commission 2011b. Commission staff working paper. Financing Natura 2000. Investing in Natura 2000: delivering benefits for nature and people. Brussels, 12 December 2011, SEC(2011) 1573 final.
European Commission 2013a. Communication from the Commission to the European Parliament, the Council, the European Economic and Social Committee and the Committee of the Regions. Regulatory Fitness and Performance (REFIT): Results and Next Steps. Brussels, 2 October 2013. COM(2013) 685 final.
European Commission 2013b. The economic benefits of the Natura 2000 network. Synthesis report. Luxembourg: Publications Office of the European Union.
European Commission 2014. Withdrawal of obsolete Commission proposals. *OJ C 153*, 21 May 2014, p. 3.
European Commission 2015. The State of Nature in the European Union. Report from the Commission to the Council and European Parliament. Brussels, 20 May 2015, COM(2015) 219 final.
European Commission 2016a. Commission refers France to the Court of Justice over its failure to protect wild birds. IP/16/4213, 8 December 2016.
European Commission 2016b. Commission refers the United Kingdom to the Court over its failure to protect marine species. IP/16/3128, 29 September 2016.
European Commission 2016c. Commission Staff Working Document. Fitness Check of the EU Nature Legislation (Birds and Habitats Directives). Brussels, 16 December 2016, SWD(2016) 472 final.
European Commission 2016d. Report from the Commission to the Council and the European Parliament under Article 18(2) of Directive 2004/35/EC on environmental liability with regard to the prevention and remedying of environmental damage, 14 April 2016, COM(2016) 204 final.
European Commission 2016e. Statement by Commissioner Vella on the orientation debate on the next steps of the Fitness Check of the Birds and Habitats Directives, 7 December 2016.
European Commission 2016f. *Statistics on environmental infringements* [Online]. Available: http://ec.europa.eu/environment/legal/law/statistics.htm [accessed 16 March 2017].
European Commission 2017a. Communication from the Commission of 28.4.2017: Commission Notice on Access to Justice in Environmental Matters, C(2017) 2616 final.
European Commission 2017b. Communication from the Commission to the European Parliament, the Council, the EESC and the Committee of the Regions. An Action Plan for nature, people and the economy. Brussels, 27 April 2017. COM(2017) 198 final.

European Commission 2017c. *The Environmental Implementation Review* [Online]. Available: http://ec.europa.eu/environment/eir/index_en.htm [accessed 26 July 2017].
European Commission 2017d. *The LIFE Programme* [Online]. Available: http://ec.europa.eu/environment/life/about/index.htm [accessed 22 June 2017].
European Commission 2017e. Natura 2000 nature and biodiversity newsletter. Number 41, February 2017.
European Commission undated. Legal Enforcement. Statistics on environmental infringements. Infringements by sector.
European Court of Auditors 2017. More efforts needed to implement the Natura 2000 network to its full potential. Special report no. 01.
European Parliament 2016. Resolution of 2 February 2016 on the mid-term review of the EU's Biodiversity Strategy (2015/2137(INI)), P8_TA(2016)0034.
European Report 1992. Council of Ministers: subsidiarity to be "the" issue of UK Presidency. *European Report*, No. 1782, 6 July.
European Report 1994a. Bruno Julien gets new position in the Environment Directorate. *European Report*, No. 1998, 3 December.
European Report 1994b. Eurocrats on the campaign trail. *European Report*, No. 1953, 28 May.
Evans, D. 2006. The habitats of the European Union Habitats Directive. *Biology and Environment: Proceedings of the Royal Irish Academy*, 106B, 167–173.
Fernández-Galiano, E. 2012. Interview with Eladio Fernández-Galiano, then head of the Secretariat of the Bern Convention, conducted by Andrew L. R. Jackson on 30 August 2012.
Fifth Environmental Action Programme 1993. Towards sustainability: A European Community programme of policy and action in relation to the environment and sustainable development. *OJ C 138*, 17 May 1993, p. 5.
Fogleman, V. 2015. The threshold for liability for ecological damage in the EU: mixing environmental and conservation law. *In*: Born, C.-H., Cliquet, A., Schoukens, H., Misonne, D. and Van Hoorick, G. (eds) *The Habitats Directive in its EU environmental law context*. Abingdon: Routledge.
Forster, A. 1999. *Britain and the Maastricht negotiations*. Basingstoke: Palgrave Macmillan.
Fraser, J. 1992. Joy as 'Grim Ripa' goes back to Rome. *The Glasgow Herald*, 29 June.
Gammell, A. 1999. The contribution of BirdLife International to the Birds Directive and the management of birds and their habitats in Europe. 20 years with the EC Birds Directive. Proceedings from a conference on the Council's Directive on the Conservation of Wild Birds, 18–19 November 1999, Elsinore, Denmark. The Danish National Forest and Nature Agency.
Goldberg, C. 2012. Interview with Claus Goldberg, Danish government official who participated in the Habitats Directive negotiations, conducted by Andrew L. R. Jackson on 31 August 2012.
Golub, J. 1996. Sovereignty and subsidiarity in EU environmental policy. *Political Studies*, 44, 686–703.
Government of Lithuania 2015. *Lithuanian position in connection with the draft Council Conclusions on the Mid-Term Review of the EU Biodiversity Strategy to 2020* [Online]. Available: www.natuurpunt.be/files/positielitouwenfitnesscheckpdf/download?token=Al5EILFx [accessed 23 March 2017].
Greenwire 1992a. EC: New enviro Commissioner takes 'cautious' approach. *Greenwire*, 8 February.
Greenwire 1992b. European Community: Delors to 'sacrifice' enviro policy. *Greenwire*, 7 July.

Groote, M. 2012. Is EU environmental legislation in decline? *ENDS Europe*.
Gunn, S. 1992. Hurd puts EC laws under scrutiny. *The Times*, 7 August.
Habitats Directive 1992. Consolidated version of Council Directive 92/43/EEC of 21 May 1992 on the conservation of natural habitats and of wild fauna and flora. *OJ L 206*, 22 July 1992, p. 7. Consolidation date 1 July 2013.
Hamer, J. 2007. Case note: Judgment of the ECJ, 14 September 2006, Case C-244/05, Bund Naturschutz Bayern E.V. *Review of European Community & International Environmental Law*, 16, 104–107.
Hardy, J. 1991. Government accused of 'tantrums' in Euro row. *Press Association*, 19 October.
Hey, C. 2005. EU environmental policies: A short history of the policy strategies. *In*: Scheuer, S. (ed.) *EU environmental policy handbook: a critical analysis of EU environmental legislation*. Dublin: International Books.
Hobolt, S. B. and De Vries, C. 2016. Turning against the Union? The impact of the crisis on the Eurosceptic vote in the 2014 European Parliament elections. *Electoral Studies*, 44, 504–514.
Jackson, A. L. R. 2014. Bund Naturschutz in Bayern Revisited: The Required Standard of Protection under the EU's Habitats Directive Prior to a Site's Adoption as a Site of Community Importance. *Journal of Environmental Law*, 26, 495–506.
Janauer, G. A., Albrecht, J. and Stratmann, L. 2015. Synergies and Conflicts Between Water Framework Directive and Natura 2000: Legal Requirements, Technical Guidance and Experiences from Practice. *Wetlands and Water Framework Directive: Protection, Management and Climate Change*, 9–29.
Jans, J. H. 2011. Who is the referee? Access to justice in a globalised legal order. *Review of European Administrative Law*, 4, 85–97.
Jordan, A. 1998. EU environmental policy at 25: the politics of multinational governance. *Environment*, 40, 14–20.
Jordan, A. and Adelle, C. 2013. Preface. *In*: Jordan, A. and Adelle, C. (eds) *Environmental policy in the EU: actors, institutions and processes*, 3rd edn. Abingdon: Routledge.
Jordan, A. and Jeppesen, T. 2000. EU environmental policy: adapting to the principle of subsidiarity? *European Environment*, 10, 64–74.
Juncker, J.-C. 2014. Mission letter dated 1 November 2014 from Jean-Claude Juncker, President of the European Commission, to Karmenu Vella, Commissioner for Environment, Maritime Affairs and Fisheries.
Khetani-Shah, S. 2016. *UK's David Cameron supports EU nature laws, says they are reason to stay in the EU*. BirdLife International, 2 June [Online]. Available: www.birdlife.org/europe-and-central-asia/news/uks-david-cameron-supports-eu-nature-laws-says-they-are-reason-stay-eu [accessed 23 March 2017].
Kingston, S., Heyvaert, V. and Čavoški, A. 2017. *European environmental law*. Cambridge: Cambridge University Press.
Knill, C. and Liefferink, D. 2013. The establishment of EU environmental policy. *In*: Jordan, A. and Adelle, C. (eds) *Environmental policy in the EU: actors, institutions and processes*, 3rd edn. Abingdon: Routledge.
Koester, V. 2012. Interview with Veit Koester, senior negotiator for the Danish government in respect of the Habitats Directive, conducted by Andrew L. R. Jackson on 22 August 2012.
Krämer, L. 2011. *EU environmental law*. London: Sweet & Maxwell.
Lasén-Díaz, C. 2001. The EC Habitats Directive approaches its tenth anniversary: an overview. *Review of European Community and International Environmental Law*, 10, 287–295.

Marine Strategy Framework Directive 2008. Directive 2008/56/EC of the European Parliament and of the Council of 17 June 2008 establishing a framework for community action in the field of marine environmental policy. *OJ L 164*, 25 June 2008, p. 19.

Marshall, A. 1992. Green vacuum after 'grim Ripa' goes. *The Independent*, 30 June.

McGillivray, D. 2015. Compensatory measures under Article 6(4) of the Habitats Directive: no net loss for Natura 2000? *In*: Born, C.-H., Cliquet, A., Schoukens, H., Misonne, D. and Van Hoorick, G. (eds) *The Habitats Directive in its EU environmental law context*. Abingdon: Routledge.

Meade, G. 1992. Green di Meana exits – in red Alfa. *Press Association*, 29 June.

Milieu and IUCN 2014. Experience gained in the application of ELD biodiversity damage. Final report for the European Commission, DG Environment.

Moules, R. 2017. Significant EU environmental cases: 2016. *Journal of Environmental Law*, 29, 177–188.

Neal, H. 2012. Interview with Hilary Neal, UK government official who participated in the Habitats Directive negotiations and early implementation, conducted by Andrew L. R. Jackson on 14 August 2012.

Neslen, A. 2016. RSPB and WWF urge backers to vote to stay in EU to protect UK wildlife. *The Guardian*, 2 June.

Ó Bríain, M. 2012. Interview with Micheál Ó Bríain of DG Environment's Nature Unit conducted by Andrew L. R. Jackson on 21 August 2012.

Ó Bríain, M. 2013. Personal communication: e-mail dated 18 March 2013 from Micheál Ó Bríain of DG Environment's Nature Unit to Andrew L. R. Jackson.

Oakley, R. 1992. Howard intent on raising green profile of Tories. *The Times*, 16 April.

Palmer, J. and White, M. 1992. EC chaos after 'no' vote; Sceptics delighted at Danes' David and Goliath rejection of Maastricht. *The Guardian*, 3 June.

Parker, G. and Hammond, G. 2017. Developers set for Brexit triumph over great crested newt. *Financial Times*, 10 February.

Peterson, J. 2017. Juncker's political European Commission and an EU in crisis. *JCMS: Journal of Common Market Studies*, 55, 349–367.

Potočnik, J. 2012. Speech – 'Living well, within the limits of the planet' (SPEECH/12/900, 4 December).

Potočnik, J. 2015. Foreword. *In*: Born, C.-H., Cliquet, A., Schoukens, H., Misonne, D. and Van Hoorick, G. (eds) *The Habitats Directive in its EU environmental law context*. Abingdon: Routledge.

Ripa Di Meana, C. 1992. Why I will stay away from the Earth Summit; Carlo Ripa di Meana, the EC environment commissioner, on what's wrong with Rio. *The Guardian*, 30 May.

Schoukens, H. 2017. Habitat restoration measures as facilitators for economic development within the context of the EU Habitats Directive: balancing no net loss with the preventive approach? *Journal of Environmental Law*, 29, 47–73.

Seventh Environmental Action Programme 2013. Decision No. 1386/2013/EU of the European Parliament and of the Council of 20 November 2013 on a General Union Environment Action Programme to 2020: "Living well, within the limits of our planet". *OJ L 354*, 28 December 2013, p. 171.

Slepcevic, R. 2009. The judicial enforcement of EU law through national courts: possibilities and limits. *Journal of European Public Policy*, 16, 378–394.

Stahl, L. 2015. The concept of 'conservation objectives' in the Habitats Directive: a need for better definition? *In*: Born, C.-H., Cliquet, A., Schoukens, H., Misonne, D. and Van Hoorick, G. (eds) *The Habitats Directive in its EU environmental law context*. Abingdon: Routledge.

Stuffmann, C. 1994. The European Community and EECONET. *In*: Bennett, G. (ed.) *Conserving Europe's natural heritage: towards a European ecological network*. London: Graham and Trotman.

Stuffmann, C. 1999. The upcoming of the Birds Directive. 20 years with the EC Birds Directive. Proceedings from a conference on the Council's Directive on the Conservation of Wild Birds, 18–19 November 1999, Elsinore, Denmark. The Danish National Forest and Nature Agency.

Taylor, M. 2017. Michael Gove as environment secretary is 'fox in charge of hen house'. *The Guardian*, 12 June.

The Guardian 1992. Lisbon summit: bold defenders of green and social issues to quit. *The Guardian*, 29 June.

Usborne, D. 1991. Environment row threatens EC Treaty, Major says. *The Independent*, 22 October.

Van der Sluis, T., Foppen, R., Gillings, S., Groen, T., Henkens, R., Hennekens, S., Huskens, K., Noble, D., Ottburg, F., Santini, L., Sierdsema, H., Van Kleunen, A., Schaminee, J., Van Swaay, C., Toxopeus, B., Wallis De Vries, M. and Jones-Walters, L. 2016. How much biodiversity is in Natura 2000?; the "umbrella effect" of the European Natura 2000 protected area network. Alterra report 2730B. Wageningen Alterra Wageningen UR (University and Research centre).

Vieten, U. M. and Poynting, S. 2016. Contemporary far-right racist populism in Europe. *Journal of Intercultural Studies*, 37, 533–540.

Walker, T. 1992. Contrite Delors talks of devolving power. *The Times*, 25 June.

Ward, S. 1997. The IGC and the current state of EU environmental policy: consolidation or roll-back? *Environmental Politics*, 6, 178–184.

Water Framework Directive 2000. Directive 2000/60/EC of the European Parliament and of the Council of 23 October 2000 establishing a framework for Community action in the field of water policy. *OJ L 327*, 22 December 2000, p. 1.

Wates, J. 2016. We need to green Juncker's 10 priorities. *Euractiv*, 18 January.

Wolf, J. 1992a. Brussels Eurocrats to divest powers. *The Guardian*, 28 June.

Wolf, J. 1992b. EC drops five British environment cases; M3 Twyford Down extension given benefit of doubt. *The Guardian*, 1 August.

Wurzel, R. 1999. The role of the European parliament: interview with Ken Collins MEP. *The Journal of Legislative Studies*, 5, 1–23.

Wurzel, R. 2002. *Environmental policy-making in Britain, Germany and the European Union: the Europeanisation of air and water pollution control*. Manchester: Manchester University Press.

Wurzel, R. 2008. Environmental policy: EU actors, leader and laggard States. *In*: Hayward, J. (ed.) *Leaderless Europe*. Oxford: Oxford University Press.

11

DISCUSSION AND CONCLUSIONS

Introduction

In this final chapter, an historical institutionalist analysis is applied to the history of the Birds and Habitats Directives to address the four central issues introduced in Chapter 1. Namely, what insights can this detailed history and the attendant historical institutionalist approach contribute to understanding:

1. Conflicts in establishing the Natura 2000 network?
2. The resolution of acute conflicts between human land uses and Natura 2000 sites?
3. The process of EU integration in the nature policy field?
4. The future of EU nature policy?

Applying an historical institutionalist analysis to Natura 2000 policy

To recap, Pierson's (1996, p. 123) historical institutionalist analysis posits, in essence, that "Losses of [Member State] control result not only from the autonomous actions of supranational organizations, but from member-state preoccupation with short-term concerns, the ubiquity of unintended consequences, and the instability of member-state policy preferences. Once gaps in control emerge, change-resistant decision rules and sunk costs associated with societal adaptations make it difficult for member states to reassert their authority".

First phase (1958–1987): foundational period

Chapters 3 and 4

Autonomous actions

It is worth recalling at the outset Meyer's (2010, p. 179) comment that the Birds Directive (1979) was negotiated and adopted "against all odds". Thus, the idea that EU institutions might obtain responsibility for nature conservation regulation could of course not be taken for granted at the time a directive on birds was first suggested.

Indeed, the mere existence of the Birds Directive represents in one sense a gap in Member State control, given certain Member States' extreme sensitivities regarding hunting regulation, and given Denmark's repeatedly expressed reluctance to allow the EU to regulate nature conservation at all; Denmark "did not want to accept any Community competence in the area of nature conservation and therefore strongly questioned the validity of the legal basis [during the Birds Directive negotiations]" (Stuffmann, 1999, p. 17).

In addressing this puzzle, one can highlight, first, the autonomous actions of the European Parliament in successfully applying pressure, together with NGOs and individuals, on the Commission and the Member States to (respectively) propose and adopt EU legislation on bird conservation. To this end, the European Parliament and others lobbied, resulting in the inclusion within the First Environmental Action Programme (1973) of a commitment to carry out a study with a view to possible harmonisation of national regulations on bird conservation.

Second, one can highlight the actions of the Commission, which more or less invited a demand to include bird conservation within the First Environmental Action Programme, as Meyer (2010) argues. That the European Parliament's actions ultimately had an important catalysing effect was confirmed by the Commission, which explained that its 1976 Birds Directive proposal was a response to the "wishes expressed repeatedly by the European Parliament in several of its Resolutions, debates and petitions . . . and to the requests of a broad section of public opinion indicated in more than 50,000 letters received by the Commission alone" (CEC, 1977a, p. 2). One might also highlight here the autonomous actions of the Commission in commissioning a scientific study by Stanley Cramp (on the advice of John Temple Lang of the Commission's Legal Service), and drafting key provisions of the Birds Directive proposal (John Temple Lang), which together moved the aims of nascent EU bird conservation policy beyond hunting control to embrace habitat conservation.

In addition to institutional factors, we see also at this stage of the history the importance of policy entrepreneurs, leaders and small networks (e.g., Stuffmann-Temple Lang-Cramp-Gammell). The role played by Alistair Gammell of the RSPB is important to emphasise, showing as it does that NGOs were directly involved in EU environmental policy development as early as the mid-to-late 1970s.

Once the Commission's Birds Directive proposal had launched, we again saw autonomous actions by the European Commission and the CJEU, which drove forward the process of EU integration by adopting a dynamic interpretation of Article 235 of the Treaty of Rome (1957),[1] whereby the Article 2 objectives of the EU were read to include environmental protection, even though environmental protection was not explicitly mentioned. In the context of the Birds Directive negotiations, this reasoning allowed the Council Legal Service (1977) to determine that Article 235 gave the EU legal competence to make legislation on bird conservation, noting that the Council had endorsed the First and Second Environmental Action Programmes, which stated that "improvement in the quality of life and the protection of the natural environment are among the fundamental tasks of the Community". This approach to Article 235 was cemented legally when the CJEU held, in 1985, that "environmental protection . . . is one of the Community's essential objectives" (CJEU, 1985, paragraph 13).

Even at the very end of the negotiation process regarding the Birds Directive, Germany and Denmark remained uneasy regarding the EU's foray into nature conservation regulation. It will be recalled that Germany prepared a statement to the effect that: "The German delegation points out that its agreement to this Directive cannot be interpreted as constituting a precedent for any further action by the European Communities in the field of nature conservancy" (CEC, 1977b, p. 30). For its part, Denmark challenged the legal basis for the Directive to the end, but – and this highlights Pierson's (1996) point about potential institutional obstacles to closing gaps in control – Denmark was willing to go along with the other Member States, even though the requirement for unanimity would have allowed Denmark to veto the Birds Directive: "While the Danish Government has not wished to refute the arguments put forward in this matter by the other eight Member States and the legal departments [i.e., the Council and Commission Legal Services], it wishes to point out that it does not consider that the adoption of this Directive can create a precedent for the adoption of further Community acts of a similar nature" (CEC, 1977b, p. 31; CEC, 1978). A similar analysis might be applied in respect of the UK: in addition to not wanting the Birds Directive to be a popular success for the EU (Stuffmann, 1999), the UK's then Secretary of State for the Environment Peter Shore wrote in May 1978 that he was opposed in principle to the EU's becoming involved in the regulation of issues such as bird protection (cited in Fairbrass, 2000, p. 22). Yet, as Fairbrass and Jordan (2001, p. 9) note, "in the dying days of the 1974–79 Labour government, [Shore's] reservations could not have been that important because in April 1979 he approved the proposal, nicely illustrating Pierson's point about politicians (even the most ardently Eurosceptic) having short time horizons".

The European Commission was not put off by Denmark's and Germany's reticence, and continued to exercise its autonomy, with the head of the Environment and Consumer Protection Service (ECPS) Michel Carpentier commenting in 1981, just two years after the Birds Directive was adopted: "Though we have to interpret the Treaty cleverly we can use it to justify conservation measures. A re-writing of the Treaty might burden us with more restrictions than we have now with our pragmatic approach. The Birds Directive, for example, was accused of being *ultra vires*, but it has been very popular . . . We're trying to extend our programme, not just to deal with pollution" (Caufield, 1981, p. 417). Moreover, Carpentier's colleague within the ECPS, Claus Stuffmann, had a strong personal commitment to European unity and nature conservation, and reportedly "always defended very much more strongly than it's ever done nowadays the competence of the [EU]. He was fighting very hard in CITES, in Bern, in Bonn, everywhere, for the Union to exert as much competence as possible" (Devillers, 2013).

Unintended consequences

It is worth recalling, at this stage, Meyer's (2011, p. 84) comment that, "[t]he Birds Directive was originally motivated by Northern European resentment against the hunting of songbirds in Southern Europe", but ended up introducing, in addition to its hunting control measures, "remarkably strict rules on habitats protection valid in all member states. It provided an important precedent for EC action in the area of nature protection and paved the way for the Habitats directive of 1992 and the Natura 2000 Programme". As Alistair Gammell reflects, "I really don't think that it was in the minds of the British or any other government that this was primarily about habitat protection. It was primarily about not killing migratory birds, and controlling hunting" (Gammell, 2012).

Whilst the site protection provisions were debated and negotiated by the Member States, their content barely changed over the course of the negotiations; as Stuffmann (1999, p. 17) reflects, "after only two years the Commission's draft of 1976 was adopted as a Council directive with very minor modifications rather reinforcing the important conservation issues than weakening them". It seems clear, in retrospect, that the Member States had little idea that the habitat protection aspects of the Birds Directive would result in a large network of, in many cases, new protected areas across Europe. One might point in this regard to the UK's comment that the government "did not think the Directive posed any major problems" (CEC, 1977a, pp. 2–7). Member States even went so far during negotiations as to supplement by 12 species the Commission's Annex I list of birds requiring the designation of SPAs.

Denmark's position is interesting: its negotiator was effectively divided between resisting EU competence over nature conservation on the one hand, yet on the other hand giving up national control "to a limited extent" to gain a stronger hand for the Ministry of Environment in respect of disputes over hunting at the national level between the Ministries of Environment and Agriculture (Koester, 2012). The unintended consequence here is that, in agreeing to the Birds Directive, Denmark would give up control over nature conservation regulation by more than just a limited extent.

Another unintended consequence of the Birds Directive, from the perspective of at least certain Member States, is revealed by the assumption implicit in a draft Resolution prepared by France during the negotiations, and of the wording finally used in that Resolution, revealing a lack of clarity regarding how the system of protected areas would operate in practice. The French text called upon the Commission to make proposals regarding the criteria to be adopted for special protection areas, saying (emphasis added), "These proposals will determine, among other things, *the proportion of the designated areas which shall be afforded special protection*"; in other words, the French draft envisaged special protection applying only to parts of a designated area, rather than to the whole area (CEC, 1977b). As it would later turn out, the CJEU interpreted the protective provisions of the Birds Directive in a different manner: that is, strictly and uniformly across sites, apparently confounding certain Member States' expectations.

Path dependence

Further highlighting the influential and even pivotal role played by certain individuals and NGOs, the final point to be highlighted in respect of the immediate aftermath of the Birds Directive negotiations is the advisory role played by John Temple Lang of the Commission's Legal Service, who assisted with the early implementation of the habitat protection parts of the Directive by advising Alistair Gammell of the RSPB to "ask for Commission funding to do these initial studies [regarding Important Bird Areas (IBAs)], which we did", Gammell (2012) explains, "and we got Commission funding to do the first [IBA] studies. So that was a very, very influential discussion, and after that John also was the one to point out, and he certainly led me through this procedure, about how you could make an official complaint to the European Union about Member States not having designated areas or not protecting areas that should be designated but . . . had some development on, so he actually advised us about how you send a formal letter of complaint to the European Commission, etc."

Thus, Temple Lang helped to kick-start a process that would effectively become path dependent when, in 1998, the CJEU cemented the legal significance of the IBA lists by holding that, although not legally binding on the Member States concerned, the IBA lists could, by reason of their acknowledged scientific value, be used by the Court as a basis of reference for assessing the extent to which a Member State has complied with its obligation to designate special protection areas (CJEU, 1998). Thereafter, an area's presence on an IBA list would indicate that it ought to be an SPA under the Birds Directive; and the production of expanded IBA lists could then of course help to drive forward the expansion of the SPA network across the EU.

In summary, the pattern that emerges from the Birds Directive negotiations, on an historical institutionalist analysis, is of EU-level regulation of bird conservation being driven forward by a series of autonomous actions on the part of EU institutions, NGOs, and policy entrepreneurs/leaders/small networks, in spite of some Member State scepticism and resistance. To a certain extent the creation of a large network of SPAs then became path dependent, in particular owing to the use of

IBA lists to establish, legally, a site's potential status as an SPA. Finally, the adoption of the Birds Directive has arguably had both intended and unintended effects from the Member State perspective – more below.

Chapter 5

Autonomous actions

Notwithstanding the concerns of (at least) Denmark and Germany in the immediate aftermath of the Birds Directive, the Commission immediately made its intentions clear regarding its desire to expand its nature conservation mandate: e.g., "The Commission hopes that certain hesitations which have become apparent as to the scope of the Community's powers which have hindered the adoption of important and urgent measures will be overcome in the future" (European Commission, 1980, pp. 36–37). Moreover, it will be recalled that as early as 1980 the Commission had a clear vision of a future Habitats Directive, accompanied by EU financing: "As far as the protection and maintenance of wild species is concerned, the [Birds] Directive of April 1979 will be supplemented by the creation and maintenance of a network of sufficiently large protected zones which will be protected and administered under common rules and principles. If this aim is to be achieved by informing and interesting the public, the Community will doubtless have to introduce legislation and provide funds" (European Commission, 1980, p. 37). Yet there must have been a degree of resistance from the Member States, because the Third Environmental Action Programme (1983, p. 13) ultimately struck an uneasy balance: "While it is recognized that local, regional and national responsibilities are decisive in this case, a Community framework is becoming essential if greater cohesion is to be given to such efforts".

As it had in the case of the Birds Directive, the European Parliament – by now directly elected – also played an important role during this period in lobbying for a Habitats Directive, led by activist MEPs such as Stanley Johnson and Hemmo Muntingh. For example, by July 1987 (by which time Johnson was back working for the Commission) – six months before the process of drafting the Habitats Directive would begin – the Parliament was calling for "land and marine parks and nature reserves to form a Community network of protected areas which is representative of the natural resources of the various geographical regions of the Community" (European Parliament, 1987, p. 122). Environment Commissioner Stanley Clinton Davis assured the Parliament that "the contents of the report have been very fully noted, both by myself and by my staff, and they will serve as an important guide for the Commission's work in this important field of nature conservation" (Clinton Davis, 1987, p. 288). This represents a clear example of the mutually reinforcing interplay between EU institutions, both of which wished to see greater EU action in the field of nature conservation.

Also in the European Parliament, Altiero Spinelli, Stanley Johnson and the rest of the Crocodile Club played an important role during this period in helping to catalyse changes to the Treaty, ultimately resulting in a specific EU environmental policy mandate via the Single European Act (SEA) (1986). And rather than the SEA merely representing a constitutionalisation of existing environmental functions, the SEA went further, giving the EU broad competences in previously contested environmental policy matters, such as nature conservation. In other words, disputes over the precise ambit of the EU's environmental remit were subsumed by the political imperative to move forward with EU integration, complete the common market, and reform the EU's decision-making procedures. It would be hard to sustain an argument that this expansion of EU environmental competences was an unintended consequence; rather, the better analysis would seem to be that the expansion of such competences was a price paid for desired progress in other areas.

On foot of the signature of the SEA, NGOs focused their attention on the Commission, keen to see what use might be made of the new environmental policy-making powers in the Treaty, particularly given the opportunity presented by the drafting of the Fourth Environmental Action Programme (1987). As Alistair Gammell explains, "Myself and one or two others . . . lobbied to put [a proposal for a directive on habitats and species protection] in [the Programme], although there were Commission officials in DG XI . . . who were . . . very easy to lobby . . . They wanted to do it and we wanted it in" (Gammell, 2012). And it was not only via the Fourth Environmental Action Programme that the EU nature conservation agenda was pushed at this time: e.g., the CORINE biotopes team within DG XI saw their work as:

> the best thing we could do. We felt that it was a way to get [something akin to a Habitats Directive] through the back door, if you like. We did not think that it was possible to get a binding instrument, but we felt that notoriety is essential, and that if we made a scientific inventory of what needs to be conserved, it would bring a strong leverage on what the Member States would have to do.
>
> *(Devillers, 2013)*

In respect of EU financing for nature conservation, there is clear evidence that a learning process had begun during this phase in the history, with the Member States and the Commission engaging in something of a ping pong match throughout the 1980s on the subject of the ambit of EU financing measures for nature conservation. Despite having its proposals to extend EU financing for nature conservation beyond birds rejected twice by the Member States – in 1984 and in 1987 – the issue remained very much on the Commission's agenda, and an extension of financing to other fauna and flora would later be secured. In the meantime, and notwithstanding these knock-backs, DG XI's Nature Unit creatively took action in respect of organisms other than birds, launching studies on the monk seal and sea turtles, for example: "we tried to go as far as we could . . . I wouldn't say it was stretching the Treaty but it was certainly using it to its fullest possibilities" (Devillers, 2013).

Unintended consequences

Unintended consequences of the Birds Directive's adoption began to play a pronounced role during this phase of the history. While there is a strong argument that legal actions brought to enforce EU law obligations (which were negotiated and agreed by the Member States) are simply part and parcel of EU membership and cannot be characterised as unintended consequences from the Member States' perspective, there was nevertheless a clear sentiment on the part of some Member States that insufficient "flexibility" was being shown by the Commission in adopting a "pernickety attitude" to its enforcement functions (DG XI, 1988a). As Fairbrass and Jordan (2001, pp. 15 and 8) argue, the UK, for example, "failed to protect its voluntary approach to nature conservation" which "implied that farmers could be encouraged (sometimes with financial inducements) to protect biodiversity, but never compelled by government fiat". Likewise, at EU level, the other Member States were also beginning to realise that they could not adopt a similar voluntary approach to EU nature conservation obligations. Commission-led legal actions under the Birds Directive during this period therefore served as something of a wake-up call for the EU's Member States.

In addition to the *Duich Moss* dispute on Islay in Scotland, a raft of other legal proceedings under the Birds Directive were brought during this phase of the history: by February 1988, DG XI's Legal Unit had launched 29 infringement proceedings against Member States seeking to force them to comply with their obligations to designate SPAs under the Birds Directive (DG XI, 1988b). This arguably kick-started a learning process on the part of the Member States, who would subsequently seek to use the opportunity presented by the Habitats Directive negotiations to address perceived gaps in their control over EU biodiversity policy.

Second phase (1987–1992): the golden era

Chapter 6

Autonomous actions; and path dependence

Having become main adviser to DG XI's Director General Laurens Jan Brinkhorst in 1987, it is clear that Stanley Johnson was in a position to "push things through" (Johnson, 2011, 2012). And yet, even after the new environmental powers in the SEA came into force in July 1987, extending the EU's reach in terms of nature conservation policy would remain a challenge – not only with the Member States, but also within the Commission.

In drafting the Fourth Environmental Action Programme (1987), Stanley Johnson included language about the need for a Habitats Directive. However, as Alistair Gammell reflects, "governments allowed it to remain in, but I'm pretty certain that they never intended to do anything about it" (Gammell, 2012).

As noted in Chapter 6, this is not impossible; after all, in the Resolution accompanying the Fourth Environmental Action Programme (1987, p. 4), the Council and Member States merely "[noted] the action programme set out in the Annex to this resolution and [approved] in general the guidelines". The UK government would later argue that "[a possible Habitats Directive] had not been one of the 22 priorities identified explicitly in the Resolution of 19 October 1987" (House of Lords, 1989, paragraph 38).

The Member States having allowed wording relating to a directive on habitats to remain in the Fourth Environmental Action Programme, there was then an element of path dependence in what followed (i.e., Johnson could argue he was simply fulfilling a mandate from the Action Programme, as well as implementing the Bern Convention), plus a considerable degree of autonomous action. DG XI had been given a mandate (it argued), via the Fourth Environmental Action Programme, and Stanley Johnson rapidly produced a draft Habitats Directive, relying for production of the first draft on John Temple Lang, who had been instrumental in drafting the habitat protection parts of the Birds Directive, and Pierre Devillers, who had been heavily involved in the early implementation of the Birds Directive, and who was at this time immersed in DG XI's CORINE biotopes project.

The Birds Directive had of course mandated the creation of a network of protected areas, and there was thus a certain degree of path dependence at this stage, with the draft Habitats Directive introducing the idea of a greatly expanded coherent Natura 2000 network, building on the network of SPAs under the Birds Directive. Despite not holding responsibility for nature conservation policy within DG XI, Johnson pressed ahead with his plan, notwithstanding some "humming and hawing [regarding] whether the Commission dared to launch something as major as this when the Birds Directive was . . . not under attack, but was in the news" (Sundseth, 2012). Johnson did not formally consult Member State governments in advance of the Commission proposing the Habitats Directive. In contrast, with the help of Alistair Gammell of the RSPB, Johnson circulated the draft instrument to a large number of NGOs (Gammell, 2012) and others, effectively letting the genie out of the bottle.

Having taken the draft this far, the entire proposal was then almost stymied by the intervention of Commission President Delors' *cabinet*, which was concerned at the length of the annexes of species and habitats and the potential reactions of Member States given their experience with the Birds Directive (Johnson, 2012). Johnson then took a tactical decision to drop the annexes from DG XI's proposal altogether, with a view to their reinsertion later in the process: "we'll introduce them or we'll somehow get them introduced at the stage of the Parliamentary discussion" (Johnson, 2012). However, the intervention of Delors' *cabinet* was not without effect, and Johnson redrafted the Explanatory Memorandum for the proposal to note that the draft directive required "an appropriate balance to be struck between conservation and economic interests and [provides] for the possibility of derogations from the general obligations in certain specific cases" (DG XI, 1988c, p. 4).

This episode is arguably an example of a reassertion of control by the Member States, albeit partial and by proxy. In other words, the Member States' reaction to the implementation of the Birds Directive arguably triggered a learning response within the *cabinet* of the Commission president, leading the proposal for the Habitats Directive to be framed in terms more conducive to the Member States' wishes. Even if the immediate effect was simply an amendment to the draft Explanatory Memorandum (not an operative part of the legislation), DG XI's card had effectively been marked by Delors' *cabinet*, arguably establishing certain parameters that would thereafter serve to frame DG XI's positions during negotiations.

Another important example of path dependence – as well as a critical juncture in the history – is provided by the intervention of DG XI's Legal Unit during the directive's drafting, emphasising the desirability of retaining the Birds Directive and the Habitats Directive as separate instruments (DG XI, 1988b). As Pierson (2004, pp. 134–135) explains, "historical institutionalists have frequently argued that institutional change typically involves a dynamic of 'punctuated equilibrium' (Krasner 1989; Collier and Collier 1991). There are brief moments in which opportunities for major institutional reforms appear, followed by long stretches of institutional stability. Junctures are 'critical' because they place institutional arrangements on paths or trajectories, which are then very difficult to alter". Amongst other things, the Legal Unit foresaw a "reduction of the level of protection provided for by [the Birds Directive] since the draft [Habitats Directive] provides for derogations and has not such a clear wording as [the Birds Directive]" (DG XI, 1988b, p. 2). The Legal Unit's fears were later realised in this regard when the Member States applied Article 6(2) to (4) of the Habitats Directive to SPAs under the Birds Directive in place of the first sentence of Article 4(4) of the Birds Directive (via Article 7 of the Habitats Directive). However, in ensuring the two instruments were retained, the Legal Unit created the opportunity for the CJEU later to revive the stricter protective regime under the Birds Directive by applying the first sentence of Article 4(4) to sites that should have been designated as SPAs but had not been (CJEU, 2000a).

In summary, the European Commission (and individuals within the Commission) can be seen to have exercised considerable autonomy and policy entrepreneurship during the production of the Habitats Directive proposal, assisted by key NGO representatives such as Alistair Gammell of the RSPB. Having secured the instrument's inclusion in the Fourth Environmental Action Programme, the Directive proposal arguably became path dependent in five important respects: first, Stanley Johnson could argue that he was simply fulfilling a political mandate (as well as implementing the Bern Convention) in producing a draft instrument; second, on the draft's being circulated to NGOs and others around Europe, the Directive proposal built up a head of steam that became hard to stop, particularly during a phase of EU environmental policy when environmental awareness and interest were high; third, by introducing the idea of the Natura 2000 network, the protected area concept introduced by the Birds Directive would be secured as a central feature of EU biodiversity policy; fourth, DG XI's Legal Unit having intervened in favour of retaining the Birds Directive and the Habitats Directive as separate instruments, the two Directives then set off on separate, albeit interlocking, paths, leading later to important differences in the protection afforded to sites under the two Directives, as a result of judgments of the CJEU; fifth, the 'appropriate assessment' impact assessment regime introduced by the Habitats Directive led over time to learning, investment and adaptation in the private and public sectors at national level – having adjusted to this important regulatory mechanism, the mechanism itself arguably became embedded and path dependent.

It was, in sum, a remarkable achievement for DG XI to have succeeded in launching the Habitats Directive proposal, given the preceding history in respect of EU nature conservation policy, and given the resistance within the Commission itself (at the most senior level), not to mention the Council.

Chapters 7 to 9

A critical juncture: closing gaps in control

This section of the history represents, in a general sense, a critical juncture, in that the Habitats Directive negotiations gave the Member States a rare opportunity to close gaps in control that they perceived to have emerged during the Birds Directive's implementation. Pierson (1996, p. 123) is right to argue that "Once gaps in control emerge, change-resistant decision rules and sunk costs associated with societal adaptations make it difficult for member states to reassert their authority". Nevertheless, from the Member States' perspective, the Habitats Directive proposal represented a fortuitous opportunity to retake control of certain aspects of EU biodiversity policy. In terms of historical institutionalist analysis, arguably we see coming to the fore during the Habitats Directive negotiations attempts by Member States to close gaps in their control of EU biodiversity policy, and in certain cases related defensive moves by the Commission. Another way to put it would be to say that it took all of the negotiating and diplomacy skills of Commission officials such as Claus Stuffmann to steer the Habitats Directive safely to shore.

Changes in government preferences played a crucial role in the history at this time. This is particularly clear in the case of the UK. As Roy Bunce comments, "under that part of the Thatcher government and under the Major government which followed it, the cabinet was, well the Tory party in fact was hopelessly split . . . [E]very time you got a new Minister, and you got a new Minister quite frequently, it was like a change of government. You know the whole policy would simply change" (Bunce, 2012).

Immediately following the Commission's proposal of the Habitats Directive it was unclear whether the Directive would gain sufficient political support to proceed at all. The situation in the UK was particularly problematic, with Secretary of State Nicholas Ridley openly hostile to the proposal, saying there would be a habitats directive "over my dead body" (Gammell, 2012). Alistair Gammell of the RSPB feared Ridley's attitude was in part attributable to the UK government's experience in the *Duich Moss* case under the Birds Directive; as Gammell stated in early 1989, "there is the risk that the new directive will have its teeth removed" (Brooke, 1989, p. 55). Thus, a learning process had clearly taken place in light of the implementation of the Birds Directive. Commissioner Stanley Clinton Davis noted "a good deal of opposition to [the Habitats Directive] in the Council. It is not very worthy opposition, it is opposition which is designed to erode the purpose of the habitats directive, based upon anxieties alleged in relation to the birds directive. I think it is a reflection of the difficulties which emerged from that directive" (Clinton Davis, 1988). The question remained whether, and in what way, the Member States might seek to retake control.

The first strategy involved seeking to undermine the Habitats Directive proposal altogether. This took several forms. First, at a ministerial-level meeting in November 1988, the UK did its best to derail the directive, joined principally by Spain, with both "contesting the necessity for EU action in the area" (DG XI, 1989a). Next, a concerted albeit ultimately failed attempt was made by certain countries, led by Denmark and the UK, to undermine the need for a Habitats Directive by seeking to refine the habitat protection provisions of the Bern Convention (Fernández-Galiano, 2012; Coffey and Richartz, 2003). At this point in time – the end of 1988 – the Directive's future remained very much in the balance. What tipped the scales in favour of the Habitats Directive was a fortuitously-timed French Presidency under Green Minister Brice Lalonde, from July 1989, and very soon thereafter an important political change in the UK. The incoming French Presidency was keen to revise the draft directive, and sought to embed a different approach to the one Member States had experienced in respect of the Birds Directive's implementation (Wildlife Link, 1989; Megret, 2013a, 2013b). Just one month into the French Presidency, a UK cabinet reshuffle saw Nicholas Ridley moved sideways, to be replaced as Environment Secretary by Chris Patten, a pro-European (White, 1989). The UK then became a positive advocate of the Directive (Bunce, 2012), and Denmark's opposition also dissipated (Koester, 2012).

Shortly before Ridley left office, an important event in the history took place. Times were changing in spring 1989: "Across Western Europe, mainstream politicians [were] scrambling to embrace the cause of the endangered environment as small ecological parties [registered] startling electoral triumphs", causing Margaret Thatcher and French President François Mitterrand to perform "sharp U-turns . . . to put environmental concerns at the top of their policy agendas" (Markham, 1989). Illustrating the complex interplay between actors and the notable role of NGOs in EU environmental policy, at this point the UK government asked the RSPB to produce a discussion paper regarding the Habitats Directive proposal, which the RSPB did in March 1989. The UK seemed to "cautiously like the solution" (RSPB, 1989). The RSPB (1989) also met with the French government at this time, which was due to take over the Presidency of the Council in July 1989, and the RSPB was due to be "making some proposals to them in the next few weeks". Amongst other things, the RSPB's discussion paper suggested introducing the idea of "favourable conservation status" as the central organising principle of the draft Habitats Directive, and raised the possibility of site protection derogations in the "overriding national interest". Both ideas were of central importance and they represent clear forerunners of provisions that would ultimately find their way into the Habitats Directive as adopted.

France's Marcel Jouve made a tour of European capitals at the beginning of the French Presidency in July 1989, and the Presidency then produced an overhauled draft of the Directive (Jouve, 2013), having assembled a broad consensus around key points, including (amongst others): a site designation process that would be more consistent with the principle of subsidiarity; the integration into the Natura 2000 network of SPAs under the Birds Directive; the principle of management plans for protected areas, supportive of ongoing human activities compatible with areas' conservation objectives. Crucially, there were very good working relations between the French Presidency and DG XI's Claus Stuffmann (Megret, 2013b; Jouve, 2013), smoothing the path for the Presidency's compromise draft. In October 1989 the French Presidency formally presented its compromise text to the other Member States (CEC, 1989b). There had been a complete overhaul in respect of the process for site selection and designation (DG XI, 1989b), and the approach and language of the Directive had shifted subtly, its title now referring to conservation rather than protection (DG XI, 1989c). Moreover, the compromise text had scrapped plans to apply the EIA Directive to Natura 2000 sites, the French Presidency having instead produced a standalone impact assessment provision for the Directive (CEC, 1989b).

In respect of site designations, a new provision introduced by the French Presidency (what ultimately became Article 5 of the Habitats Directive) proposed enabling the Commission to propose sites to the Council for designation, where the Commission believed a site ought to have been proposed (CEC, 1989a). This would later be restricted to priority sites and amended to require a bilateral consultation process with the relevant Member State where the Commission believed that a site is "essential for the maintenance of [a] priority natural habitat type or for the survival of [a] priority species", and should have been submitted for adoption as a Natura 2000 site. If the dispute remained unresolved after six months of consultation, the Commission could then submit a proposal to the Council relating to the selection of the site as a Natura 2000 site. This provision and the French Presidency's original version of it were aimed squarely at avoiding the sorts of legal disputes experienced under the Birds Directive; i.e., the Member States and the Commission had high hopes that this provision would allow site designation disputes (at least in respect of priority sites) to be resolved at the level of the Council rather than ending up before the CJEU. This is therefore not quite an example of the Member States retaking complete control, since the Commission would also be involved, but rather of their trying to sort matters out politically, without engaging the CJEU.

An example of (internal) resistance by the Commission came in early 1990 when Commission President Delors intervened in the process for a second time, again with a view to blocking the Directive proposal. Stanley Johnson again helped to rebuff this move. Later that year, in October 1990, the Commission stressed that the Habitats Directive proposal risked reaching a stalemate, in part owing to certain Member States' ongoing issues regarding implementation of the Birds Directive (DG XI, 1990). As at 15 March 1990, the Commission had commenced 37 cases against Member States for failure to comply with their site designation and site protection obligations under Article 4 of the Birds Directive (CJEU, 1990), and decisions kept coming. As a result, the Habitats Directive proposal was facing a "strong blockage", led by Spain but supported by other Mediterranean countries and France to a certain extent, owing to those countries' experience with the implementation of the Birds Directive (Troya, 2012). These countries hoped to address, via the Habitats Directive, the "dysfunctionality" they perceived in the implementation of the Birds Directive, which was impacting on projects funded under the Structural and Regional Funds, for example (Troya, 2012).

At this point, in February 1991, the CJEU (1991) handed down its decision in the *Leybucht Dykes* case under the Birds Directive, ruling that socio-economic considerations could not be relied upon to justify reducing the size of an SPA (e.g., to accommodate a construction project). The Member States thus decided to "retaliate" (Cichowski, 2007, p. 157), via the then ongoing Habitats Directive negotiations, by amending the Birds Directive to (in effect) overrule the *Leybucht Dykes* decision. They did this by applying a derogation – Article 6(4) of the Habitats Directive – to allow damaging projects to go ahead in or near SPAs under the Birds Directive as well as sites under the Habitats Directive, for reasons of overriding public interest, including for social or economic reasons (Krämer, 2009; Baldock, 1992; Holder, 1997).

Where this derogation was to be applied in respect of a priority site, a special rule was to apply: "Where the site concerned hosts a priority natural habitat type and/or a priority species, the only considerations which may be raised are those relating to human health or public safety or to beneficial consequences of primary importance for the environment" (CEC, 1991). When a group of Member States agitated for this special rule to be deleted from the draft Directive, the Commission proposed a compromise, thereby resisting this attempt by the Member States to assert greater control: rather than deleting the provision altogether, the application of the derogation in respect of priority species/habitats would be made the subject in certain cases of an opinion from the Commission (DG XI, 1991). As with the Article 5 site designation process, which sought to provide for disputes to be resolved in the Council rather than before the CJEU, here we see a similar idea: a positive opinion from the Commission on the issue could avoid the need to appear before the CJEU (subject of course to a challenge at the national level and a preliminary reference to the CJEU). Again, Member States did not retake control completely in this way, but one major route to the CJEU (via Commission-led infringement proceedings) was nevertheless thereby diminished. This also, of course, had the effect of increasing the Commission's role while reducing the CJEU's involvement.

A final point to note here is that, in the UK at least, the restricted time horizons of politicians played a role during the Habitats Directive negotiations, as Hilary Neal records:

> We [the UK] prided ourselves on having all the bases covered and knowing exactly what we were talking about, and being able to see through to implementation, but even . . . we were subject to the political pressures of Ministers wanting the directive adopted . . . [T]hey wanted to be able to go back to the environmental NGOs back home and say, we've negotiated this directive, isn't it terrific, and so we were having to, by virtue of that pressure, we were having to compromise where colleagues perhaps in [the Department of Trade and Industry], in the Treasury might not have wanted us to compromise, and certainly the Scottish Office might not have wanted [to compromise].
>
> *(Neal, 2012)*

In summary, the negotiation of the Habitats Directive can be seen as a balancing act from DG XI's perspective, with its Directive proposal facing attacks from powerful opponents both within and outside the Commission. As such, for the most part, DG XI engaged during negotiations in forcefully resisting attempts to undermine the Directive, while adopting a pragmatic approach in respect of Member State wishes to close certain gaps in their control over EU biodiversity policy, such as the one opened by the *Leybucht Dykes* case. As the UK's Roy Bunce wrote to Claus Stuffmann following a meeting to discuss that judgment (emphasis added), "You kindly indicated that the Commission were *not unsympathetic* to the UK's concerns but that you were concerned to ensure that the protection offered to these sites protected them from all but overriding issues" (Department of the Environment, 1991).

While Pierson (1996) makes a convincing case regarding the opening up of gaps in Member State control and subsequent difficulties in closing them, the Habitats Directive nevertheless presented Member States with an obvious opportunity to reassert control. As Cichowski (2007, p. 158) argues, "This legislative act [the Habitats Directive] shifts the control over national conservation measures back into the hands of member states".

Third phase (1992–present): weakening environmental policy dynamics

Chapter 10

Unintended consequences

If the adoption of the Article 6(4) derogation and the application (via Article 7) of Article 6(2) to (4) of the Habitats Directive to SPAs under the Birds Directive arguably shifted control back to the Member States, decisions of the CJEU post the adoption of the Habitats Directive have clearly rebalanced matters somewhat in the direction of supranational control, and in the direction of unintended consequences from the Member States' perspective. This has arguably created a feedback effect in terms of conflicts at the national level during the establishment of the Natura 2000 network, with Member States seemingly reluctant to propose sites for designation when they know those sites will be strictly protected by EU law. Moreover, strict legal interpretations of the Habitats and Birds Directives have arguably forced more and more plans and projects down the Article 6(4) derogation route (in cases where countries engage seriously with their EU legal obligations).

It is unnecessary to rehearse the legal judgments in detail here, since they are discussed in Chapter 10, but it is worth flagging a few key decisions, which include:

- The CJEU's ruling that the Habitats Directive applies beyond territorial waters in the marine area (CJEU, 2005);
- The CJEU's rulings to the effect that socio-economic considerations cannot be taken into account when selecting Natura 2000 sites and their boundaries (CJEU, 1996; CJEU, 2000b);
- The CJEU's precautionary take on the circumstances in which an Article 6(3) appropriate assessment must be carried out, and the limited circumstances in which a plan or project can be approved in light of the outcome of that assessment (CJEU, 2004) – this ruling in principle forces many plans/projects down the Article 6(4) derogation route;
- The CJEU's decision that sites that should have been designated as SPAs but have not been so designated remain subject to protection under the first sentence of Article 4(4) of the Birds Directive, meaning that the Article 6(4) derogation is not available in respect of such sites (CJEU, 2000a);
- A separate decision of the CJEU that arguably has a similar effect in respect of pSCIs under the Habitats Directive (CJEU, 2006);
- The CJEU's decision in the *Sweetman* case, offering a very high level of protection to sites for priority habitats, which will in principle have the effect of forcing additional plans/projects down the Article 6(4) derogation route (CJEU, 2013);
- The CJEU's narrow definition of mitigation (versus compensation) in the *Briels* line of judgments, again with the effect that more plans/projects must meet the tests of the Article 6(4) derogation if they are to proceed (CJEU, 2014, 2016b, 2017); and
- The judgment in the *LZ II* case, in which the CJEU (2016a) held that the public participation requirements of Article 6(1)(b) of the Aarhus Convention, and hence also the access to justice requirements of Article 9(2) and 9(4) of the Convention, apply in the context of appropriate assessment under Article 6(3) of the Habitats Directive.

Given the painstaking efforts of certain Member States to retake control of aspects of EU biodiversity policy during the Habitats Directive negotiations, many of these CJEU decisions must be regarded as having had unintended consequences from the Member States' perspective. Strikingly, most of the judgments referred to above came about as a result of preliminary references from national courts and not as a result of Commission-led infringement action against Member States (cf. Krämer, 2015).

Since 1992, CJEU decisions have thus served to strengthen the legal protections afforded to Natura 2000 sites, yet these decisions have played out against a relatively difficult political backdrop. DG Environment has therefore faced the tricky task during this third phase of defending the Habitats and Birds Directives from renegotiation and weakening.

A critical juncture, changing government preferences, and path dependence

The REFIT exercise in respect of the Birds and Habitats Directives from 2013–2016 clearly represented a critical juncture in the history of EU nature policy. Would the directives remain in place or would they be renegotiated and merged to create a "more modern instrument", fulfilling Commission President Juncker's (2014) wishes? While it would appear that criticism from certain Member States and others helped to trigger the REFIT exercise in the first place, once the process was underway and a massive show of public support was mobilised by NGOs in response to the Commission's consultation, many Member States came out in support of retaining the directives in their current form. The UK ultimately came around to this view (Khetani-Shah, 2016), with David Cameron no doubt looking for all the support he could get in advance of the Brexit referendum, emphasising Pierson's (1996) point about changing government policy preferences.

In terms of path dependence, an interesting feature of the REFIT evaluation report is its discussion of evidence regarding the legal uncertainty that would be caused by a renegotiation: e.g., some infrastructure developers "argue that the Directives provide a clear legal framework and that, in their absence, a loss of legal certainty would be expected to increase administrative burdens" (European Commission, 2016a, p. 58). This echoes a point made by then Environment Commissioner Potočnik (2015) in seeking to argue why there should not be a REFIT exercise in the first place: "In these times of uncertainty, we must also not forget that today, economic operators benefit from a stable and predictable legislative framework". As the years have gone by, the regulatory regime has become embedded, leading to a degree of path dependence. This path dependence has been strengthened by the European Commission's (2016c) formal decision not to reopen the directives, and the Council's 2017 conclusions regarding the Action Plan for nature, people and the economy (CEU, 2017a).

Turning now to the central issues identified in Chapter 1.

Conflicts in establishing the Natura 2000 network

As the history revealed in this book shows, in light of the difficulties and litigation that had been a feature of the early implementation of the Birds Directive, the Member States and the Commission together sought to embed a different site designation process in the Habitats Directive. This was to be a process in which the European Commission had a greater role, and that seemed to provide the potential for disagreements to be settled politically rather than before the CJEU. A clear example of this is the Article 5 process under which the Commission could, in exceptional cases, after a bilateral consultation with the relevant Member State, propose a priority site for designation, with a unanimous Council decision required. However, this carefully crafted provision has never in practice been used, as discussed in Chapter 10.[2]

In addition to the factors discussed in Chapters 1 and 10, a factor that has therefore arguably contributed to conflicts during the establishment of Natura 2000 has been the Member States' and the Commission's failure to pursue site designations in the spirit intended by those who negotiated by the directive, using all the tools available in that regard. That said, where Member States were evidently falling far short of their site designation obligations, the Commission must have felt it finally had no option other than to pursue infringement actions before the CJEU. In turn, as argued in Chapter 10, judgments of the CJEU following the adoption of the Habitats Directive have arguably created a feedback loop: strict and sometimes unexpected interpretations of the Directive by the CJEU, with unintended consequences from the Member States' perspective, have arguably made Member States reluctant to propose new sites for designation, knowing that those sites will then be strictly protected by EU law.

The impact of Commission-led infringement actions may have dissipated somewhat given the Commission's apparent de-prioritisation of this work in the environmental sector in recent years (see Chapter 10). However, there is always the possibility of legal action at the national level followed in appropriate cases by a preliminary reference to the CJEU, although, given the difficulties inherent in pursuing such a solution, "one should not make the mistake of considering it as a panacea for the enforcement of EU law" (Slepcevic, 2009, p. 391). In the years ahead we can therefore expect to see continuing slow-but-steady growth in the development of the Natura 2000 network, particularly in the marine area where much work remains to be done. A key difference between the future and past periods may be the extent to which Commission-led litigation plays a role in speeding up the establishment of the network.

Acute conflicts in conserving Natura 2000 sites

Acute conflicts pitting nature conservation against other societal goals clearly present serious difficulties for EU policymakers. In the context of the Habitats Directive, these often play out under Article 6, including the derogation provision, Article 6(4).

In recent years the Commission faced an unenviable task: if EU law was strictly enforced in every case, the Commission arguably ran the risk of triggering a renegotiation of the Habitats and Birds Directives, potentially leading to substantially weaker biodiversity protection provisions. This risk seems to have subsided given that the Habitats and Birds Directives passed through the REFIT check unscathed, and it is very unlikely there will be political will to review the directives again any time soon. Nevertheless, if the Commission engages in politicised enforcement, the rule of law will be compromised and with it important environmental protections.

Schools of thought on Article 6(4)

Of the 20 published cases that have come before the European Commission for an opinion regarding the existence or absence of imperative reasons of overriding public interest, all but one has been approved by the Commission (European Commission, 2016b; CJEU, 2012d).[3] Having reviewed the then 11 Commission opinions delivered under Article 6(4) (ten positive and one negative), Krämer (2009, p. 84) concluded that, "In my opinion, not one of the positive Commission Opinions would, with the reasoning made, successfully survive scrutiny by the Court of Justice".[4] His opinion has not changed since, with Krämer (2015, p. 236) adding that such opinions tend to be issued "without too much checking of the facts and the arguments by the Commission itself", without consulting parties other than the Member State, and without checking whether promised compensatory measures are actually provided.

"Rather unexpectedly", comment Clutten and Tafur (2012, p. 172), "the [CJEU] does not appear, in the near-two decades since the Directive was adopted, to have had directly to express a view on the question of what is meant by [imperative reasons of overriding public interest] under Article 6(4)". This changed with the CJEU's 2012 decision in the *Solvay* case – a case brought before the CJEU by way of a preliminary reference from a Belgian court, and not by the Commission – in which the CJEU held that: "the mere construction of infrastructure designed to accommodate a management centre cannot constitute an imperative reason of overriding public interest within the meaning of Article 6(4) of the Habitats Directive" (CJEU, 2012c, paragraphs 77–78).[5]

There would currently appear to be two leading schools of thought regarding Article 6(4), the first of which is exemplified by Krämer (2009, pp. 84–85), who writes:

> the Commission would not lose, but rather win credibility, if it were to take a position that was more along the line of the 'rule of law' – in other words if it were to strictly apply Article 6(4) of the Habitats Directive. How can local, regional and national authorities, private investors and economic operators be persuaded to apply Article 6 and follow the Commission's guidelines, if the Commission itself does not set an example and apply this provision? The rule of law is for everybody; and letters, visits or telephone calls from prime ministers or members of national governments are nothing more than protectionist lobbyism.[6]

The second school of thought – which would arguably appear to have been DG Environment's approach – regards the Article 6(4) derogation as a "safety valve" that provides legal "flexibility" and at times serves to protect the Habitats and Birds Directives from those who would like to see them renegotiated/weakened. Note in this regard the following comment from the European Commission's (2017b, p. 2) Action Plan for nature: "those who implement the Directives, particularly at regional and local level, are sometimes not sufficiently aware of their requirements or of the flexibility and opportunities [the directives] offer. This can lead to tensions between nature protection and economic activity".

DG Environment could point to the fact that the institutional balance established within the EU is such that the Commission has discretion as to whether or not to pursue infringement proceedings (CJEU, 1989). DG Environment might further argue that Natura 2000 comprises more than 27,500 sites, yet only 20 opinions have ever been published under Article 6(4) (all bar one being positive, as above). In the context of such an enormous network of protected areas, it is in the greater biodiversity good – the argument might run – that damage is permitted to a small number of Natura 2000 sites in exchange for compensatory measures (i.e., no net loss), in order to protect the continuing existence of the Habitats and Birds Directives.

In response one might highlight that the aim of EU biodiversity policy is "to halt the loss of biodiversity and the degradation of ecosystem services in the EU by 2020, restore them in so far as feasible, while stepping up the EU contribution to averting global biodiversity loss" (Environment Council of the EU, 2010, p. 2; European Commission, 2011). It is hard to see how this can be achieved in practice when, in then Commission President Barroso's words, "There is no instance of the EU Nature Directives having prevented significant economic development in [any Member State]" (Barroso, 2009). Surely, one might argue, a proper application of Article 6(4) would and should have stopped at least some significant economic development in the EU. If Barroso's statement is factually correct, the Commission would appear to place great faith indeed in the ability of compensatory measures to halt the loss of biodiversity in the EU by 2020.

Consider, further, that there is no clear picture of what is happening to non-priority Natura 2000 sites across the EU via Article 6(4), since there is no need for an opinion from the Commission to be sought in such cases, and it would appear that not all countries notify the Commission of the compensatory measures they have adopted, as they are obliged to do.[7] One explanation for this failure to notify – at least in respect of certain countries – may be that Article 6(4) is not always being applied when it should be: as Sundseth and Roth (2013, p. 63) comment, "Some countries seem to try to push everything through under Article 6.3 to avoid having to use 6.4", on the basis that the Article 6(4) route may increase costs and result in considerable delays. But if certain Member States are permitting plans/projects via Article 6(3) in this way when the relevant legal test cannot be met, the European Commission should step in and enforce EU law. In this regard, the passage of the Commission's 2017 Action Plan cited above, emphasising "the flexibility and opportunities [the directives] offer" (European Commission, 2017b, p. 2), may potentially be seen by some as an invitation to Member States to rely on the Article 6(4) derogation process instead of pushing plans and projects through unlawfully under Article 6(3). The Article 6(4) derogation may thus in future be used more frequently than is the case now.

Similarly, there is limited visibility of what is happening to priority Natura 2000 sites via Article 6(4) where the imperative reason relied upon is human health or public safety or beneficial consequences of primary importance for the environment, since here, again, no opinion from the Commission need be sought. So the Commission would arguably seem to be taking the conservation of the Natura 2000 network on faith to a certain extent, whilst failing to enforce the law in certain acute conflict situations.[8] However, compromising the rule of law – one of the foundational principles of the EU (TEU, 2012, Article 2) – by failing properly to enforce the Habitats Directive could of course lead to an overall weakening of respect for the rule of law regarding the protection of Natura 2000 and more generally.

A third view of Article 6(4) might hold that there is scope for further legal development of Article 6(4) by reference to the principle of proportionality (as set down in Article 5 of the TEU (2012)). That is, the "safety valve" view of Article 6(4) arguably gives too much away in terms of the conservation of Natura 2000 sites. A better approach might be to seek to further refine the application of the Article 6(4) derogation to provide for a weighing up of the claimed "imperative reasons of overriding public interest" (IROPI) and the precise impacts foreseen (i.e., scale and type). There is some evidence that DG Environment is moving in this direction: e.g., see the European Commission's (2010, paragraph 107) written observations in the *Solvay* case. While this would serve to further calibrate the approach in respect of the question of IROPI, it would not of course address the issue of projects proceeding under Article 6(4) where the compensatory measures appear legally insufficient or where there would appear to be alternatives.[9] This third view would therefore go some way to addressing concerns with the second, but would still leave certain problematic issues untouched.

So which of these three views is to be preferred? The historical institutionalist analysis presented here provides valuable insights to inform this debate. The analysis reveals that the Birds and Habitats Directives were forged when EU environmental policy dynamics were much more positive than they are now, and that even at the time of their negotiation it was a major achievement to obtain agreement regarding the Directives. Further, decisions of the CJEU since the Directives were adopted have arguably had unintended consequences from the Member States' perspective, in various ways. If a strict application of the Article 6(4) derogation were in future to tip the balance in favour of a renegotiation – as a strict interpretation of Article 4(4) of the Birds Directive effectively did in the *Leybucht Dykes* case in 1991 (CJEU, 1991) – such a renegotiation would play out not during a golden era for EU environmental policy, but instead during the present phase of EU environmental policy, during which the environmental policy dynamics have significantly weakened. In such circumstances, what hope would there be for a positive outcome for nature conservation?

In response to this, one might argue that the rule of law is too important a principle to compromise and that the Commission should not seek to second-guess the outcome of applying the law, particularly since, over time, adopting a "flexible" approach to Article 6(4) of the Habitats Directive could result in serious adverse impacts on Europe's most threatened species and habitats. DG Environment might counter here by saying that it always seeks to secure the best possible outcome for Natura 2000 via Article 6(4) by way of compensatory measures (i.e., 'no net loss'). However, the ability of compensatory measures to halt the loss of biodiversity may prove rather uncertain, even where those measures are properly implemented on the ground, which is by no means guaranteed.

These are very difficult issues that go to the heart of EU nature policy, and it is important to recognise that their resolution may not be entirely within the Commission's control. That is, while the Commission controls the EU infringements process by bringing (or not bringing) Member States before the CJEU for alleged breaches of EU law, it is clear that the primary responsibility for implementing EU law lies with authorities in the Member States (e.g., see European Commission, 2006). Thus, enforcement of the Habitats and Birds Directives at national level by national authorities could result in a strict application of the Directives, even if this seems unlikely in practice.

More likely than this would be the enforcement of the Directives at national level by way of legal actions brought by individuals and NGOs, followed by preliminary references on points of legal interpretation from national courts under Article 267 of the TFEU (2010). As Cichowski (2001, p. 26) argues in the context of the Member States' decision to enact the Article 6(4) derogation in response to the *Leybucht Dykes* judgment: "in the area of nature conservation, Member State governments have diminished or at least delayed the impact of expansive [CJEU] rulings. While such intervention has minimized the immediate policy consequences of these rulings and perhaps slowed the process of integration, the general trajectory is not easily reversed once the rules are created. The cycle begins again. Through the action of individuals, national courts, and the CJEU, these EU rules can be changed". One could cite in this regard the fact that most of the recent important Habitats Directive judgments came before the CJEU via references from national courts. For those who would like to see improved implementation of the Habitats and Birds Directives, a central task, in parallel to applying pressure on Member States and the Commission to take effective action, will be the bringing of strategic legal cases at the national level by individuals and NGOs, and thereafter seeking preliminary references to the CJEU.

Slepcevic (2009) argues convincingly that four variables determine the potential success of such a litigation strategy: the organisational capacity of potential litigants; their ability to access the courts (e.g., questions of legal standing and prohibitive expense may arise); the interpretation by national courts (e.g., a national court may decline to make a reference to the CJEU); and the reaction of the competent authorities on the receiving end of any such litigation. Since these variables vary considerably across the EU, the potential for enforcing EU environmental law via national courts varies accordingly (Slepcevic, 2009). Such enforcement may therefore prove to be something of a long game, since it may require first working on one or more of these variables (e.g., by way of communications to the Aarhus Convention's Compliance Committee in respect of access to justice considerations) before legal actions under the Habitats and Birds Directives can be successfully pursued by a wide range of public interest litigants. Nevertheless, it is clear that such litigation can have a positive impact in the context of Natura 2000 (Slepcevic, 2009), and historically litigation has been an important driver of EU nature policy.[10]

The question for individuals and NGOs will ultimately be the same one the European Commission faces: when and in what cases should legal action be pursued? The crucial point here is that litigation pursued at the national level by citizens and NGOs serves to democratise enforcement. It is for this reason that providing more effective access to justice at the national level remains such a key issue. The Commission's Interpretative Communication "bringing together all the substantial existing CJEU case-law, and . . . drawing careful inferences from it" (European Commission, 2017a, p. 5) falls far short of the harmonising effect of a directive but is useful nonetheless. Whether the Commission will demonstrate its commitment to improving access to justice at the national level by following up on its Communication with strategic infringement actions in appropriate cases remains to be seen.

Ultimately, viewing Article 6(4) of the Habitats Directive through the lens of a long-term historical perspective tells us this: in circumstances where the Commission appears to be making less extensive use of its infringement powers, and in the absence of direct access to the CJEU,[11] environmentalists may need to focus on improving access to justice at the national level, with a view to bringing appropriate cases before the CJEU by way of preliminary references. In this way, conflicts regarding Natura 2000 may be resolved in full compliance with EU law, and we may continue to see progressive interpretations of the Birds and Habitats Directives from the CJEU.

EU integration in the field of nature conservation

As noted in Chapter 1, there appears to be a consensus around the idea that the Member States in some sense lost control of EU nature policy, and that the policy has had unintended consequences from the (or at least from *certain*) Member States' perspective (e.g., see Fairbrass and Jordan, 2001).

Supranationalism

As noted in Chapter 2, Sandholtz and Stone Sweet's (2012, pp. 18–19) supranational theory generates a number of testable hypotheses including, of particular relevance here: "EU organizations like the Commission and Court of Justice will routinely produce supranational outcomes that the Member States would not have produced on their own"; and "Supranational governance will routinely produce outcomes that conflict with the revealed preferences of the most powerful states".

Stone Sweet (2004, p. 235) has in the past used EU nature law as an example in support of his theory, arguing that it "shows that the activities of supranational organizations, such as the European Court, routinely produce 'unintended consequences', from the perspective of those who have designed and redesigned the EC". Stone Sweet (2004) relied in this regard on the following snapshot of the history detailed in this book: first, the *Leybucht Dykes* judgment, in which the CJEU (1991) interpreted the Birds Directive strictly, holding that economic interests are insufficient to justify reducing the size of an SPA. Next, as detailed in Chapter 9, the Member States decided to "retaliate" (Cichowski, 2007, p. 157), via the then ongoing Habitats Directive negotiations, by crafting the Article 6(4) derogation and amending the Birds Directive to in effect overrule *Leybucht Dykes*, applying Article 6(4) to SPAs under the Birds Directive as well as to sites under the Habitats Directive (Krämer, 2009; Baldock, 1992; Holder, 1997).

The final step in Stone Sweet's (2004) timeline is the CJEU's (1996) decision in the *Lappel Bank* case, in which the UK government, supported by the French government, argued that in selecting SPAs and their boundaries, Member States should be allowed to take economic considerations into account. The CJEU disagreed, holding that, in selecting SPAs and defining their boundaries, Member States must act solely on the basis of the ornithological criteria set out in Article 4(1) and (2) of the Birds Directive (CJEU, 1996). "The Court's forthright approach to the part to be played by economic considerations in the designation of land might be better understood", argues Holder (1997, p. 1479), in the context of the Member States' earlier decision to overrule the CJEU's judgment in the *Leybucht Dykes* case regarding site protection post-designation. Stone Sweet (2004, p. 229) comments, "Statutory review of EC directives typically serves to reduce national autonomy in the domain. In *Lappel Bank* . . . the Court imposed its interpretation of secondary legislation over the objections of a consortium of powerful Member States. The Court did so even after the legislation had been revised to 'reverse' a prior [CJEU] ruling".

While one can endorse Stone Sweet's (2004) analysis insofar as it goes, the analysis arguably suffers from the weakness that it focuses on a brief snapshot from a lengthy history – a snapshot that supports Stone Sweet's argument that a supranational organization has acted to reduce the domain of national autonomy, etc. In other words, using Holder's (1997, p. 1479) characterisation, the period considered by Stone Sweet (2004) was arguably a dialogue of "legislation and counter-legislation" between the CJEU and the Member States. Stone Sweet (2004) bases his analysis on what might appear to have been, but was not, the last word in that dialogue: i.e., the CJEU's (1996) decision in the *Lappel Bank* case, holding that Member States cannot take socio-economic considerations into account in selecting SPAs and their boundaries. Had Stone Sweet (2004) chosen, in addition, to trace the post-adoption history of the Member States' part of the dialogue – i.e., the operation of the Article 6(4) derogation – he might potentially have uncovered "counter-forces" (Alter, 2009, p. 16), at odds with his supranational account of EU integration.

In this regard, Article 6(4) is a very clear example of the Member States retaining a significant degree of control over EU biodiversity policy. Moreover, the history recorded in this book reveals that the provision was designed with this precisely in mind, as a reaction to the loss of control Member States perceived in respect of the Birds Directive's implementation. The UK went so far as to let the Commission know that it would withdraw its support for the Habitats Directive if action was not taken to overrule *Leybucht Dykes* (see Chapter 9). The UK rallied support amongst other Member States for this view, but Spain then saw a tactical opportunity to withhold its support from the UK with a view to gaining leverage in its demands for financing for Natura 2000 to be addressed via the text of the Directive. There were thus powerful intergovernmental forces at work during the negotiation of the Article 6(4) derogation – these forces did not cease operating after the CJEU's decision in *Lappel Bank*. Additional evidence of the importance of intergovernmental dynamics during the Habitats Directive's history can be found, for example, in the crucial roles played by the presidencies of France, Luxembourg and the Netherlands, compared with other less effectual presidencies. These are not examples of what Fairbrass and Jordan (2001, p. 1) describe as "firm state control in the short-term". Rather, the cited presidencies set (and changed) the agenda, and in carefully crafting the text of the Article 6(4) derogation in response to *Leybucht Dykes*, the Member States regained a significant degree of control over land use planning decisions, on a continuing basis.

The present historical institutionalist analysis reveals a "counter-force" to supranational EU integration in the biodiversity field (Alter, 2009, p. 16). Regarding Sandholtz and Stone Sweet's (2012, pp. 18–19) hypotheses: first, "EU organizations like the Commission and Court of Justice will routinely produce supranational outcomes that the Member States would not have produced on their own". The history here certainly supports this hypothesis insofar as the CJEU is concerned. The Commission's position is more complex. We can certainly endorse the hypothesis in respect of the Commission during much of the history discussed here (and certainly until 1992). However, in cases where Member States assert imperative reasons of overriding public interest for proceeding with a damaging plan or project, the Commission has not, in the third phase of the history described here, routinely produced supranational outcomes in respect of site protection that the Member States would not have produced on their own.

This links to the second supranational hypothesis we set out to test: "Supranational governance will routinely produce outcomes that conflict with the revealed preferences of the most powerful states". The present analysis supports this second hypothesis as a general principle, with the important caveat that the Member States' negotiation of the Article 6(4) derogation under the Habitats Directive allows Member States to permit damaging and potentially damaging plans/projects by positing the existence of imperative reasons of overriding public interest,[12] and these decisions have tended not to be challenged by the European Commission. Thus, the "revealed preferences of the most powerful states" would appear to have a good chance of holding sway in respect of the protection (or otherwise) of Natura 2000 sites in such cases.

This analysis supports Alter's (2009, p. 16) argument that "Historical institutionalism . . . offers a more open framework [than neo-functionalism and] deals better with contingency, and . . . captures important contextual elements and counter-forces that scholars who are looking only for forces supporting integration will miss. By focusing on counter-forces and surprising puzzles, we may well reveal dynamics that ultimately shape long term change, and thus that better account for important systemic transformations we have observed in Europe".

New intergovernmentalism

From the new intergovernmentalist perspective, we can usefully test the six hypotheses set out by Bickerton et al. (2015) in respect of the third phase of the history discussed in this book (i.e., since 1992).

Hypothesis 1: deliberation and consensus have become the guiding norms of day-to-day decision-making at all levels

Certainly deliberation and consensus have gained ground at the expense of Commission-led infringement actions in the field of nature conservation, with the number of infringement cases falling by 58% between 2007 and 2015 (European Commission, 2016d, undated). This is complicated to some extent by the fact that the infringements procedure under Articles 258/260 TFEU itself provides for deliberation throughout the process, allowing for cases to be settled (cf. Čavoški, 2016). Nevertheless, the fall in the number of infringement actions is striking. The Commission's Action Plan for nature is the epitome of a deliberative approach, with only fleeting mention of the adversarial "assuring compliance" (European Commission, 2017b). More generally, the Environmental Implementation Review process and the EU Pilot process represent further moves in the direction of deliberation and consensus, intended as they are to resolve issues politically to avoid the need for legal action (European Commission, 2017c; Čavoški, 2016).

Hypothesis 2: supranational institutions are not hard-wired to seek ever-closer union

While the CJEU appears to have acted as an engine for deeper integration throughout the entire history of EU nature policy, some support for this hypothesis can be found in the history presented here insofar as the Commission is concerned. Evidence for this can be found even before 1992, albeit close to this date. For example, in 1991 the Commission indicated that it was "*not unsympathetic* to the

UK's concerns" regarding the CJEU's *Leybucht Dykes* judgment (Department of the Environment, 1991), and accommodated these concerns within the Article 6(4) derogation, allowing Natura 2000 sites to be damaged in the case of imperative reasons of overriding public interest (IROPI), including those of a social or economic nature, provided the other conditions of Article 6(4) are also met. This derogation test, and IROPI in particular, is broad enough to provide for national autonomy in a manner that can act as a counter-force to integration. This has arguably proven the case, with the Commission (as noted above) having issued a negative opinion under Article 6(4) only once, of the few published cases where it has been asked for such an opinion (European Commission, 2016b).

While the Commission actively and creatively pursued ever-closer union in the field of nature conservation for the first two phases of the history discussed here and for much of the third phase, in recent years the Commission's role appears to have changed. As discussed in Chapter 10: power has become centralised in the Commission president, Secretariat General, and Vice-Presidents; the Commission appears more directly linked to national capitals; and DG Environment is relying less on infringement proceedings. Arguably we see in the REFIT exercise an example of the Commission President acting as the proxy of certain Member States' wishes to see the Habitats and Birds Directives renegotiated in favour of a "more modern" instrument, only for this approach to be rejected by the European Parliament, the Council, environmental NGOs, a vocal public, and ultimately the College of Commissioners. If the directives *had* been renegotiated, it seems highly unlikely that the final outcome would have represented deeper integration.

Becker et al. (2016, p. 1013) challenge the new intergovernmentalist analysis of changes in the Commission's behaviour since 1992, arguing, for example, that the reduction in the Commission's policy activism is due to factors internal to the Commission rather than to inter-institutional dynamics. That is, they argue that "[Commission President] Barroso after 2010 and Jean-Claude Juncker, his successor, have used the enhanced powers and resources available to the office [of President] to take personal control over the Commission agenda and to end the expansionist policy activism that historically characterised the organisation". There is certainly support for such a view in the environmental sector. For example, Becker et al. (2016, p. 1016) emphasise Barroso's conversion of the Secretariat General, which acts as a central coordinating body, "into a personal service of the Commission Presidency", extending "the Commission President's reach into the administration and grip over policy". This resonates with reports from 2013 that then Secretary General, Catherine Day, had been actively blocking environmental proposals, believing that the Commission's work should be focused on "growth and the euro" (*ENDS Europe*, 2013).

Becker et al. (2016, p. 1027) argue that the financial crisis was a *catalyst* not a *cause* of changes in the Commission, since the process of change had already been underway for some years, and that the reduced legislative output of the EU was "due to the ability and choice of a strong president to focus the attention on crisis-related areas". Becker et al. (2016) view the Commission's actions as strategic use of a variety of methods to *eventually* further integration, arguing that the Commission remains an engine of integration but in a different guise. Perhaps, but this does not appear to explore fully the potential *causal* factors at play in the Commission's changing role nor fully the inter-institutional dynamics. Arguably, we see in the nature conservation field the Commission (with some internal resistance) functioning almost as part of the 'intergovernmental' process.

The argument here is not that the Council and Commission form a simple 'principal–agent' relationship. Rather, it is that the Commission appears itself to have become more 'intergovernmental' in nature, as Peterson (2008) argues, in a way that suggests a need to redefine the meaning of 'intergovernmentalism' itself. Peterson (2017, p. 364) makes the point clearly:

> A paradigm shift in the way the EU works may well have occurred post-Maastricht. But the Commission often has seemed complicit in it. Not only by gauging what the political traffic will bear (as under Barroso), but (under Juncker) being 'more connected to the political process' (in the words of one of his advisors) and influencing debates in the European Council, where Junker's experience far out-stripped that of Barroso. One indicator of this connection was that no fewer than 90 per cent of the Juncker College were members of their state's governing parties, the highest share in the Commission's history. None of this is to deny that the Commission and European Council remain rivals in a kind of 'competitive cooperation', but one in which they are 'joint agenda setters' (Bocquillon and Dobbels, 2014). What is new about the new intergovernmentalism is that simple binaries about who wins or loses when European integration advances are no longer helpful.

From the EU nature policy case we can see that the Commission has arguably been complicit in this regard since before the Maastricht Treaty was signed. One could cite in this regard the Commission's role in negotiating provisions of the Habitats Directive that, on paper, enhanced the Commission's and Member States' roles in policy processes at the expense of the CJEU's: e.g., the Commission's role in giving opinions to Member States under Article 6(4); its role in the Natura 2000 site designation process, including Article 5, which (as discussed above) sought to enable political settlement rather than legal determination. Equally, while the Commission has made liberal use of its infringement powers over the years to bring Member States before the CJEU, in recent years it has chosen to use these powers less frequently – arguably a result of the Commission's view of what the political traffic will bear.

All this makes it clear that there is no simple binary between the Council (and individual Member States) on one hand, and the Commission as one of the traditional supranational institutions on the other. The inter-institutional dynamics are much more nuanced than this. In recent years it would appear that parts of the Commission took the view that it would be in the EU's interests to amend EU nature legislation, arguably to weaken integration in this field. The centralised power of the Commission President and those closest to him (*cabinet*, Secretariat General, Vice-Presidents) appear to have held the balance of power within the Commission in this regard, while the influence of DG Environment appears to have been diminished compared with earlier periods in the history described here. But what *caused* the Commission President et al. to adopt this position? In announcing the REFIT process of the Habitats and Birds Directives in 2013, was Commission President Barroso sensitive to the views of certain Member States? The Netherlands, for example, had explicitly called for the Directives to be renegotiated in 2009, and in the UK the Tory government had attacked the Habitats Directive in 2011, commenced a 'balance of competences' review in 2012, before in January 2013 committing to renegotiate the UK's relationship with the EU followed by an 'in-out' Brexit referendum (see Chapter 1).

The recent history of EU nature policy arguably shows the Commission President taking a decision on the future direction of that policy in light of certain Member States' concerns, only to have this decision rejected by a majority of Member States (amongst others). In this sense, the Commission appears to have functioned not so much as an engine of integration but rather as an 'intergovernmental' weather-vane that misread the wind.

Hypothesis 3: where delegation occurs, governments and traditional supranational actors support the creation and empowerment of de novo institutions

This has not particularly been a feature of EU nature policy.

Hypothesis 4: problems in domestic preference formation have become standalone inputs into the European integration process

Nor does this appear to have been a particular feature of EU nature policy.

Hypothesis 5: the differences between high and low politics have become blurred

Bickerton et al. (2015) are not making the obvious point here that what constitutes high and low politics can vary over time. Instead they are arguing that there appears now to be close monitoring by Member State governments of EU decision-making and a wariness of the independent powers of traditional supranational actors across a range of policy areas that might not previously have obtained such rigorous scrutiny: "it has become difficult to say what constitutes an issue of heightened national concern and what may be considered a purely technocratic issue which could be left to expert decision-making" (Bickerton et al., 2015, p. 35).

Given its relatively late start – from a base of zero in the 1970s – for a while EU environmental policy as a whole was seen as a 'flanking' policy to the single market. But EU nature policy has proven itself of relatively high political salience over the years, and increasingly so. For example, its influence on land use decision-making at national level and its focus on wildlife conservation have the potential to capture the attention of politicians and public alike. Its operation has attracted the attention of heads of State (e.g., Balkenende in the Netherlands, Cameron in the UK, Trump in the US – see Chapter 1). Its potential weakening resulted in the largest ever public response to an EU consultation exercise (see Chapter 10). Etc. This has all occurred in the post-Maastricht period, but this is almost an inevitability given that the Habitats Directive was adopted after the Maastricht Treaty was signed and the network of SPAs and SACs has been established largely since 1992. While the present author would not describe EU nature policy as 'low politics' (it has certainly never been a purely technocratic issue), the policy area has certainly grown in prominence over the years – and particularly post-Maastricht – and now receives a degree of national scrutiny from Member State governments that was certainly not the case when, for example, the Birds Directive was proposed and negotiated. To that extent, this hypothesis can therefore be tentatively endorsed.

Hypothesis 6: the EU is in a state of disequilibrium

This is perhaps true viewed in the round, though cf. Bulmer's (2015) point about disequilibrium being in the eye of the beholder. However, EU nature policy itself has arguably passed through a period of disequilibrium (the REFIT exercise from 2013–2016) to reach a plateau of stability via the European Commission's (2017b) Action Plan and its endorsement by the Council (CEU, 2017a). This is a resilient area of policy.

Summary

The apparent consensus in the literature emphasising supranational integration in the field of EU nature policy and losses of control by the Member States requires updating to reflect important intergovernmental dynamics, including the changing role of the European Commission. These dynamics:

- are clear from the key role played by certain Council presidencies (France, Luxembourg, Netherlands) in the negotiation of the Habitats Directive and the insignificant role played by others;
- resulted in provisions (e.g., Article 6(4), Article 5) being deliberately woven into the fabric of the Habitats Directive at the time of its negotiation enhancing the Commission's and Member States' roles and arguably diminishing the CJEU's;
- have long-term as well as short-term effects on nature policy and land use planning;
- played a key role in the rejection of Commission President Juncker's plan to merge the Habitats and Birds Directives to create a "more modern" instrument (Juncker, 2014).

EU nature policy is not as supranational as it is often portrayed. While supranationalism has undoubtedly played an important role in the history of EU nature policy, it would be a simplification to seek to place this policy on Stone Sweet and Sandholtz's (1998) intergovernmental-supranational continuum. There is arguably no simple binary currently between the Council (and Member States) on one hand and the Commission, one of the EU's traditionally supranational organisations, on the other. The Commission is a complex institution with diverse preferences and power bases (particularly when granularised to the level of President, individual DGs, etc.). To fully understand and accurately characterise EU nature policy requires unpacking these dynamics in a clear light.

While the new intergovernmentalism has identified important dynamics in the post-Maastricht era – some of which have been at play in respect of EU nature policy – this pre-Maastricht policy area does not bear out all of the new intergovernmentalist hypotheses.

The future of EU nature policy

In looking to the future of EU nature policy we can usefully reflect on themes from its past.

First, while the enlargement of the EU and the enlargement and centralisation of the Commission may well have made it more difficult for officials to make their mark in the development and implementation of EU policy, it is clear that policy entrepreneurs, policy leaders and small networks (bridging, for example, the European Parliament, national governments, NGOs, the Commission) played a central role in the history of the Birds and Habitats Directives. There is no reason why this need not be the case in the future, though much will depend on the extent to which the Commission frees or constrains individual DGs and officials. The current centralising tendency with a narrow focus on Juncker's ten political priorities is not conducive to ambitious EU environmental policy. Addressing this may ultimately require a restructured Commission: as Becker et al. (2016, p. 1015) comment, "Strong presidentialism is no longer an exception that is based on personal qualities, such as in the case of Delors, but underpinned by formal organisation".

Second, NGOs have played a prominent role – in myriad ways – in the history of EU nature policy: e.g., lobbying for the need for EU legislation on birds and habitats; informally assisting the Commission with advice and drafting; formally assisting the Commission (cf. Ian Hepburn's role on secondment – see Chapter 7); helping to develop Important Bird Area lists; the RSPB's influential 1989 discussion paper in the context of the Habitats Directive; communicating breaches of the Directives to the Commission; lobbying for improved implementation; litigating at the national level; and (in more recent times) organising the Nature Alert campaign and mobilising hundreds of thousands to respond to the Commission's REFIT consultation, etc. Time and again Europe's NGOs have shown themselves remarkably able to coordinate their activities in support of strong EU nature conservation policy. If the EU is to maintain and develop its nature policy, Europe's NGOs will need to maintain this high standard of engagement.

Third, the role of individual Member States in shaping the history of EU nature policy is clear, with France, Luxembourg and the Netherlands having driven forward and shaped the negotiation and ultimate adoption of the Habitats Directive during their presidencies. When the Habitats Directive came under threat from the Barroso and Juncker Commissions, individual Member States coordinated and resisted a revision of the Birds and Habitats Directives. Following Brexit the Council dynamics will undoubtedly change and it will be interesting to see the coalition possibilities in favour of strong EU nature policy, particularly as Europe emerges from the financial crisis.

Fourth, public opinion and a degree of luck played evident roles in the history of EU nature policy. For example, public opinion in favour of strong environmental policy – as was the case at key moments in the 1970s, 1980s and early 1990s – appears to have been a necessary precondition for adopting ambitious EU nature policy "against all odds" (Meyer, 2010, p. 179). Similarly, without fortuitous constellations of facts (driven to some extent by prevailing public opinion) – including a French presidency of the Council led by Green minister Brice Lalonde, the almost simultaneous departure of the UK's obstructive Eurosceptic Minister Nicholas Ridley, and Denmark's position softening – it is hard to see how the Habitats Directive proposal would have succeeded. It seems clear that EU nature conservation policy will progress further and faster in future with a knowledgeable and supportive public, which will require greatly improved engagement and public participation. Political will is required for this but is often lacking at national and local levels (Kati et al., 2015). This is something of a 'chicken and egg' problem in that political will seems likely to be influenced by levels of public interest and awareness. NGOs could therefore play a vital role here – both in terms of public awareness and political will – if adequately resourced for the tasks.

Fifth, the centrality of adequate funding to effective EU nature policy has been a consistent and consistently unheeded refrain. Member States were right to insist during the Habitats Directive negotiations on the need for proper financing, but this has never in practice been forthcoming – at national or EU levels. Increased levels of funding could be used to tackle key issues such as improving site management, carrying out better and more frequent monitoring, and improving the knowledge base, amongst other things (EEA, 2015). While the Commission's 2017 Action Plan has made strengthening investment in Natura 2000 a priority, this needs to translate into environmental policies obtaining a much greater share of the EU's budget than is currently the case. Negotiations regarding future multiannual financial frameworks will therefore be key.

Sixth, the need for effective policy integration is another consistently unheeded refrain. Again, the Commission's Action Plan targets this, citing the common agricultural policy (CAP) and the common fisheries policy, amongst others. Many half-hearted attempts at policy integration have been made and have failed. While the priorities of the current Commission may not seem conducive to ambitious policy changes of this sort, it will be important for policy entrepreneurs and leaders to continue pushing for fundamental and ambitious reforms of the CAP and other key EU policy sectors.

Seventh, litigation has played a crucial role in the development of EU nature policy. The CJEU has played a consistently progressive role in this regard but it is empowered to do so only by virtue of the cases that are brought before it. The Commission should make more effective use of the infringements procedure (Čavoški, 2016) but for now seems unlikely to do so. Since the Commission's activity has dropped (for now) in this area, it will fall on others to pick up the slack in terms of using the law as an instrument of progressive change. This points to the importance of improving access to justice at the national level, and of using the preliminary reference procedure to ensure strategically important cases are determined at CJEU level.

Finally, while climate change did not really feature during the negotiations of the Birds and Habitats Directives, its impacts on Natura 2000 are of course expected to increase over the coming decades (EEA, 2015). Claims that EU nature policy is climate proof (e.g., Trouwborst, 2015) will undoubtedly be tested in the coming years, and many authors have proposed measures to future proof Natura 2000 (see EEA, 2015, pp. 141–143). The connectivity of the Natura 2000 network and protection of the wider countryside seem likely to be issues of central concern – and increasingly so as the establishment phase of Natura 2000 nears completion in the coming years. Considering that around 18% of the EU's terrestrial area, for example, is within Natura 2000, some 82% lies outside. Neither the Commission nor players at national level have made full use of the possibilities under the Birds and Habitats Directives in this regard. For example, the obligation in the Birds Directive to "strive to avoid pollution or deterioration of habitats" outside SPAs has proven itself enforceable before the CJEU (2007, paragraphs 176–193) but has not been pursued systematically. The idea of protecting areas outside the Natura 2000 network, including corridors and stepping stones, was heavily debated during the Habitats Directive negotiations, yet ultimately the provisions agreed were very weak (see Article 10). Given that this was the outcome during a period more conducive to EU environmental policy development than now, one might justifiably be sceptical of the chances of so-called green infrastructure policy succeeding where the Habitats Directive negotiations failed. However, it is encouraging in this regard that the Council, in its conclusions regarding the Commission's 2017 Action Plan, reiterated its call on the Commission to put forward a proposal for a trans-European network for green infrastructure (TEN-G) (CEU, 2017a), equivalent to existing networks for energy, telecommunications, and transport. The creation of such a network would represent grand policy-making akin to the creation of the Natura 2000 network itself.

On the 20th anniversary of the Birds Directive's adoption, Alistair Gammell commented:

> [Tomorrow's Europeans] will be grateful that a handful of people in the early 1970s had the foresight to act to try to protect these areas and species and that a movement of European citizens . . . fought, in some cases successfully, to conserve them . . . Tomorrow's Europeans will thank us for that.
>
> *(Gammell, 1999, p. 12)*

Tomorrow's Europeans deserve at least the same levels of ambition and commitment from the current generation.

Notes

1 This is perhaps only semi-autonomous on the part of the Commission given that the Paris Summit (1972, paragraph 14) declaration stated that the Heads of Government "agreed that in order to accomplish the tasks laid out in the different action programmes, it was advisable to use as widely as possible all the provisions of the Treaties including Article 235 of the EEC Treaty".

2 Why the Article 5 process has never been used is unclear. There are several possible explanations. One would be that the Nature Unit lost Claus Stuffmann, Henriette Bastrup-Birk and Catherine Yourassowsky very soon after the Directive was adopted, such that there was perhaps a loss of institutional memory regarding the intended purpose of Article 5. A second potential explanation would be that Article 5 presupposes the existence of a scientific disagreement between the Commission and the Member State in question (this was the Commission's argument in CJEU, 2012b), rather than simply a dispute in principle over whether a particular site should be designated or not.

3 The proposed creation of a new industrial and commercial area in Germany was the subject of a negative opinion (European Commission, 2003).

4 Winter's analysis in Harte and Winter (1997) provides some support for Krämer's view, at least in respect of certain of the Commission's opinions.

5 Also see the CJEU's decision in Case C-43/10 (CJEU, 2012a, paragraph 8 of operative part).

6 This view is effectively shared by Clutten and Tafur (2012), who write: "In spite of the rhetoric of conserving the EU's natural heritage at the centre of the Habitats Directive, the ease with which conservation interests can be overridden casts serious doubt on its effectiveness as a conservation tool. Neither the CJEU nor the European Commission, with responsibility for ensuring compliance with European law, are taking measures to counteract the broadening of the exception under Article 6(4). Member States appear similarly at ease with a wide interpretation of IROPI [imperative reasons of overriding public interest]. In answer to the question posed in the title of this paper, it seems that the purpose of the Habitats Directive really is imperilled by the interpretation being accorded in practice to the phrase 'imperative reasons of overriding public interest'". To be fair to the CJEU, it can only "counteract the broadening of the exception under Article 6(4)" where relevant cases are brought before it, as happened in the *Solvay* (CJEU, 2012c) and *Nomarchiaki* (CJEU, 2012a) cases (both preliminary references from a national court, so not brought by the Commission). The CJEU held that imperative reasons of overriding public interest were absent in the *Solvay* case – it is worth noting that the European Commission (2010) was of the same view as the CJEU on this point.

7 The European Commission (2008, 2012) has released two reports on the operation of sub-paragraph 1 of Article 6(4), covering the years 2004–2006 and 2007–2011. The reports reveal that numerous Member States either expressly have not applied the provision or have not provided information regarding compensatory measures to the Commission.

8 E.g., see Jackson (2011).
9 Cf. the case of the Sabor dam in Portugal in Jackson (2011).
10 Though cf. Beunen and Duineveld's (2010, pp. 324–325) claim that "Nature conservationists also do not like the high number of lawsuits because they believe these lawsuits and the ensuing negative publicity about nature conservation laws undermine public support for nature conservation". There are no doubt a range of views on this issue amongst nature conservationists.
11 The EU's treatment of this issue continues to cause considerable controversy: e.g., see the Council's decision to merely "take note of" rather than "endorse" findings of the Aarhus Compliance Committee against the EU in this regard (CEU, 2017b).
12 Alongside "no alternatives" and the adoption of compensatory measures.

References

Alter, K. J. 2009. *The European Court's political power.* Oxford: Oxford University Press.

Baldock, D. 1992. The status of Special Protection Areas for the protection of wild birds. *Journal of Environmental Law*, 4, 139–144.

Barroso, J. M. 2009. Letter dated 26 October 2009 from Commission President Barroso to Jan Peter Balkenende, Prime Minister of the Netherlands, attaching a response to the Dutch position paper on Natura 2000.

Becker, S., Bauer, M. W., Connolly, S. and Kassim, H. 2016. The Commission: boxed in and constrained, but still an engine of integration. *West European Politics*, 39, 1011–1031.

Beunen, R. and Duineveld, M. 2010. Divergence and convergence in policy meanings of European environmental policies: The case of the Birds and Habitats Directives. *International Planning Studies*, 15, 321–333.

Bickerton, C., Hodson, D. and Puetter, U. 2015. The new intergovernmentalism and the study of European integration. *In:* Bickerton, C., Hodson, D. and Puetter, U. (eds) *The new intergovernmentalism: states and supranational actors in the post-Maastricht era.* Oxford: Oxford University Press.

Birds Directive 1979. Council Directive 79/409/EEC of 2 April 1979 on the conservation of wild birds, *OJ L 103*, 25 April 1979, p. 1, since codified as Directive 2009/147/EC of the European Parliament and of the Council of 30 November 2009 on the conservation of wild birds. *OJ L 20*, 26 January 2010, p. 7.

Brooke, M. 1989. A birthday for the birds. *New Scientist* (4 February), 54–57.

Bulmer, S. 2015. Understanding the new intergovernmentalism: pre- and post-Maastricht EU studies. *In:* Bickerton, C., Hodson, D. and Puetter, U. (eds) *The new intergovernmentalism: states and supranational actors in the post-Maastricht era.* Oxford: Oxford University Press.

Bunce, R. 2012. Interview with Roy Bunce, senior negotiator for the UK government in respect of the Habitats Directive, conducted by Andrew L. R. Jackson on 13 August 2012.

Caufield, C. 1981. Environment: Britain lags, Europe leads. *New Scientist* (13 August), 416–417.

Čavoški, A. 2016. An assessment of compliance strategies in the environmental policy area. *European Law Review*, 41, 252–274.

CEC 1977a. Note: Proposal for a Council Directive on bird conservation [R/3265/76 (ENV 145)] – Outcome of the meeting of the Working Party on the Environment on Friday 11 March 1977. Council of the European Communities. Brussels, 22 March 1977 (29.03). R/633/77 (ENV 46) (AGRI 174).

CEC 1977b. Note: Subject: Proposal for a Council Directive on bird conservation [R/3265/76 (ENV 145)] – Report from the Permanent Representatives Committee to the Council (document concerning the Articles). Council of the European Communities. Brussels, 4 December 1977 (08.12). R/3077/77 (ENV 170) (AGRI 823).

CEC 1978. Note: Subject: Proposal for a Council Directive on bird conservation [R/3265/76 (ENV 145)] – Report by the Permanent Representatives Committee to the Council. Council of the European Communities. Brussels, 22 May 1978 (26.05). R/1214/78 (ENV 74) (AGRI 345).

CEC 1989a. Background note from the Secretariat to the Committee of Permanent Representatives dated 13 November 1989, attaching revised draft directive. Brussels, 9847/89, ENV 205. Council of the European Communities.

CEC 1989b. Cover note from General Secretariat to Delegations. Subject: Proposal for a Council Directive on the protection of natural and semi-natural habitats and of wild fauna and flora. Attaching French Presidency compromise text of Directive. Brussels, 13 October 1989, 9204/89, ENV 186. Council of the European Communities.

CEC 1991. Outcome of proceedings of Working Party on the Environment on 4 September 1991. Brussels, 10 September 1991, 7992/91, ENV 250. Council of the European Communities.

CEC 2017a. Council conclusions on EU Action Plan for nature, people and the economy. Press release 387/17, 19 June 2017.

CEC 2017b. Council Decision (EU) 2017/1346 of 17 July 2017 on the position to be adopted, on behalf of the European Union, at the sixth session of the Meeting of the Parties to the Aarhus Convention as regards compliance case ACCC/C/2008/32. *OJ L 186*, 19 July 2017, p. 15.

Cichowski, R. A. 2001. Litigation, compliance and European integration: the preliminary ruling procedure and EU nature conservation policy. Annual meeting of the European Communities Studies Association, 31 May–2 June 2001, Madison, Wisconsin.

Cichowski, R. A. 2007. *The European Court and civil society: litigation, mobilization, and governance*. Cambridge: Cambridge University Press.

CJEU 1985. Judgment of the Court of Justice of the EU of 7 February 1985, *ADBHU*, Case 240/83, EU:C:1985:59.

CJEU 1989. Judgment of the Court of Justice of the EU of 14 February 1989, *Star Fruit Company SA* v. *European Commission*, Case 247/87, EU:C:1989:58.

CJEU 1990. Opinion of Advocate General Van Gerven of 5 December 1990, *Commission* v. *Germany*, C-57/89, EU:C:1990:432.

CJEU 1991. Judgment of the Court of Justice of the EU of 28 February 1991, *Leybucht Dykes, Commission* v. *Germany*, C-57/89, EU:C:1991:89.

CJEU 1996. Judgment of the Court of Justice of the EU of 11 July 1996, *Lappel Bank*, C-44/95, EU:C:1996:297.

CJEU 1998. Judgment of the Court of Justice of the EU of 19 May 1998, *Commission* v. *Netherlands*, C-3/96, EU:C:1998:238.

CJEU 2000a. Judgment of the Court of Justice of the EU of 7 December 2000, *Commission* v. *France*, C-374/98, EU:C:2000:670.

CJEU 2000b. Judgment of the Court of Justice of the EU of 7 November 2000, *First Corporate Shipping*, C-371/98, EU:C:2000:600.

CJEU 2004. Judgment of the Court of Justice of the EU of 7 September 2004, *Waddenzee*, C-127/02, EU:C:2004:482.

CJEU 2005. Judgment of the Court of Justice of the EU of 20 October 2005, *Commission* v. *United Kingdom*, C-6/04, EU:C:2005:626.

CJEU 2006. Judgment of the Court of Justice of the EU of 14 September 2006, *Bund Naturschutz*, C-244/05, EU:C:2006:579.
CJEU 2007. Judgment of the Court of Justice of the EU of 13 December 2007, *Commission* v. *Ireland*, C-418/04, EU:C:2007:780.
CJEU 2012a. Judgment of the Court of Justice of the EU of 11 September 2012, *Nomarchiaki*, C-43/10, EU:C:2012:560.
CJEU 2012b. Judgment of the Court of Justice of the EU of 15 March 2012, *Commission* v. *Cyprus*, C-340/10, EU:C:2012:143.
CJEU 2012c. Judgment of the Court of Justice of the EU of 16 February 2012, *Solvay*, C-182/10, EU:C:2012:82.
CJEU 2012d. Opinion of Advocate General Sharpston of 22 November 2012, *Sweetman*, C-258/11, EU:C:2012:743.
CJEU 2013. Judgment of the Court of Justice of the EU of 11 April 2013, *Sweetman*, C-258/11, EU:C:2013:220.
CJEU 2014. Judgment of the Court of Justice of the EU of 15 May 2014, *Briels*, C-521/12, EU:C:2014:330.
CJEU 2016a. Judgment of the Court of Justice of the EU of 8 November 2016, *LZ II*, C-243/15, EU:C:2016:838.
CJEU 2016b. Judgment of the Court of Justice of the EU of 21 July 2016, *Orleans*, C-387/15, EU:C:2016:583.
CJEU 2017. Judgment of the Court of Justice of the EU of 26 April 2017, *Commission* v. *Germany*, C-142/16, EU:C:2017:301.
Clinton Davis, S. 1987. European Parliament, Verbatim report of proceedings, 9–10 July 1987, Strasbourg.
Clinton Davis, S. 1988. Extract from the European Parliament's Intergroup on Animal Welfare bulletin, 52nd session, 13 December 1988.
Clutten, R. and Tafur, I. 2012. Are imperative reasons imperiling the Habitats Directive? An assessment of Article 6(4) and the IROPI exception. *In:* Jones, G. (ed.) *The Habitats Directive: a developer's obstacle course?* Oxford: Hart Publishing.
Coffey, C. and Richartz, S. 2003. *The EU Habitats Directive: generating strong responses*. Berlin: Institute for European Environmental Policy.
Council Legal Service 1977. Note: Subject: Proposal for a Council Directive on bird conservation (R/3265/76 (ENV. 145)) (Brussels, 25 July 1977, R/1867/77 (JUR 95) (ENV 106)).
Department of the Environment 1991. Letter dated 10 June 1991 from the UK government's Department of the Environment to the head of the Nature Unit of DG XI.
Devillers, P. 2013. Interview with Pierre Devillers of Belgium's IRSNB conducted by Andrew L. R. Jackson on 25 April 2013.
DG XI 1988a. Draft headed "Speaking notes a l'attention de M. Brinkhorst. Object: reunion de hauts fonctionnaires 'protection de la nature' le 29 Septembre 1988 de 11h30 a 13h00".
DG XI 1988b. Internal DG XI note dated 18 February 1988 from the Legal Unit to the Director of Directorate B, the adviser to the Director General, the Nature Unit, and the Economics Unit. Subject: Draft of a directive on habitats protection. Ref: Version of FFH directive of 15.2.1988.
DG XI 1988c. Projet de directive concernant la protection des habitats naturels et seminaturels ainsi que de la faune et de la flore sauvages. Communication de M. Clinton Davis. Bruxelles, 15 July 1988, COM(88) 381/5. Corrigendum: replacement des pages 1 à 4. On the agenda of the 928th meeting of the Commission, on Tuesday 19 July 1988.

DG XI 1989a. DG XI (unit B3) file note dated 23 January 1989, headed "Object: Entretien du Commissaire Ripa di Meana avec le Président du Conseil Environnement le 24.1.89".

DG XI 1989b. DG XI note headed "Conseil Environnement, le 28 novembre 1989, Point 'Habitat', Summary Background Note".

DG XI 1989c. Note dated 14 November 1989 from the head of the Nature Unit (B3) of DG XI to the Director of Directorate B, headed "Subject: Item: 'Draft Directive FFH' on the Coreper agenda of 15–16/11/1989 – Preparation of the Environment Council of 28/11/1989".

DG XI 1990. Annex headed "Explications des points defensifs out questions supplementaires", annexed to a background note, in turn attached to an internal DG XI briefing package dated 16 October 1990, headed "Discussion du rapport de H. Muntingh sur la proposition de directive concernant la protection des habitats naturels et semi-naturels ainsi que de la faune et de la flore sauvages".

DG XI 1991. DG XI document headed "Points defensifs, Conseil Environnement, 1er octobre 1991".

EEA 2015. *State of nature in the EU: results from reporting under the nature directives 2007–2012*. European Environment Agency. EEA Technical report No 2/2015.

ENDS Europe 2013. Green measures blocked by top EU official – sources (1 March 2013). *ENDS Europe*.

Environment Council of the EU 2010. Environment Council conclusions of 15 March 2010, endorsed by the European Council conclusions of 25/26 March 2010.

European Commission 1980. Progress made in connection with the environment action programme and assessment of the work done to implement it. Communication from the Commission to the Council. COM(80) 222 final, Brussels, 7 May 1980.

European Commission 2003. Opinion of the European Commission of 24 April 2003 delivered upon request of Germany according to Art. 6(4) Sub Par. 2 of the Habitats Directive, concerning the creation of a new industrial and commercial area "Siegerland" within the former military training area Trupbach near Siegen/Freudenberg (North Rhine-Westfalia).

European Commission 2006. Seventh Annual Survey on the implementation and enforcement of Community environmental law – 2005. Brussels, 8 September 2006, SEC(2006) 1143.

European Commission 2008. Implementation of Article 6(4), first subparagraph, of Council Directive 92/43/EEC during the period 2004–2006. Summary report, 30 June 2008.

European Commission 2010. Commission's written observations of 30 July 2010 in Case C-182/10, *Solvay*, JURM(10) 8097. Obtained by author following access to information request.

European Commission 2011. Our life insurance, our natural capital: an EU biodiversity strategy to 2020 (3.05.2011). COM(2011) 244 final.

European Commission 2012. Implementation of Article 6(4), first subparagraph, of Council Directive 92/43/EEC (Habitat Directive). Period 2007–2011. Summary report, March 2012.

European Commission 2016a. Commission Staff Working Document. Fitness Check of the EU Nature Legislation (Birds and Habitats Directives). Brussels, 16 December 2016, SWD(2016) 472 final.

European Commission 2016b. *European Commission Opinions relevant to Article 6(4) of the Habitats Directive* [Online]. Available: http://ec.europa.eu/environment/nature/natura2000/management/opinion_en.htm [accessed 28 March 2017].

European Commission 2016c. Statement by Commissioner Vella on the orientation debate on the next steps of the Fitness Check of the Birds and Habitats Directives, 7 December 2016.

European Commission 2016d. *Statistics on environmental infringements* [Online]. Available: http://ec.europa.eu/environment/legal/law/statistics.htm [accessed 16 March 2017].

European Commission 2017a. Communication from the Commission of 28.4.2017: Commission Notice on Access to Justice in Environmental Matters, C(2017) 2616 final.

European Commission 2017b. Communication from the Commission to the European Parliament, the Council, the EESC and the Committee of the Regions. An Action Plan for nature, people and the economy. Brussels, 27 April 2017. COM(2017) 198 final.

European Commission 2017c. *The Environmental Implementation Review* [Online]. Available: http://ec.europa.eu/environment/eir/index_en.htm [accessed 26 July 2017].

European Commission undated. Legal Enforcement. Statistics on environmental infringements. Infringements by sector.

European Parliament 1987. Resolution on the establishment and conservation of Community nature reserves (10 July 1987). *OJ C 246*, 14 September 1987, p. 121.

Fairbrass, J. 2000. EU and British biodiversity policy: ambiguity and errors of judgement. CSERGE Working Paper GEC 2000–4: Centre for Social and Economic Research on the Global Environment, University of East Anglia and University College London.

Fairbrass, J. and Jordan, A. 2001. European Union environmental policy and the UK government: a passive observer or a strategic manager? *Environmental Politics*, 10, 1–21.

Fernández-Galiano, E. 2012. Interview with Eladio Fernández-Galiano, then head of the Secretariat of the Bern Convention, conducted by Andrew L. R. Jackson on 30 August 2012.

First Environmental Action Programme 1973. Declaration of the Council and of the European Communities and of the representatives of the Governments of the Member States meeting in the Council of 22 November 1973 on the programme of action of the European Communities on the environment. *OJ C 112*, 20 December 1973, p. 1.

Fourth Environmental Action Programme 1987. Resolution of the Council of the European Communities and of the Representatives of the Governments of the Member States meeting within the Council of 19 October 1987 on the continuation and implementation of a European Community policy and action programme on the environment (1987–1992). *OJ C 328*, 7 December 1987, p. 1.

Gammell, A. 1999. The contribution of BirdLife International to the Birds Directive and the management of birds and their habitats in Europe. 20 years with the EC Birds Directive. Proceedings from a conference on the Council's Directive on the Conservation of Wild Birds, 18–19 November 1999 Elsinore, Denmark. The Danish National Forest and Nature Agency.

Gammell, A. 2012. Interview with Alistair Gammell, formerly of the RSPB, conducted by Andrew L. R. Jackson on 31 July 2012.

Harte, J. D. C. and Winter, G. 1997. Nature conservation: the rule of law in European Community environmental protection. *Journal of Environmental Law*, 9, 139–180.

Holder, J. 1997. Case C-44/95, *R.* v. *Secretary of State for the Environment*, ex parte Royal Society for the Protection of Birds, judgment of the Court of Justice 11 July 1996, [1996] ECR 1-3805. *Common Market Law Review*, 34, 1469–1480.

House of Lords 1989. House of Lords' Select Committee on the European Communities. 15th report: Habitat and Species Protection. Ordered to be printed 18 July 1989.

Jackson, A. L. R. 2011. Renewable energy vs. biodiversity: Policy conflicts and the future of nature conservation. *Global Environmental Change*, 21, 1195–1208.

Johnson, S. 2011. Interview with Stanley Johnson conducted by Christian van de Velde on 18 October 2011. Ref: INT185. HISTCOM.2: Histoire interne de la Commission européenne 1973–1986. Project coordinator: Université catholique de Louvain.

Johnson, S. 2012. Interview with Stanley Johnson, former European Commission official and former MEP, conducted by Andrew L. R. Jackson on 16 July 2012.

Jouve, M. 2013. Personal communication: e-mail dated 8 April 2013 from Marcel Jouve, ex head of the International and European Affairs division of the Nature Protection Directorate at France's Department of the Environment, to Andrew L. R. Jackson.

Juncker, J.-C. 2014. Mission letter dated 1 November 2014 from Jean-Claude Juncker, President of the European Commission, to Karmenu Vella, Commissioner for Environment, Maritime Affairs and Fisheries.

Kati, V., Hovardas, T., Dieterich, M., Ibisch, P. L., Mihok, B. and Selva, N. 2015. The challenge of implementing the European network of protected areas Natura 2000. *Conservation Biology*, 29, 260–270.

Khetani-Shah, S. 2016. *UK's David Cameron supports EU nature laws, says they are reason to stay in the EU.* BirdLife International, 2 June [Online]. Available: www.birdlife.org/europe-and-central-asia/news/uks-david-cameron-supports-eu-nature-laws-says-they-are-reason-stay-eu [accessed 23 March 2017].

Koester, V. 2012. Interview with Veit Koester, senior negotiator for the Danish government in respect of the Habitats Directive, conducted by Andrew L. R. Jackson on 22 August 2012.

Krämer, L. 2009. The European Commission's Opinions under Article 6(4) of the Habitats Directive. *Journal of Environmental Law*, 21, 59–85.

Krämer, L. 2015. Implementation and enforcement of the Habitats Directive. *In:* Born, C.-H., Cliquet, A., Schoukens, H., Misonne, D. and Van Hoorick, G. (eds) *The Habitats Directive in its EU environmental law context*. Abingdon: Routledge.

Markham, J. M. 1989. Europe's politicians thinking green as concern grows on environment. *The New York Times*, 12 April.

Megret, A. 2013a. Personal communication: e-mail dated 23 March 2013 from Alain Megret, deputy director of the Nature Protection directorate of the French government's Department of the Environment at the time of the Habitats Directive negotiations, to Andrew L. R. Jackson.

Megret, A. 2013b. Personal communication: written comments from Alain Megret, deputy director of the Nature Protection directorate of the French government's Department of the Environment at the time of the Habitats Directive negotiations, attached to e-mail dated 1 April 2013 from Alain Megret to Andrew L. R. Jackson.

Meyer, J. H. 2010. Saving migrants: a transnational network supporting supranational bird protection policy in the 1970s. *In:* Kaiser, W., Gehler, M. and Leucht, B. (eds) *Transnational networks in regional integration: informal governance in Europe 1945–83*. Basingstoke: Palgrave Macmillan.

Meyer, J. H. 2011. Green activism: the European Parliament's Environmental Committee promoting a European environmental policy in the 1970s. *Journal of European Integration History*, 17, 73–85.

Neal, H. 2012. Interview with Hilary Neal, UK government official who participated in the Habitats Directive negotiations and early implementation, conducted by Andrew L. R. Jackson on 14 August 2012.

Paris Summit 1972. Declaration of the first summit conference of the enlarged Community: Meetings of the Heads of State or Government, Paris, 19–21 October 1972. French (VO) and English versions. *Bulletin of the European Communities*, No. 10.

Peterson, J. 2008. Enlargement, reform and the European Commission. Weathering a perfect storm? *Journal of European Public Policy*, 15, 761–780.

Peterson, J. 2017. Juncker's political European Commission and an EU in crisis. *JCMS: Journal of Common Market Studies*, 55, 349–367.

Pierson, P. 1996. The path to European integration. *Comparative Political Studies*, 29, 123–163.

Pierson, P. 2004. *Politics in time: history, institutions, and social analysis*. Princeton, NJ: Princeton University Press.

Potočnik, J. 2015. Foreword. *In:* Born, C.-H., Cliquet, A., Schoukens, H., Misonne, D. and Van Hoorick, G. (eds) *The Habitats Directive in its EU environmental law context*. Abingdon: Routledge.

RSPB 1989. Note dated 15 March 1989, prepared by the RSPB at the request of the UK government's Department of Environment, with a handwritten, undated covering note to DG XI.

Sandholtz, W. and Stone Sweet, A. 2012. Neo-functionalism and supranational governance. Unabridged version, available at http://works.bepress.com/alec_stone_sweet/. Abridged version. *In:* Jones, E., Menon, A. and Weatherill, S. (eds) *The Oxford handbook of the European Union*. Oxford: Oxford University Press.

Single European Act 1986. Single European Act. *OJ L 169*, 29 June 1987, p. 1.

Slepcevic, R. 2009. The judicial enforcement of EU law through national courts: possibilities and limits. *Journal of European Public Policy*, 16, 378–394.

Stone Sweet, A. 2004. *The judicial construction of Europe*. Oxford: Oxford University Press.

Stone Sweet, A. and Sandholtz, W. 1998. Integration, supranational governance, and the institutionalization of the European polity. *In:* Sandholtz, W. and Stone Sweet, A. (eds) *European integration and supranational governance*. Oxford: Oxford University Press.

Stuffmann, C. J. 1999. The upcoming of the Birds Directive. 20 years with the EC Birds Directive. Proceedings from a conference on the Council's Directive on the Conservation of Wild Birds, 18–19 November 1999, Elsinore, Denmark. The Danish National Forest and Nature Agency.

Sundseth, K. 2012. Interview with Kerstin Sundseth, environmental consultant and former DG XI *stagiaire*, conducted by Andrew L. R. Jackson on 13 September 2012.

Sundseth, K. and Roth, P. 2013. Study on evaluating and improving the Article 6.3 permit procedure for Natura 2000 sites. Final report, November 2013.

TEU 2012. Treaty on European Union. Consolidated version. *OJ C 326*, 26 October 2012, p. 13.

TFEU 2010. Treaty on the Functioning of the European Union. Consolidated version. *OJ C 83*, 30 March 2010, p. 47.

Third Environmental Action Programme 1983. Resolution of the Council of the European Communities and of the representatives of the Governments of the Member States, meeting within the Council, of 7 February 1983 on the continuation and implementation of a European Community policy and action programme on the environment (1982 to 1986). *OJ C 46*, 17 February 1983, p. 1.

Treaty of Rome 1957. Treaty establishing the European Economic Community. United Nations Treaty Series.

Trouwborst, A. 2015. The Habitats Directive and climate change: is the law climate-proof? *In:* Born, C.-H., Cliquet, A., Schoukens, H., Misonne, D. and Van Hoorick, G. (eds) *The Habitats Directive in its EU environmental law context*. Abingdon: Routledge.

Troya, A. 2012. Interview with Antonio Troya, Spanish government official seconded to the European Commission for the Habitats Directive negotiations, conducted by Andrew L. R. Jackson on 5 October 2012.

White, M. 1989. Major leaps up cabinet ladder: Howe is made deputy PM. *The Guardian* 25 July.

Wildlife Link 1989. Internal Wildlife Link memorandum (LINK/MEM/89/21) dating from some time before 14 July 1989.

INDEX

FSC
www.fsc.org
MIX
Papier | Fördert
gute Waldnutzung
FSC® C083411

Kolibri 360 GmbH
Ferdinand-Jühlke-Straße 7
99095 Erfurt, Deutschland
produktsicherheit@kolibri360.de